42D CONGRESS, 1st Session. } HOUSE OF REPRESENTATIVES. { EX. DOC. No. 19.

FINAL REPORT

OF THE

UNITED STATES GEOLOGICAL SURVEY

OF

NEBRASKA

AND PORTIONS OF

THE ADJACENT TERRITORIES,

MADE UNDER THE DIRECTION OF

THE COMMISSIONER OF THE GENERAL LAND OFFICE.

BY

F. V. HAYDEN,

UNITED STATES GEOLOGIST.

MARCH 23, 1871.—Ordered to be printed.

WASHINGTON:
GOVERNMENT PRINTING OFFICE.
1872.

LETTER TO THE COMMISSIONER.*

WASHINGTON, D. C., *March* 1, 1868.

SIR: I have the honor to transmit my final report of the United States Geological Survey of Nebraska and adjacent Territories.* The observations here recorded were made with some care at the time, and, inasmuch as they will not probably be repeated for years to come, may prove of value to the state. The history of the origin and purposes of the survey have been so clearly set forth in your annual report of the General Land Office for 1867, that I beg permission to introduce it in this connection:

In the second section of the act of Congress, approved March 2, 1867, making appropriations and to supply deficiencies, it is declared "that the unexpended balances of the appropriations heretofore made for defraying the expenses of the legislative assembly to the Territory of Nebraska, shall be diverted and set aside for the purpose of procuring a geological survey of Nebraska, to be prosecuted under the direction of the Commissioner of the General Land Office."

It has been estimated that the unexpended balance applicable to the geological survey is $5,000. Under that authority Dr. F. V. Hayden was appointed on the 29th April, 1867, to make a geological examination and survey, with compensation of $2,000 per annum. With the limited means provided, he was allowed an assistant geologist and paleontologist, at the rate of $1,000; three collectors and laborers at not exceeding $700, the sum of $300 having been set apart for chemistry and natural history; while the sum of $1,000, or the residue of the means, was designed for general expenses of outfit and incidentals in the service, which was restricted to one year from the date of the appointment.

The geologist was directed to proceed as soon as necessary arrangements could be made to the sphere of his operations. He was instructed to ascertain the order of succession, arrangement, relative position, dip, and comparative thickness of the several strata and geological formations in the State, to search for and examine all the beds, veins, and other deposits of ores, coals, clays, marls, peat, and other like mineral substances, as well as the fossil remains of the various formations; to obtain chemical analyses of such of those substances, and of the different varieties of soil, whereof it may be deemed desirable to ascertain the elementary constituents. He was required also to determine by careful barometrical observations the relative elevations and depressions of the different parts of the State of Nebraska, and to gather in the field of his explorations collections in geology, mineralogy, and paleontology, to illustrate the notes taken in the field. In order to enable the Commissioner to present to Congress the results of the geological survey, it was stipulated that a preliminary report should be made of the progress of the work, accompanied by such maps, sections, and drawings as might be considered requisite to illustrate the report, it having been ordered that the final report under the appropriation should embody the results of the entire survey, and be accompanied by a geological map, with carefully-prepared sections and diagrams, showing by different colors and other marks and characters the principal

* It is but just to me to be permitted to state that this report was completed in its present condition three years ago, and the manuscript delivered to the Commissioner. It is now printed without any opportunity for revision. Several chapters which have appeared in other publications are omitted.

localities and geographical range of the various geological formations of the district explored, and by drawings and descriptions of the characteristic fossil remains of the different groups of strata—advance data having been called for at short intervals, in order that the Department might know the progress of the work.

It was required in our instructions that the region of Nebraska south of the Platte River should be first examined, it being occupied by the limestones of the true coal measures, and that a careful search should be instituted for the localities, depths, and extent of deposits of that most valuable mineral. It was deemed important to extend the explorations and examinations along the Missouri to Sioux City, as it had been reported that there was a bed of coal outcropping from rocks of the chalk formation near the Omaha reserve, then under survey for the accommodation of the Omaha and Winnebago Indians. It was desired that the geologist, who was furnished with a map of public surveys, should locate geological formations by townships and ranges of the sixth or governing principal meridian in Nebraska. As the unsurveyed region also includes settled portions of the State, it was required that the explorations should also be there directed to determine the location and extent of natural resources in coal, metallic ores, hydraulic and common limestone, fire-clays, freestone, flagstone, and marbles, properly belonging to the various formations there existing, and which would be of immediate use to the people. As the predominating interest in the State is farming, his attention was directed to the examination of its soils and subsoils, to their adaptability to particular crops, as well as to the best method of preserving and increasing their fertility.

Information was also called for in regard to the introduction of suitable forest trees, in order to promote the growth of timber.

* * * * * * * *

Attention should likewise be given to the materials for the construction of roads, houses, bridges, such as building stones, limestones for the manufacture of quick-lime, sand, clays for making bricks and tiles, as well as for potters' use. Particular attention should be given to the various soils and subsoils, and their adaptability to the growth of different kinds of crops, fruit and ornamental as well as forest trees. If the district is hilly or mountainous, barometrical observations should be made to determine the heights of the elevations above the sea and the principal streams, and attention should likewise be given to the climatology of the country. Full sets of collections of all the different kinds of rocks, soils, ores, minerals, and mineral waters of every description, as well as of the various organic remains characterizing the different formations, should be carefully collected and preserved for study and analysis. These collections to be arranged and permanently preserved in the Department. It is proposed that specimens of every kind be transported to the Department for careful investigation, in order that final and more detailed reports may be made out, illustrated by maps, sections, diagrams, and drawings of the various fossil remains, characteristic of the different rocks.

Authority should be given for the publication of the final reports in a suitable form, and in such manner as to be creditable to the country. A few such reports, properly prepared by competent and reliable authorities, with full statistics of our resources, would, if distributed abroad, have a tendency to stimulate immigration, and cause the rapid settlement of our vast unoccupied public domain, thus increasing the national wealth and power, and relieving the burdens of general taxation.

In accordance with your instructions, the survey was prosecuted with all the energy in my power, and I was enabled during the season to examine the entire eastern portion of the State, and some portions farther west. Mr. F. B. Meek, the very able paleontologist of the survey, passed through the State of Iowa, from the Mississippi at Iowa City to the Missouri at Nebraska City, in company with Mr. C. A. White, the State geologist of Iowa, in order that he might connect the geological formations of that State with those of Nebraska, and also trace out the coal-beds in their western extension. His report, herewith transmitted, will show the success and thoroughness with which he performed his labors.

The entire area of the State of Nebraska belongs to what is called the plain or prairie country of the West. The strata are entirely horizontal, and are seldom exposed to the scrutiny of the geologist, except along the banks of streams. The greater portion of the State is covered with a great thickness of yellow marl, which oftentimes conceals the underlying formations, but it gives to the soil the wonderful and inexhaustible fertility which the entire eastern portion possesses. As an agricultural and grazing State Nebraska must eventually take a very high rank.

Nebraska, the youngest State in the American Union, extends from the Missouri westward to the Rocky Mountains, with an extreme length of 412 miles, decreasing to 310 miles on the southern border, its extreme width being 208 miles, diminishing to 138 miles on the west. Its area is 75,905 square miles, or 48,636,800 acres.—(Report Commissioner General Land Office, 1867.)

But three of the principal geological formations are represented in the State, Carboniferous, Cretaceous, and Tertiary. Of the Carboniferous strata only the upper members occur, and these seem to thin out in their western extension, until they almost disappear in the region of the Rocky Mountains.

Much attention was given to the study of the coal-beds in the southeastern portion of the State, and the conclusion was reached that no coal-beds would be found more than from 12 to 30 inches in thickness. In a country which is so destitute of timber these beds may be wrought with advantage. The subject of tree-planting in Nebraska cannot be too strongly impressed upon the settlers in that State. No labor or expense should be spared and no delay permitted in this direction. Not many years will elapse before fine forests of young timber will cover much of this fertile region. The influence of these forests upon the climate and the soil has been discussed in the report. A careful examination of the geological formations, and the physical geography of the State, will show at once that its mineral resources must be very limited. No ranges of mountains come within its borders. Some iron ores occur, but there is not sufficient fuel to utilize them. Gold and silver in paying quantities will not probably be found.

The supply of building material has been shown to be inexhaustible, and the skill and industry of the inhabitants is now fast turning it to practical use.

The Union Pacific Railroad, passing directly across the State from east to west, has given a wonderful impulse to its material prosperity, and must continue to do so for all time to come. The demand for branch roads in every direction has already been made by successful agriculture, and the ease with which they can be built, especially along the bottoms of streams, is quite remarkable.

With the exception of the Missouri River, there are no navigable streams in Nebraska. The Platte, although very broad, is always too shallow, is full of sand-bars, and quicksands, and at certain seasons of the year it is nearly dry. The supply of water for milling and manufacturing purposes, however, seems to be abundant. In conclusion I would say that my examination of the resources of the State has greatly increased my already favorable impression in regard to them. With its broad, fertile bottoms bordering a navigable river like the Missouri for over three hundred miles; with the rich soil of the uplands, and their almost unlimited supply of all the agricultural productions adapted to that climate; with its railroads, water privileges, and the marked energy and industry of its inhabitants, Nebraska is certainly destined to take a high position among the States of our Union.

It remains now for me to acknowledge my indebtedness to many friends, for aid and sympathy during the progress of the survey. The appropriation was so small that it would have been impossible for me to have brought the survey to a successful termination without the generous assistance of the United States military authorities. Armed with a letter from General Grant to the military commanders in the West, requesting them to afford me all the aid in their power not inconsistent with the public service, I was everywhere received by them with the utmost courtesy. General William Myers, chief quartermaster Department of the Platte, supplied me with a complete outfit for traveling, horses, equipments, &c.; General Barriger, the chief commissary, supplied me with commissary stores at officers' prices. From Colonel E. B. Carling, at Fort D. A. Russell, I obtained an outfit, which enabled me to make an examination of the coal-fields along the base of the Rocky Mountains, near the close of the season. To the citizens of Nebraska, and the press generally, I am indebted for the most cordial sympathy and aid, and the mere list of their names would occupy pages of this report. The officers of the Union Pacific Railroad supplied me with every facility in their power, as well as free transportation for party and freight along the road.

To the Hon. P. W. Hitchcock, representative of the Territory in Congress, the people of Nebraska are indebted for the appropriation which enabled the Geologist to make the survey. I wish to express my obligations for many favors of great value to Dr. G. L. Miller, Captain William Wilcox, and Hon. A. S. Paddock of Omaha.

The reports of F. B. Meek, paleontologist, and Mr. S. H. Scudder, on certain insects, injurious to vegetation, are of the highest value. The report of Mr. Scudder was prepared without any expense to the Government.

I take pleasure in extending my cordial thanks to my assistants in the field, Mr. James Stevenson and Edward Chase, who labored with zeal to advance the interests of the survey.

Very respectfully, your obedient servant,

F. V. HAYDEN,

United States Geologist.

To the *Commissioner General Land Office, Washington, D. C.*

REPORT.

CHAPTER I.

DETAILED REPORT OF DOUGLAS AND SARPY COUNTIES.

Detailed reports of counties must necessarily be attended with more or less repetition. Yet in presenting a report on the geology of a State it would appear to be important to the inhabitants of each county that the local geology be dwelt upon with as much minuteness as possible. It is for this reason that I have called this portion of the report the local geology, much of which will no doubt be more interesting to the inhabitants of the State than the general geology. I will, therefore, commence with Douglas and Sarpy Counties. The basis rocks of these two counties are for the most part of the age of the Upper Coal-Measures.

If the Permo-Carboniferous and the Permian were ever deposited over this area, they were swept away by erosion prior to the deposition of the Cretaceous rocks. If we follow the valley of the Platte westward on the northern side, we shall see the junction of the two great periods, Carboniferous and Cretaceous, and we shall find that the beds of the Dakota Group, or what we suppose to be the Lower Cretaceous beds of the West rest directly down on the limestones of the Upper Coal-Measures. As an illustration of this statement, we find near the Old Otoe Village, eight miles above the mouth of the Platte, a good exposure of the sandstone resting conformably on the Carboniferous limestone.

In Douglas County there are very few exposures of the underlying rocks. Indeed, the only quarries of any importance in the county are near Omaha, on the Missouri River. But along the Platte the limestones jut out in massive beds, which supply all that region with excellent building stone, and most of that used in Omaha is transported from these quarries. The exposure of limestones near Omaha is not great, only about eight or ten feet above the water's edge, and over these limestones there are from 150 to 200 feet of drift and yellow marl, which must be stripped off before the rocky layers below can be made available. These difficulties will always render this quarry an expensive one. These limestones have also been seen in one or two localities farther up the river. At Rockport, about ten miles above Omaha, they form the bed of the river, and on this account this locality was formerly known as Rock Bottom, and attracted the attention of railroad engineers as a suitable crossing for a line of road. I was also informed that at low water, limestones are seen at Fort Calhoun and DeSoto, and it is probable that at the latter locality, the Carboniferous beds disappear beneath the river, not to be seen again in the valley of the Missouri, until disclosed by the uplifting of the Rocky Mountain ranges. Passing westward from Omaha there is no important exposure of rocky beds to be observed. The scenery is very monotonous—a rolling prairie, with scarcely a tree to be seen.

Near Bellevue, Sarpy County, there is an exposure of the Upper Coal-Measure rocks at low water, a very careful section of which has been given in Mr. Meek's report. This quarry has been wrought more or less for years, and the limestones used for building purposes, and burned into lime. But the most important quarry in this country north of the Platte is that belonging to Mr. Watson, and located on the Pappillion, three miles west of Bellevue. This quarry has been worked for many years, and contains several layers of valuable rocks for building purposes. It is a source of considerable revenue to the owners, and the materials are taken to Bellevue and Omaha in great quantities, where a ready market is found. The following is a section of the beds in descending order:

6. Vegetable soil, two to four feet thick, with a few stray water-worn rocks.
5. A bed, like No. 3, with fragments of fossils, capped with loose layers of limestone 18 inches to 2 feet thick.
4. Three inches of light-yellow clay, a hard layer.
3. Yellow indurated calcareous clay, full of shells, *Chonetes granulifera*, *Spirifer plano-convexus*, &c., 10 inches.
2. Several layers of hard limestone, very compact, with crinoids, corals, *Chonetes granulifera*, *Athyris subtilita*, *Spirifer cameratus*, *Productus semi-reticulatus*, &c., 6 feet.
1. Greenish yellow clay, underneath the most valuable and massive bed of limestone, 20 inches thick. Below the clay, but only exposed by removing the earth, is a layer of yellow limestone 18 inches thick.

Bed 2 in the above section produces the best rock for building purposes, and the organic remains indicate its geological position to be in the Upper Coal-Measures.

I then visited the quarries along the north bank of the Platte with Messrs. Clark and Watson, of Bellevue. The first point examined was Duclos's quarry, about four miles above the mouth of the Platte. Here are some most excellent limestones for building purposes, and blocks of large size are taken out for the Omaha market. Upon the principal layer rests a bed of yellow clay, filled with water-worn pebbles of small size, usually 2 or 3 inches in diameter, 3 to 4 feet thick. Above this is yellow marl of indefinite thickness. The surface of the rock seems to have been planed smooth, probably by glacial action, as if by the hand of art, so that as far as exposed by the stripping off of the superincumbent clay and marl beds, no portion of the surface seems to be higher than another. The main scratches or grooves are 27° east of north; but there are some exceptions, 19°, 23°, 27° east of north. There is one groove crossing the main scratches nearly north and south. The observations were taken with great care with a large surveyor's compass, belonging to Mr. Watson. Mr. W. regarded the variation at this point as 11¾°. The entire process seems to have been rather a smoothing one, with a few small pebbles in the bottom of the glaciers. A section at this quarry, in descending order, reveals the following beds:

6. Yellow marl. Loess.
5. Yellow clay, full of white lumps of magnesia, with small pebbles.
4. A layer of limestone varying from 10 to 15 inches in thickness, which makes most excellent lime and building blocks. *Athyris subtilita*, *Fusulina cylindrica*, and *Spirifer cameratus*.
3. Slope, doubtless intercalations of clay and thin beds of rock, 30 feet.
2. Massive layers of limestone, yellowish-white, full of organic remains. *Spirifer cameratus*, *Athyris subtilita*. *Productus semi-reticulatus*, *Fusulina cylindrica*. The fusulina very abundant in the middle layers, 10 to 15 feet.
1. Slope, same as bed 3, 25 feet above the Platte.

About two miles above Duclos's quarry, the little streams cut deep valleys along the region of the Platte, exposing some beds which doubtless hold a lower position than those last mentioned. Some very fine springs issue from these rocks, which are regarded with great favor. In the vicinity of these springs there is usually a large accumulation of tufaceous deposits, sometimes forming large masses like thick layers of porous or spongy limestone. All along the Platte the beds are concealed except near the foot of the bluffs, and the following section is disclosed by artificial cuttings:

5. Greenish-yellow and brown clays with irregular hard layers of limestone, 2 to 6 inches thick, 10 to 15 feet.
4. Reddish clay, 2 feet.
3. Heavy, bedded, bluish limestone, very hard, 6 to 8 feet.
2. Yellowish laminated clay, 2 feet.
1. Bluish, hard, argillaceous limestone, 12 to 18 inches.

The Carlisle quarries are about ten miles above the mouth of the Platte. The rock here is a blue limestone in layers 2 to 12 inches thick, with intercalations of clay 1 or 2 inches thick. It breaks into fine blocks for caps and sills, and is much sought after. Near Fair View, just below the mouth of Buffalo Creek, a section was taken:

8. Yellowish calcareous clay, full of *Fusalina cylindrica*, 2 to 4 feet.
7. Yellowish indurated clay, 18 inches.
6. Rather porous, impure limestone, 12 inches.
5. Yellow and ash-colored clay, 2 feet.
4. Massive yellow limestone, very good, full of organic remains, *Allorisma*, *Athyris*, *Productus*, &c., 3 to 4 feet.
3. Slope.
2. Limestone.
1. Slope.

Farther up the Platte bed 4 appears again; and always, wherever it occurs, forms most valuable quarries for the inhabitants. All these layers of limestone continue nearly to the mouth of the Elkhorn, gradually passing by a gentle dip beneath the Cretaceous sandstone. Near the mouth of the Elkhorn and along that stream are several bluffs of the yellow sandstone, soft, and yielding readily to atmospheric influences; but the mural faces are often covered with hieroglyphics, so that they seem to have been of some service to the Indian on which to record his history.

In summing up the geology of these two counties we can at a glance see the simplicity of its character. A few beds of the Upper Coal-Measures and the rusty sandstones of the lowest division of the Cretaceous series, the Dakota Group, are all that occur. We find at one or two localities the Cretaceous and Carboniferous beds in apposition; and though the eye can observe no apparent want of conformity in these beds, yet we can readily imagine the tremendous effects of the erosion prior to the deposition of the sandstone, from the fact that hundreds of feet of clays and limestones must have been swept away.

The first interesting feature in the geology of this district is the vast deposit of yellow marl or Loess, which covers the whole of Eastern Nebraska. Although I shall allude to this deposit quite often in another portion of this report, yet it will be as well to define it as clearly as possible in this connection. Above the basis rocks there is generally a considerable thickness of a deposit which goes under the general name of Drift. Near Omaha it is 40 or 50 feet in thickness composed of loose sand, water-worn pebbles, &c., the layers of deposition inclining at all angles. Indeed, there is every evidence of turbulent waters during its deposition. Mingled with the water-worn pebbles are fragments of iron ore, and often thin layers of rusty sand, which shows that a large portion of the materials of this drift is derived from the washing away of the Cretaceous rocks. The coarse gravel gradually passes up into fine sand, then clay, and not unfrequently the clay graduates up into the yellow marl. In most instances the bluffs of yellow marl present a massive front, without any signs of stratification. Sometimes there are indistinct traces of stratification, but, as a general rule, the yellow marl indicates a deposition in very quiet waters, probably in a lake. This yellow marl is sometimes called the bluff formation, from the fact that it forms the greater portion of the high hills bordering the Missouri River on either side. Indeed, it is this formation that gives the contour to the hills and character to the entire surface of the country forming Eastern Nebraska. But its most powerful influence upon this region lies in the wonderful fertility which it has imparted to all soils, lowland or upland. It varies from 5 to 150 feet in thickness in different places; yet, throughout its entire thickness, it is filled with nutritious matter for the growth of vegetation. It is owing to the universal distribution of this marl that the underlying basis rocks are so rarely revealed to the eye of the geologist. Passing westward from Omaha, as far as the eye can reach, nothing can be seen but one series after another of rounded hills or long sloping ridges. The yielding nature of this marl is such that it wears away into gentle slopes, so that traveling is not difficult; yet there is no more variation in the form of the surface than in that of the sea when agitated by the winds. This, too, is a great advantage to the country, for on this account the drainage is perfect. Sometimes the farmers complain that their farms are rough and not as desirable, but this feature will be regarded at some future period as most important in the cultivation of certain crops. This marl was called, many years ago, Loess by Mr. Lyell, who examined it along the banks of the Mississippi. It forms the characteristic features of the hills along the Missouri and the Mississippi from the Great Bend to the Gulf of Mexico. But through the State of Missouri, so far as I have observed it, it contains more clay. The hills upon which the village of St. Joseph is located are fine examples of the yellow marl. Here very deep artificial excavations have been made. Scattered through it everywhere is an abundance of land and fresh-water shells which all belong, so far as we know, to recent species; a great variety of the genus *Helix*, also *Limnea*, *Physa*, *Planorbis Pupa*, *Succinnea*, valves of *Cyclas*, and some others. These shells, of course, illustrate its purely fresh-water character. This marl is also often filled with whitish calcareous concretions or nodules, which are sometimes hollow. These nodules are especially abundant at the foot of the lofty marl bluffs which form the precipitous range of hills along the Missouri River below the mouth of the Big Sioux. Not unfrequently the bones of extinct animals, mingled with those existing at the present time, are found in this marl. The bones and skull of the buffalo often occur. A most excellent skull of the buffalo was recently found in the marl bluffs near Dakota City, 30 feet below the surface. Near Plattsmouth Rev. J. G. Miller obtained the bones and skull of a rodent, enveloped in one of these calcareous concretions, which on examination proved to belong to the *Geomys bursarius*, pouched or pocket gopher. This animal is still very abundant all over the rock bottoms of the Missouri from mouth to source, and yet we determine from this specimen that it extended back in time to the period of the Loess and was a cotemporary with the mastodon and elephant. In a large number of localities the bones and teeth of mastodons and of the beaver have been found. The teeth of the *Mastodon giganteus* and *Elephas primigenius* are not uncommon. This carries us far back into the past, when the great valley of the Missouri, or a large portion of it, was one vast fresh-water lake, with its myriads of streams pouring in from every side, on the shores of which tramped the huge mastodon and elephant, along with the smaller quadrupeds that now inhabit this region. The beaver also built his dams along the streams, and cut the timber that grew along the banks, while the gopher burrowed, as now, under the rich soil of the bottoms. As I have before remarked, this yellow-marl formation has sometimes been called the Loess from a similar deposit which exists in the valley of the Rhine, and which has given this valley a celebrity in song and story all over the world. The vine-clad hills of the Rhine, and the delicious wine, have long been associated with this district; and if grape-vines that will withstand the severity of the climate can be found, these hills along the Missouri River will rival those along the Rhine in the abundance of the grapes and the income of their vineyards.

There is another feature in the scenery along the Missouri which should not be overlooked here, from the fact that it is quite conspicuous, and contains within itself a most important truth, throwing much light on the history of the great West. The terraces along the Missouri and many of its tributaries are not confined to the district where the yellow marl is deposited, for I have seen these terraces even more numerous and conspicuous about the sources of the Missouri and Yellowstone, and they are described as occurring in the valley of the Columbia and other rivers, flowing into the Pacific. They are doubtless instructive as showing the pauses that occurred in the upward rising of the country during the elevation of the Rocky Mountain range. There is generally the uncertain sandy bottom next to the river, then an abrupt terrace, 4 to 15 feet high, which forms the proper river bottom; then comes the second terrace, which also varies much in height above the river bed of from a few feet to 100 or 150 feet. The city of Omaha is located on one of these high terraces, which is almost entirely composed of yellow marl. This terrace slopes gently up to the foot of the marl hills, when the ascent becomes more abrupt.

It is not often that the terrace is worn away into the rounded hills, as shown near Council Bluffs. Sometimes there are three or four of these terraces, but usually not more than two conspicuous ones. When the terraces are well shown on one side of a river they seem to be wanting on the opposite side. Bellevue is also located on a high terrace overlooking a vast extent of country, and so fine is the location and so beautiful the prospect that it has called forth the involuntary admiration of travelers from Lewis and Clark to those of the present time. In a purely economic view this deposit is of very great value to the settlers. The erosion of the limestones and clays of the Upper Coal-Measures alone could never have given such wonderful fertility to the soil of Nebraska; indeed, it is almost or entirely independent of it. Indeed, it is composed of a mixture of the eroded materials of all the formations in the Missouri Valley, and particularly of the Cretaceous and Tertiary formations and the soft, yielding, chalky limestone of No. 3 Cretaceous enters largely into its composition. I made collections of the soils and the yellow marls from various localities. They are now in the possession of Professor S. W. Johnson, of Yale College, for analysis, and no report has yet been made. In the excellent report of Professor Swallow on the geology of Missouri, 1855, several analyses of yellow marl from different points are given as conducted by Professor Litton. Professor L.

also gives a table of analyses of the Loess of the Rhine, and I will quote it here to show how closely related these two deposits seem to be in their chemical composition. It is not to be supposed that they are necessarily synchronous in their deposition, as some of the older formations are supposed to be, but merely this, that similar causes have produced similar results. The Loess of the Rhine may have been deposited in a quiet lake, and the materials may have been composed of the eroded débris of the rocks of a similar chemical composition. It will be seen also that those chemical constituents predominate in the yellow marl that are useful for the growth of the different cereals. In one hundred parts of this marl, dried, from near Hannibal, Missouri, analyzed by Dr. Litton, there were, from two analyses—

	1.	2.
Silica	76.98	77.02
Alumina and peroxide of iron	11.54	12.10
Lime	3.87	3.25
Magnesia	1.68	1.63
Carbonic acid	Not determined.	2.83
Water	2.01	2.43
	96.17	99.26

Brick made from this formation, from the same locality, gave the following result:

Silica	79.52
Alumina and peroxide of iron	12.80
Lime	3.45
Magnesia	1.95
Carbonic acid and water	1.11
	98.83

Bischoff's analyses of the Loess of the Rhine.

Number of analysis.	1st.	2d.	3d.	4th.	5th.
Silicic acid	58.97	79.53	78.61	62.43	81.04
Alumina	9.97	13.45 }	15.26	7.51	9.75
Peroxide of iron	4.25	4.81 }		5.14	6.67
Lime	0.02	0.02	....	....	....
Magnesia	0.04	0.06	0.91	0.21	0.27
Potash	0.11	1.05 }	3.31	1.75	2.27
Soda	0.84	1.14 }			
Carbonate of lime	20.16	....	....	11.63	....
Carbonate of magnesia	4.21	....	....	3.02	....
Loss by ignition	1.37	....	1.89	2.31	....

The table of five analyses is taken from Bischoff's Chemical Geology, and it will be seen at once that the two deposits are not essentially different from each other chemically. This deposit is found everywhere along the Missouri River to the Great Bend, and for a considerable distance up all the tributaries of that river. It gradually thins out as we proceed westward until it disappears beyond the mouth of Loup Fork. This shows clearly that the deposit was made after all the great water-courses in the West were marked out, and was one of the latest of the important geological events in this region.

CHAPTER II.

CASS, OTOE, AND NEMAHA COUNTIES.

Cass County.—This county lies immediately south of the Platte River, and borders on the Missouri. It is one of the most fertile and productive counties in the State. The Weeping Water, a considerable stream, passes nearly diagonally through it and flows into the Missouri near the little town of Wyoming. The Platte forms the north limit and the Missouri the east, so that it seems to be most favorably located so far as water is concerned. But while this proximity to streams renders a large portion of the surface very rolling, or rough, as the farmers term it, there is undoubtedly more wood-land on that account. Along the Missouri there is a good supply of timber, and a young growth of trees is continually aggressing upon the prairie portion. If not removed by the ax or destroyed by fire, these hills and valleys will, in a few years, be clothed with a thick growth of valuable timber. Among the ravines of the Platte are a great number of patches of oak, sufficient for fuel, on every farm. Then again, along the valley of the Weeping Water there is a good deal of excellent wood-land, and an abundance of limestone for all kinds of building purposes, and some excellent mill-sites. Nearly all the farms are occupied by actual settlers, and the crops of wheat, during the autumn of 1867, were unsurpassed by any other portion of the United States. I should judge also that this county is settled by a thrifty and intelligent class of farmers, for we find not only well-cultivated farms, with neat farm-houses, but also neat churches and school-houses. The entire county is covered with an immense thickness of the yellow marl which conceals all the basis rocks except along the streams. As in the counties already described, only the Upper Coal-Measures and the Cretaceous rocks are found. Along the Platte, about twelve miles west of Plattsmouth, and from thence to the mouth of Salt Creek, there are numerous quarries of the rusty sandstones of the Dakota Group. In these sandstones are numerous impressions of dicotyledonous leaves similar to those found at Blackbird Hills, Decatur, and other localities to the northward. Near Plattsmouth is a series of beds of clays, sandstones, and limestones of the Carboniferous period, and corresponding with those already noticed on the north side of the Platte. Near the summit of the hill is a layer of limestone four to six feet in thickness, which has been much wrought for building materials. As the superincumbent clays and marls are stripped off, a smooth surface is revealed, with indistinct scratches, the same as seen on the opposite side of the Platte. How extensive this planing operation was carried on in this region it is impossible to tell, yet, according to the investigations of Dr. White in Iowa, this glacial action must have extended over a large area, and perhaps over all the Northwest. In a very interesting article by Dr. White, published in the American Journal of Science for May, 1867, entitled "Observations upon the Drift Phenomena in Southern Iowa," he states that as far back as 1858 he discovered distinct glacial scratches on an exposed layer of the upper Burlington limestone. During the past season he discovered other traces of this action on limestones of the Upper Coal-Measures in Mills County. In Nebraska these scratches have been observed at Omaha, on the Platte, and along the Missouri below the mouth of the Platte, and the inference is a just one that, if the great thickness of drift and yellow marl could be removed from the surface of the limestones, we should find that this glacial action was universal. This is undoubtedly the beginning of a series of events which continued up into the present period. I hope to be able to discuss this matter more fully in another part of this report. Through the great kindness of the Rev. J. G. Miller, of Plattsmouth, I was able to examine much more of this county than I could otherwise have done, and by his aid I secured all the essential points. On our way to Weeping Water from Plattsmouth we pass over a rolling prairie country underlaid with yellow marl, but having a surface soil of almost unexampled fertility. Near Eight-Mile Grove, at Mr. Austin's, we could ascertain, approximately, the thickness of the yellow marl at that distance from the river. A well dug 30 feet revealed no rocky layer on high ground and away from the main water-courses. The water is excellent. Near Mr. A.'s house there is a peat-bed that will undoubtedly be of some value when wrought. The peat appears to be two or three feet thick. There are several of these peat marshes in this neighborhood. In the valley of Weeping Water there is a bed of light-yellow limestone—a few layers compact and good for building purposes. It is mostly brittle, breaking into irregular fragments, with no true regular cleavage. A fine large mill and several houses are constructed of this limestone in the valley. The fossils are quite rare. The grassy slopes are so numerous everywhere that I found it difficult to secure a full section of the rocks.

7. Limestone, hard, whitish, and yellowish-white; cropping out at the summits of the hills, and lying on the slopes in large masses 8 to 10 feet thick.
6. Slope, 20 to 30 feet.
5. Layers of bluish limestone, brittle and cherty; full of flint, 6 to 8 inches thick, and breaking into square blocks 6 to 8 feet thick.
4. Slope, 20 to 30 feet.
3. Shelving limestone, yellow, not very useful for buildings; underneath this is a small bed of clay which forms a slope.
2. Yellowish-white, brittle limestone, with much sulphuret of iron in cavities.
1. Shale or black slate.

These limestones are often filled with the comminuted fragments of fossils, as corals, bryozoa, and stems of crinoids, and not unfrequently thick layers are made up of an aggregate of the *Fusulina*, so that when the rock disintegrates the surface looks sometimes as if it was covered to a considerable thickness with grains of wheat. The crinoids are extremely abundant in their fragmentary condition, but it is very seldom that a specimen can be found sufficiently perfect to be characterized. Near the mouth of Stone Creek, section 12, range 10, township 10, indications of coal have been observed. Mr. E. L. Reed, residing at the little village of Weeping Water, sunk a shaft through the following beds:

9. Sandstones, which form the bed of the creek, 10 feet.
8. Slate and clay, 3 feet.
7. Coal, 9 inches.
6. Whitish, fire clay, 3 feet.
5. Crystalline quartz, 3 inches.
4. Bluish clay, 4 feet.
3. Whitish, fire clay, 6 feet.
2. Red clay, 3 feet.
1. Soft white limestone, —.

The coal above, although so thin a seam as to render it unprofitable for working, is of good quality, and is useful to the blacksmiths in the vicinity. These alternate beds of limestones, sands, and clays give to the surface of the country bordering on the Weeping Water an unusually rugged character. The bottoms of the little streams are narrow, the soil is good, water excellent, and the valley is well settled and is prosperous.

Near the residence of Dr. Childs, about fourteen miles west of Plattsmouth, is a bed of red earth under the Cretaceous sandstones which looks like red ocher. There is along the Platte, in this county, connected with the Cretaceous sandstones, quite a series of the variegated sands and clays of the Dakota Group, which must, at some future time, be made very useful for economical purposes. Still at this time there is so much excellent limestone in this county that the wants of the settlers are fully supplied. There are many beautiful houses built of limestone. This rock seems to work easily into any desirable shape. I have estimated that the quarries of this county would supply the whole State with building material for many years to come; and where I examined the clays and sands they are quite inexhaustible. The demand for fuel will never be as good in this county as in some of the others farther west, on account of the supply of wood-land; but I would ask the attention of the farmers to the importance of peat as an article of fuel at some future time, or as an article of trade with the counties west. I am confident that a moderate supply at least can be found in this county, and that it will soon become an earnest object of pursuit and a source of great profit. But, like all the eastern portion of Nebraska, the great wealth of this county lies in its inexhaustible soil, which yields the most abundant crops. The wheat crops will average 30 to 35 bushels per acre; oats, 40 to 50; and corn, 60 to 75, while for grazing and stock-raising the whole country is unsurpassed by any portion of the globe. With all these facts before us, it is very easy to predict for eastern Nebraska a remarkable destiny in the future.

Otoe and Nemaha Counties.—The geology of all the counties bordering on the Missouri south of the Platte has been so ably and minutely discussed by Mr. Meek, in his report, that I shall allude to it only in a general way, although the geological structure of this district seems very simple. Yet, to trace out the details of the beds, so as to be satisfied with the perfect accuracy of the work, is very difficult. The Carboniferous rocks of Nebraska are made up of a series of layers of limestones, sandstones, sands, and clays of all colors and composition, and, unlike the rocky beds of nearly all the older Paleozoic periods, there seems to be no persistency in the lithology over even small areas. No two sections of strata not eighteen miles apart are alike, and the same bed changes apparently in short distances. If a very careful section of the beds, as revealed in some natural exposure along the Missouri River, is taken, it will fail to correspond with the results of a boring farther into the interior. Sections of the natural exposures at Plattsmouth, Nebraska City, Brownsville, Rulo, or White Cloud, on the Missouri, will fail to correspond, except in some few particulars, and the vast thickness of yellow marl, which conceals the underlying rocks over such a great extent, leaves considerable intervals continually, over which it is impossible to trace out the beds. These Upper Coal-Measure rocks seem to be as changeable in their lithological characters as those of the Tertiary period. What adds to the difficulty, also, is the fact that the same species of fossils, with few exceptions, run through all the beds. Many of the same species run through all the beds that are exposed by the Missouri, from Leavenworth City to Omaha. As has been shown by Mr. Meek, both these counties are underlaid by rocks of the Upper Coal-

Measures, or Permo-Carboniferous. The Cretaceous sandstones have not been observed on the western borders, and it is probable that the Cretaceous belt extends westward beyond the limits. If the western portions of these two counties could be stripped of this thick covering of marl, it is very probable that true Permo-Carboniferous beds, and, quite possibly, true Permian would be disclosed at some points. The efforts in search of coal in Nebraska have so far been unsuccessful, but so important does it seem to be to the inhabitants that a source of mineral fuel should be found somewhere within the limits of the State, that its existence or non-existence becomes the most important problem for solution connected with the survey. In succeeding portions of this report, I shall allude frequently to outcroppings and seams of coal in various portions of the State; but, inasmuch as so great interest has been taken in this question by gentlemen at Nebraska City, it will not be out of place to discuss the matter briefly now. Mr. Meek, in his report, has discussed the probabilities of the existence of a bed of coal within accessible distance of the surface within the limits of the State with his usual care, and he has given the details of the sections shown by the borings at different localities. His opinion seems to be unfavorable to success anywhere within the limits of the State. As far back as 1862, I published a paragraph in a memoir read before the American Philosophical Society in Philadelphia, referring to some indications of coal along the banks of the Platte, which had attracted the attention of settlers:

A seam of carbonaceous shale, 12 to 18 inches in thickness, crops out occasionally near the water's edge, and is regarded by the inhabitants as a sure indication of coal. The great scarcity of timber throughout this region would render such a discovery of the highest importance, but I am inclined to the opinion that it is a geological impossibility for a workable seam of coal to be found within the limits of the Territory of Nebraska. The limestones of southeastern Nebraska belong to the Upper Coal-Measures, and form the extreme northwestern rim of the great coal-basin, and, inasmuch as the strata dip toward the northwest, in ascending the river, at least one foot to the mile, there must be from 600 to 1,000 feet of clays, shales, and limestones over the first seam of coal two feet in thickness, in any part of the country near the mouth of the Platte. A bed of coal, to be really valuable for economical purposes, should be at least three feet in thickness, and even then it would not prove profitable if a large amount of labor were required in opening the mine.

It had always been my opinion that the Coal-Measure rocks of Nebraska were located on the western rim of the great coal basin of the West, and that a profitable bed of coal would never be found within a workable distance of the surface. During the survey many facts bearing on that point were secured which we had never known previously. The fact that the Union Pacific Railroad Company had bored 400 feet at Omaha; that Mr. Croxton had bored down nearly the same distance at Nebraska City, without passing through any important seam of coal, seemed discouraging. Now, if we go still lower down the river, where the rocks hold a position several hundred feet lower in the geological scale than at Nebraska City, at Atchison, St. Joseph, Leavenworth, and some other localities, we shall find that borings have been made 400 or 500 feet with no better success. Again, if we examine a small but valuable memoir on the geology of Northern Missouri, by Mr. Broadhead, we shall find that in this paper he gives a continuous section of the beds as they are seen in their natural exposures, of 2,000 feet or more in thickness, including the Upper Coal-Measures and a part or all the Middle Coal-Measures, and yet he does not find a seam of coal over 2 or 2½ feet in thickness. Another fact seems to me to possess considerable weight. In the valley of the Des Moines, in the Lower Coal-Measures, are found several beds of coal varying in thickness from 1 to 7 feet. The fact that these seams of coal were not more than one hundred or one hundred and thirty miles east of Nebraska City gave encouragement that the same seams would be penetrated at the latter place at a reasonable depth. But it is probable that in the western extension there is a thinning out of the beds of coal and shale and a thickening of the limestones, clays, and sands. It is now well known that the Carboniferous limestones disclosed by the elevation of the Rocky Mountains are of the same age as those along the Missouri in Nebraska, Iowa, and Kansas. We believe that these beds pass beneath the more recent deposits in Nebraska only to reappear again along the margins of the mountains, and we have no reason to suppose that the continuity of Carboniferous rocks from the Missouri River to the mountains is interrupted underneath the Tertiary, Cretaceous, or other deposits. Now all along the margins of the mountains, from the north line to the Arkansas, there are no black shales or indications of coal in the Carboniferous rocks that would deceive the most hopeful. The beds are for the most part massive limestones, or fine, compact sandstones, and the entire absence of irregular beds of sands, clays, seams of shale or coal, would seem to indicate that the Carboniferous rocks passed through these changes gradually in their extension westward. These facts point to the conclusion, as I have before mentioned, that Nebraska lies on the western border of the coal-basin, and however deep borings may be carried anywhere along the Missouri River no seams of coal over 2 or 2½ feet will ever be penetrated. As a matter of course it is not possible to decide this point positively except by actual boring, and it certainly would be a good plan to extend a boring down a distance of 1,000 or 1,200 feet to settle the question for all time to come. This disposition of the question in Nebraska settles it for a vast area along the Missouri River of the finest land in the west.

A large portion of Kansas, Northern Missouri, Iowa, Nebraska, and Dakota Territory will be found destitute of mineral fuel. Mr. J. Sterling Morton sunk a shaft on his farm near Nebraska City, and bored down about 100 feet more, but with no indications of coal. There is a seam of coal near the river at this place, which has been wrought by drifting in a distance of 300 yards, and the seam was about 8 inches in thickness, and several thousands of bushels of coal have been taken from it. At Otoe City, eight miles below Nebraska City, there is a bed of slate and coal about 8 inches thick, which has been worked to some extent, and the coal used in blacksmith shops. Again, at Brownsville there is a seam of coal accompanied by some of the plants peculiar to the Coal-Measures. There is from 4 to 6 inches of coal, and the whole bed, coal and slate, is about 12 inches in thickness. There is also a fine quarry of limestone at this point, which is of a very superior quality for building purposes, but it contains too much sand and clay to make a good article of lime. There is also a bed of fine-grained, micaceous sandstone, which cleaves naturally into excellent flagstones. Indeed, the rock quarries in this county are of great value to the inhabitants. The materials for making brick abound everywhere in this region; clays, sands, and marls are abundant and of excellent quality. At Aspinwall, in Nemaha County, we discovered the most favorable exhibition of coal yet observed in the State. The general dip of the beds seems to be up the Missouri, or nearly north or northwest. It is difficult to determine this point with precision. The rocks at Aspinwall are all geologically at a little lower horizon than the Nebraska City beds, and mostly beneath the Brownsville beds. Two seams of coal are met with at

Aspinwall; one crops out near the river, about 15 feet above the water, 24 inches in thickness—very good quality. A few feet above this seam is a second seam—six inches of good coal. Some English miners are sinking a shaft here, with full confidence that the thickest bed can be made profitable, and I am inclined to think that, with the present scarcity of fuel, they will succeed well. Coal commands a ready sale at from 40 cents to 80 cents per bushel; and even at 80 cents a bushel coal is cheaper than wood. The miners have already sunk the shaft about 40 feet, have passed through the 6-inch seam, and are confident of soon reaching the 24-inch bed, when the work of drifting in various directions will commence and the coal be taken out for market. The beds hold such a position here that, if these miners are successful, this effort determines the existence of a workable bed of coal for Nemaha, Richardson, Pawnee, and Johnson Counties, which will be a most important matter for the whole State. We have very abundant notes in detail, and many specimens to illustrate the geology of the river counties. I am informed that excellent hydraulic lime for cement exists in Nemaha County, section 9, township 6, range 14, but I have not been able yet to make a personal examination of the locality. In both counties there are many indications of extensive peat-beds, which must attract the attention due them before many years. Mr. McPherson, of Brownsville, informed me that just twelve miles west of that town he had observed an extensive area, which appeared to be a peat-bog, into which he thrust a pole to the depth of 10 or 12 feet. It is not to be supposed that in this dry climate any very extensive peat-bogs, like those in New England or other portions of the Atlantic coast, but a bog with peat 2 feet in depth even, would be of great value. The counties of Otoe, Nemaha, and Richardson contain more timber than any others in the State.

CHAPTER III.

RICHARDSON AND PAWNEE COUNTIES.

The two counties above named are in some respects the finest in the State. Forming as they do the southeastern boundary of the State they are located geographically in that portion in which the extreme cold does not have so powerful an influence. All the fruits and cereals peculiar to northern temperate zones can be raised here in perfection. Peaches, pears, apples and grapes do well; while north of the Platte it is doubtful even yet whether the cultivatlon of fruits is an entire success. In the northern parts of the State several kinds of fruits must fail on account of the severity of the climate during the winter months.

There is more woodland in Richardson County than in any other portion of the State, and on this account very little attention has been paid to the planting of trees. While there are many excellent farmers here and there, and the county is quite thickly settled, there is not that thrift that is found in Nemaha and Pawnee Counties. The greater part of the land, however, has been taken up by actual settlers, who are now devoting themselves to the improvement of their farms, and to the raising of large crops. There is a ready market for all kinds of produce at the highest price. Although nearly all the settlers came into the county poor, many without any money at all, nearly all are becoming moderately rich, and every man with industry and prudence may become independent in a few years. This country may certainly be called the poor man's paradise. There is scarcely a foot of land in the whole county that is not susceptible of cultivation. I have never known a region where there is so little waste land. The underlying rocks of the whole county belong to the age of the Upper Coal-measures, and are composed of alternate beds of limestones, sandstones, and clays of almost all colors, textures and compositions. There are several localities along the Missouri River and the larger streams where there are good natural exposures of the rocks, but as a rule the beds are concealed by the superficial covering of yellow marl or Loess, which gives the beautiful, undulating outline to the surface, the gentle slopes, with only now and then an exposure of the basis rocks. This aids in rendering the investigation of the geological structure of the county more complicated and difficult. The river counties present better exposures of the rocks than any other counties in the State, and it is partly on this account that I have given them my first attention. Even these exposures are by no means good.

In my last communication I spoke of the coal-seam at Aspinwall, Nemaha County; that about 16 feet above the water level of the Missouri a bed of coal 22 to 24 inches in thickness was observed cropping out from the bluff; and a few feet above this, in the same range of hills, was a second seam, 6 inches in thickness. These beds do not appear again for considerable distance down the river, until we come to Rulo, except at one or two localities near St. Stephens. At Arago I saw no outcroppings of coal at all, and could not hear that any had been observed; but there are some good quarries of limestone, beds of clay, sands, &c. The next marked exhibition of coal is at Rulo and its neighborhood, about two miles above Rulo, on land belonging to Mr. S. F. Nuckolls, of Nebraska City. At this locality Mr. N. has drifted into the bank 100 feet or more, and taken thence over 200 bushels of coal, which has been used by blacksmiths with success. The outcrop was about 5 inches in thickness, but increased. as the drift was extended in the bank, to 11 inches, and again suddenly diminished to 1 inch of good coal, the remainder being composed of impurities, or niddy coal, as the miner called it. The coal which has been thus far taken from this mine sells readily for 35 to 40 cents per bushel. The abrupt termination of the coal-seam, or "fault," is undoubtedly due to the sliding down toward the river of the superincumbent beds—a phenomenon which is very common everywhere along the Missouri. Still the irregularity in the thickness of this coal-seam is everywhere apparent, vibrating between 4 and 20 inches, thus alternately exalting and depressing the hopes and prospects of the miner. On the farm belonging to Mr. St. Louis, about one and one-fourth miles below Rulo, the same bed of coal has been worked with some success by drifting, and a considerable quantity of coal taken out. Mr. St. Louis unwisely sunk a shaft at a higher point on the hill, thinking to cut the coal-seam at a more favorable spot, the expense attending it, exhausting his means at 45 feet. He sunk a drill, however, into the bed of coal and found it 12 feet below the position at the outcrop, showing an extensive inclination of the beds from the river, or toward the west.

This dip may be readily accounted for by the extensive erosion of the rock prior to the deposition of the yellow marl and drift deposits, which erosion has given rise to many perplexing local inclinations of strata. These local dips will not interfere with the miner so much, farther in the interior of the county. The thickness of the coal-bed at this locality is 10 to 12 inches, increasing in one instance to 17 inches. On the Iowa reserve, along the Great Nemaha River, the same bed again crops out in the ravines or banks of little streams, and has been wrought with some success, several hundred bushels of the coal having been taken out from time to time for several years past. The country along the Nemaha is quite rugged, or "rough," as it is termed by the settlers, owing to the several beds of sandstone, and the overlying or cap rock of the coal-bed, which prevents the water from forming gentle slopes, as in the case of the more yielding clays or marl-beds. This bed of coal is probably the equivalent of the 2-foot bed seen at Aspinwall, while the upper 6-inch bed is not exposed at all. The rocks in contact with the coal are as follows:

Underlying the coal, a bed of light-gray fire-clay, full of fragments of plants, as fern-leaves, *Neuropterisc Loschei*, and *N. hirsuta*, stems of rushes, calamites, &c., the same as occur in the underlying clays in Ohio and Illinois coal-fields. Above the coal there is about four feet of very hard laminated or shaly clay, varying from black to dark-ash color, all of which must be removed with great labor before the bed of limestone, or cap rock, as it is called, can afford suitable protection to the miner as he drifts into the bank. Thus the small amount of coal is obtained with great labor, and it is only the great scarcity of fuel that will warrant any labor being expended upon it at all.

We passed over the almost treeless prairie, from Rulo to Falls City, the county seat, about nine miles distant. Some beds of limestone crop out from the hills occasionally, but usually all the basis rocks are concealed from view, and the surface is gently and beautifully undulating. The fertility of the soil is everywhere shown by the luxuriance of the crops. Falls City is located upon high ground overlooking the valley of the Nemaha. There is not a native shrub or tree of any size growing within a mile of the town. Although the same coal-bearing beds formed the underlying basis rocks about Falls City, yet not an outcropping of coal could be found in the vicinity. Some good quarries, however, were examined. Having heard that a boring had been made at Hiawatha, the county seat of Brown County, Kansas, ten miles south of Falls City, I visited that place to ascertain the result. I was informed that a company had bored near that place 240 feet without success, and that the project had been abandoned; and as the strata in all this region are very nearly horizontal, the same result would follow any attempt at boring at Falls City, to that depth at least. About nine miles southeast of Hiawatha a bed of coal is worked with considerable success, and many hundred bushels of coal are taken out of the mines and sold annually. Mr. Laycock, a lawyer at Hiawatha, informed me that during the past winter he used about 130 bushels of coal, for which he paid 50 cents per bushel; and he found it cheaper than wood, even at that price. He spoke highly of its qualities as fuel. I am disposed to believe that it is the same bed seen along the Missouri, in Nemaha and Richardson Counties, although I did not examine it in person. Continuing our course westward to Salem, we observed no marked change in the country; indeed, there is a remarkable uniformity in the character of the country over a large area. The changes that take place are usually the result of some change in the underlying geological formations, and are, therefore, quite gradual. No outcroppings of coal could be found at Salem or vicinity, and it is quite possible that none will be found exposed to the surface in that portion of the county, except along the Missouri River. I am convinced, however, that boring at a moderate depth at almost any point would penetrate the thin bed seen at Rulo. The quarries of limestone for building purposes, &c., are much finer at Salem than at any other point observed in the county. The town is located upon an elevation on the point of the wedge of land between the two forks of the Nemaha. Forming a part of the town site is a high hill, with two beds of limestone, both of which form large quarries, which yield an abundance of stone for all economical purposes. All along the Nemaha and its numerous branches are quite well-wooded tracts of land, which are held at a high price, though no portion of the country would be called well timbered in any of the States east of the Mississippi.

Although there is less wood-land in Pawnee County, it is equally fertile with Richardson, and the latter possesses only the geographical advantage of bordering on the great navigable river Missouri. Its surface is more rolling or undulating, the slopes are more gentle, and to the eye it seems even more desirable for farming purposes. Both counties are remarkably well drained by nature; so there is scarcely a foot of land in either that is not susceptible of cultivation. I cannot ascertain that one county produces better crops than the other. Pawnee County is remarkably well watered. The numerous branches of the North and South Nemaha, circulating all over it, render the land very attractive to the settler and speculator, who have already absorbed every acre of it. Both counties are underlaid by rocks of the Upper Coal-measures, which give a very great uniformity of character to their surface. The rocks are composed of alternate beds of clays, sandstones, and limestones, with some thin beds of coal. At Salem the succession of the beds is about as follows:

10. Yellowish-gray soft limestone, 3 feet.
9. Slope.
8. Very compact white limestone, brittle, about 12 inches thick, caps the hills.
7. Slope, 50 feet.
6. A very porous, yellow limestone, 2 feet.
5. A layer of laminated, impure, rather rotten limestone, dirty yellow, with *Spirifer planoconvexus*, *S. Kentuckensis*, *Pleurophorus occidentalis*, *Hemipronetes crassus*, *Productus semireticulatus*.
4. Dark, laminated, arenaceous clay, mostly fine sand, 5 feet.
3. A very irregular bed of laminated fine sandstone; sometimes the bed is 2 feet, gradually increasing to 4 feet.
2. Dark, ash-colored, indurated clay, unusually hard, 5 feet.
1. Bluish, very hard clay, becoming yellow on exposure, 2 feet above water.

Up one of the little branches of the Nemaha, in Salem, I found about 4 feet thickness of black shale, or slate, which was not exposed at the mill. The cap rock, above the shale, is full of fossils. In the shale the *Lingula*, probably *L. Scotica* and *Orbiculoidea*, occur in considerable quantities. On Contrary Creek there is a small exposure of the basis rocks of this region:

4. Yellowish laminated shale, with a thin layer of deep-yellow, fine clay, 4 feet.
3. Blue indurated clay, 3 feet.
2. Rather hard layer of limestone, 6 inches; *Chonetes granulifera*, *Productus Prattenanus*, *P. Nebrascensis*, *Spirifer planoconvexus*.
1. Yellow indurated clay.

On Isaac Trigg's farm, seven miles west of Salem, I obtained some fossils from a well. The surface was about 20 feet above the Nemaha, and dry. Twenty-five feet before reaching water, near the bottom of the well, is a bed of shale, in which were species of *Myalina* and *Aviculopecten*.

At Miles's ranche, about 20 feet above the bed of the Nemaha, is a dark-bluish layer of fine, argillaceous limestone, which cleaves easily but makes good stone for buildings, about 10 inches thick. Above is a layer of yellow shale; then limestone, very brittle, with *Productus*, *Myalina subquadrata*, &c. The succession of the beds is thus shown:

8. Bluish, close-grained, argillaceous limestone.
7. Slope.
6. Gray arenaceous limestone, fine-grained fragments of organic remains or shells, and fish teeth, 4 feet.
5. Dark, ash-colored, laminated, indurated clay, somewhat variable in character, 30 feet.
4. Impure fibrous gypsum, and fine, light-gray, indurated, slaty clay, 4 inches.
3. Bluish indurated clay shale, 11 inches.
2. Light, ash-colored clay slate, 12 inches.
1. Bluish-black indurated clay, 6 feet above water.

Bed 6 of the above section would yield most excellent building-stone, and may be quarried out in immense blocks. All these beds seem to change their character over short areas. This bed has much the appearance of one seen at Plattsmouth, though the evidence is not positive. The hills, for 200 feet in height in this region, have little layers of limestone cropping out. The whole country along the Nemaha is much broken—more than usual—yet the fertility of the soil is very great. On Mr. Wheeler's farm, section 31, township 2, range 14, the basis rocks crop out again: 1st, shale; 2d, yellow clay, 2 feet; 3d, impure limestone, intercalated with clay, 3 feet; 4th, blue clay, 6 feet. In the third layer occur shales with *Aviculopecten*, *Productus*, *Spirifer planoconvexus*, &c. The valley of the Nemaha seems nowhere in this region to be destitute of good building-stone. As we proceed westward, even as far as Salem, the yellow marl, which is so conspicuous a feature along the Missouri, begins to grow thinner, so that in digging wells the basis rock is soon reached. This marl is somewhat unequally distributed over the country, but where it is found it renders the soil very productive. The bottom lands of the Nemaha are quite broad here, covered with a luxuriant growth of grass yielding two or three tons of hay to the acre. Water is reached in wells, at a depth of 15 to 25 feet, without passing through the alluvial soil. On Four-mile Creek, ten miles southwest of Salem, a bed of rock crops out, which is the same as the upper bed at Salem, and might be called fusulina limestone. It caps the hills on section 12, township 1, range 13. Other fossils occur, as *Aviculopecten*, *Spirifer planoconvexus*, *Pleurophorus occidentalis*, &c.

On Turner's Branch, a small stream flowing into the Nemaha, there is a seam of coal that has been worked with much labor. The principal drifts are on school section, township 1, range 12, one and a-half mile northeast of Freize's mills:

4. Massive yellow limestone; upper part cleaving.
3. Dark, ash-colored laminated clay, with two or three black bands, 2 to 4 inches thick. The whole mass is filled with shells—*Productus Nebrascensis*, *P. longispinus*, *Chonetes granulifera*, *Spirifer planoconvexus*, *Retzia punctulifera*, *Spriferina Kentuckensis*, *Hemipronites crassus*, *Bryozoa*, &c.
2. Coal, 4 to 16 inches thick, very variable in thickness and quality.
1. Yellow plastic clay, passing up into a hard blue clay, upon which the coal lies as if pressed down, 20 feet thick.

No rocks below bed 1 are seen in this immediate vicinity. The coal seemed to be packed closely down on to the clay beneath, like masses of flat rock, as if it had been originally deposited there like a layer of clay or sand. The clay below is quite hard and filled with fragments of fern leaves, stems of the rush-like calamites, like the clay underneath the coal seams in Ohio or Pennsylvania. The under surface of the coal seems to be composed of stems, like grasses, as if the vegetable débris began upon a densely grass-covered surface. The vegetable impressions do not go down into the clay more than an inch or two, and above the seam, where the coal ceases, all traces of vegetable matter disappear, and the clay is charged with a variety of molluscous remains. The clay above the coal is very hard, and yields with difficulty to the pick, and the coal is extracted with great labor. Several hundred bushels have been taken out and sold, and the bank of the creek reveals fifteen or twenty openings like that shown by the illustration. This shows the coal seam at the base, the bed of indurated clay above, which is generally 3 to 4 feet thick, all of which has to be removed, and the heavy-bedded limestone forms an excellent cap-rock above. At Frieze's Mill, still further on, this same bed of coal is again wrought with some success.

On Mr. Boston's farm, township 1, range 12, section 34, several openings have been made; and here the coal seam increases in thickness to 16 inches. Mr. B. has taken out 900 bushels of coal here. He finds a ready market for it at the mine at 30 cents per bushel. This coal seam averages a bushel of coal to a square foot of surface. I have collected abundant specimens of this coal at different localities, and they will be properly investigated at some future time.

This seam is also worked on Lee's Branch and on Miner's Creek, so that it is now wrought, more or less, over an area of ten miles square, at least. The coal seems to have been worked with more system, industry, and success than in any other portion of the State.

Near Pawnee City there is another small seam of coal holding a higher geological position, which has attracted some attention. I made a careful examination of all the localities, and found it not more than 4 inches in thickness generally. On Mr. Jordan's farm, at the water-level of Turkey Creek, a branch of South Nemaha, this seam increased to 8 inches, but is so impure and full of sulphuret of iron as to be quite unfit for use. Near Pawnee City, on a little branch of Turkey Creek, about 300 yards south of the city line, the beds succeed each other as follows:

5. Yellow, fine-grained sandstone, 20 feet.
4. A close-grained, bluish, ash-colored argillaceous limestone.
3. Two to four inches of coal, with large masses of sulphuret of iron.
2. Plastic fire-clay, nearly white, passing down into yellow clay and sand, 12 inches.
1. Yellow, arenaceous laminated clay, mostly sand, and quite firm in some parts.

At numerous localities in this county sections like those that are given above might be written, but as they teach nothing more than this, that the rocks all belong to the Upper Coal-measures, and that they are very variable in their character, it is not worth while to report them. We learn, also, that the seams of coal, so far as they have been opened, are very thin, and, although we suppose that the seam on Turner's Branch and at Frieze's Mill is the same as the one exposed at Pawnee City, the long intervals over which the rocks are concealed by superficial marl renders it impossible to trace all these beds out in their continuity, but as the inclination of the strata seems to be west or northwest, higher and higher beds must be constantly making their appearance. Throughout this county the uniformity in the specific character of the fossils running through all the beds is another obstacle. If certain species were restricted to certain beds we could then trace them with more certainty; but our hasty examination always reveals from any one bed, or series of beds, certain fossils identical in species; and if at any one locality any species seem to be wanting that occur at another, a closer examination generally reveals them all. It is somewhat doubtful whether in the whole of Pawnee County a single species of shell was found that is not also found along the bluffs of the Missouri. In this county, however, several beds of limestone make their appearance on the high lands which have not been seen east, showing the direction of the dip. On a piece of land belonging to Governor Butler there is the best quarry of limestone I have seen in the county. The bed crops out near the edge of a hill bordering a small stream about eight miles west of Pawnee City. It is soft, cream-colored rock, full of small cavities, by the decaying out of a small shell. *Fusulina cylindrica*, *Spirifer cametus* also occurs. It is a true fusulina limestone, and is a great favorite with masons for building purposes. It is easily wrought into any desirable shape, is very tenacious in texture, and durable. It seems to hold a position about 100 feet above the water level of Turkey Creek, and belongs to the age of the "Permo Carboniferous," or intermediate between the Upper Coal-measures and the Permian series, the general inclination of the beds being toward the west and northwest. New and more recent beds are continually making their appearance as we proceed toward the west, and this choice bed of limestone has made its appearance here for the first time. It will doubtless be found to extend over considerable area in a southeasterly direction. There is still another bed of bluish limestone cropping out of the hills, which, though useful, is not regarded with the favor bestowed on that just mentioned. It does not dress as nicely, is not as handsome for caps or sills. It is equally durable with the other. There are several beds in the county which are employed, to a greater or less extent, for various economical purposes.

About three miles north of Pawnee City, about 50 feet above a little branch of the North Nemaha, is a ledge of yellowish-gray limestone cropping out of the hills, filled with a very large variety of the *Fusulina*. It is much used for building purposes. At another locality in the same region, at about the same level, is another quarry in which the rocky layers seem to be made up of an aggregate of small masses, apparently organic and like *Fusulina*. The bed is 4 to 6 feet in thickness, and although the rock has a loose, porous texture, it is very tenacious, and is very useful to the settlers. This bed, though holding nearly the same geological position as that mentioned above as occurring on Governor Butler's farm, is, I think, underneath, and both belong to the series designated in this report as *Permo-Carboniferous*. As we proceed farther west of this point we shall find the beds more and more recent and approaching gradually the character and texture of the true Permian rocks, as shown near Junction City, in Kansas. Peat-beds occur in various portions of the county, sufficient, I think, to attract attention at some future time. Near Pawnee City there is a small peat-bog, on which one may stand and jar the ground for considerable distance around. I estimated the peat here at 10 or 12 feet in thickness. These low wet places are covered with water a large portion of the year, and give rise to a luxuriant growth of the large rushes and reed grasses of the country. This vegetable dies down every year; from its decay grows a still more luxuriant growth, which furnishes an abode for the muskrat, which builds its conical houses. Water is abundant all over the county. Scarcely a section of land can be found in the two eastern tiers of counties without a running stream or a flowing spring. Water is obtained everywhere, by digging, at moderate depths. Near the little streams it is only necessary to go down a little below the river bed and the best of water flows in abundantly, and on the high land water is usually found at a moderate depth. Nearly all the wells have a continual supply from 6 to 10 feet. In the limestone region the water is usually hard on account of the carbonate of lime, but it is cool and clear as crystal and very pleasant to the taste. I cannot well conceive of a healthier region either for man or beast. As a grazing country Nebraska presents unusual facilities, and the time must come when it will be unsurpassed by any State in the Union. Having described most of the counties underlaid by the Upper Carboniferous rocks, I might say a few words here in regard to the possible outline of the surface of them prior to the deposition of the yellow marl. We see, all along the Missouri River, that this comparatively recent deposit attained there a great thickness, varying at different localities from 50 to 150 feet. As we go westward from the river this deposit gradually grows thinner until it finally disappears. The question now arises as to the influence it had on the character of the scenery of this region. Even if this yellow marl did not exist here, there must have been some superficial deposit of drift or alluvial to conceal, in part at least, the underlying or basis rocks, and the character of the surface would depend somewhat on the thickness and compactness of this superficial material. We could safely infer, from the evidence we have obtained, that prior to the deposition of the yellow marl the outline of the surface underlaid with the Upper Coal-measure rocks was exceedingly rugged. Sometimes the foundation of a house dug on a high hill will rest upon the true limestones; again, a well dug will pass through the clays, sands, or limestones of the basis formations. Then, in other localities, wells are dug from 30 to 100 feet through the yellow marl without ever reaching the bottom of the alluvial or drift material; but it is especially along the bottoms of streams that the superficial deposits appear to be so deep. If we can imagine the superficial materials entirely removed from the county, the surface, as it seems to me, would be rugged in the extreme. The valleys of the streams would be much deeper and wider, and the massive piles of limestones would stand around here and there, scattered over the surface like gigantic ruins. Evidences of glacial action also would be visible, as well as striking examples of erosive action—numerous valleys with almost perpendicular rocky sides, high conical hills capped with ledges of limestones; indeed, so rugged would be the surface that it would be quite uninhabitable. The distribution over it of the thick deposit of soft-yielding marl has softened down all the slopes, so that not only is nearly every foot of land susceptible of cultivation, but the drainage is complete. To these causes are due the almost unparalleled agricultural resources. The source of all this material may be somewhat difficult to determine, but in another portion of this report I will bring together all the information that I can secure in relation to it.

CHAPTER IV.

GAGE AND JEFFERSON COUNTIES.

Gage County.—After having examined Pawnee County with considerable care, with the kind aid of many excellent friends, we took a course nearly southeast across the open, high prairie, passing over the divide between the valley of the Nemaha and that of the Big Blue. Very few exposures of rock were to be seen, the surface is rolling, covered with a luxuriant growth of grass that would yield two or three tons to the acre. Here I saw the first long stretch of treeless, waterless prairie, reminding one somewhat of the prairies farther west. There was no living water, and not a house along the road for several miles. The soil, however, is extremely rich, a thick deposit of yellow marl covering the whole surface. The compass plant, *Silphium laciniatum*, is still seen in great abundance on highland and lowland, attesting by its presence the fertility of the soil. About seven miles northeast of the Otoe reservation we see the first outcrop of limestone from the hills, forming a sort of terrace about 50 feet above the beds of the streams. The gentle slopes and the entire absence of outcrops of rock over so long a distance from Pawnee City, are doubtless due to beds of soft yielding clay and sand prevailing, which readily yield to atmospheric agencies; while the beds of limestone, yielding less readily, form a sort of cap or floor protecting the softer beds below. This border rock always gives to the surface a more abruptly rugged character, the little branches have steeper banks, and there is greater variety to the scenery. At the Otoe agency this same bed of limestone is again exposed. It is quite cherty, breaking into small fragments. There are one or two layers, 6 to 12 inches in thickness, that would furnish good materials for building purposes. The following section was obtained at several localities within two miles of the agency buildings, and gives a fair view of the succession of the beds on the Indian reservation:

7. Superficial deposits of soil and yellow marl.
6. Yellowish-white limestone, rather soft, yielding readily to atmospheric influences, 2 feet.
5. Slope same as No. 3, below.
4. Yellow, fine-grained, arenaceous limestone, 18 inches.
3. Slope supposed to be laminated clay, but covered with grass, 20 feet.
2. Yellow and gray limestone filled with seams and nodules of schist or flint. A large species of *Orbiculoidea* occurs here.
1. Bluish-gray, laminated, calcareous clay, with numerous fragments of fossils, crinoids, corals—*Productus semi-reticulatus*, *Meekella striato-costata*, &c., 30 feet above water's edge.

The Otoe reserve is located on the Big Blue River, mostly in the southern portion of Gage County, but extending into Jefferson County. It occupies a surface 10 by 24 square miles = 153,600 acres of the finest land in Southern Nebraska. The Big Blue, one of the most beautiful of the inland streams, with several of its most important branches, passes through it. Like all other portions of the State, there is comparatively little timber, yet as much as on other streams. Some of the branches have the most desirable farms bordering on them. The Otoe Indians occupy a small village bordering on the Blue and are not distributed over the reserve. The land is not divided out to them, but they are all aggregated together in a village of mud huts. They seem to have no idea of individual independence, but have all things in common, as it were. The bed of cherty limestone extends beyond Blue Spring and forms the same bluff-like bench along all the streams; it then passes beneath the water level of the Blue. Near Blue Spring this bed presents much the appearance of mason-work, the cherty material forming the cement between the blocks of limestone. These bluffs are about 10 or 15 feet high, and as they are cut through by the myriads of little streams pouring down from the hills, they present a rather rugged appearance. A few fossils were obtained here, as *Athyris subtilita*, *Hemipronetes crassus*, *Syntrilasma hemiplicata*, with bryozoa and crinoidal remains. Near the ford at Blue Spring there is an excellent mill site, and the settlers here have quarried the rock down to the water's edge, for the purpose of building a dam and laying the foundations of a mill. The succession of the beds here is as follows. All of them are below the cherty bed spoken of:

3. About 10 inches soil, 8 feet yellow marl, 2 feet worn pebbles and sand. Roots of trees pass all through this bed into the joints of the limestone.
2. Layers of cherty nodules, variable in thickness, with intercalations of fine gray sand, 2 to 2½ feet, *Productus semireticulatus*, *Hemipronetes crassus.*
1. Bluish, ash-colored argillaceous limestone, easily decomposing in water, but hardening on exposure.

This bed is made up of layers varying from a few inches to two feet in thickness, separated by thin layers of clay. It makes quite good building-stone—*Productus*, *Pinna peracuta*, *Orbiculoidea*, *Myalina subquadrata*. Underneath the argillaceous limestone there is a bed, about a foot above water's edge, of ash-colored clay, breaking into small angular fragments, containing in it irregular seams of agillaceous limestone. Beds similar to these are seen on the Nemaha, but it is impossible to say that they are identical. It is probable that they are not, but simply a portion of the series coming to view continually as we proceed eastward. On the west side of the Blue River the slopes are all very gentle, the ascent being very gradual as far back as the eye can reach. There is no sign of the cherty bed which causes the bluffs and terrace on the east side. The bottoms are everywhere very rich and black, but the hills are covered with a heavy deposit of yellow marl, so deep yellow that I suspect it to be composed mostly of the eroded materials of the Cretaceous and White River Tertiary formations. On the bottom of the Blue Mr. Tylor dug a well 25 feet deep, reaching water near the level of the bottom of the river. At the Blue Spring Village a well was dug on high ground 55 feet deep, through clays and quicksands, without reaching the basis rocks. At a depth of 54 feet a large bone was found, which probably belonged to the mastodon. The water came in in great quantities, and is of the finest quality. On John Hagen's farm a well was dug 44½ feet through alluvial marl and gravel. The water rushed in with great power, and is now 8 feet deep. The wells show the depth of the superficial deposits, and we are enabled to ascertain to some extent their influence on the county. We know that in the valley of the Big Blue the soil is inexhaustible, and that the water is of the best quality, and abundant everywhere. About five miles west of Blue Spring, on the road to Beatrice, there are layers of yellow limestone, that are used for building purposes. Two miles farther there is a bed of yellow magnesia limestone, with *Aviculopecten*, *Syntrilasma hemiplicata.* The rock is 15 to 20 feet thick, arranged in layers 6 inches to 2 feet in thickness, and is full of geode-like cavities, with the same white crystalline lining "calc-spar" inside as seen at Blue Spring. Beatrice, the county seat of Gage County, is very pleasantly located on the Big Blue, with many natural advantages for becoming a prosperous town. The inhabitants are energetic and thriving; the land is fertile, and an abundance of excellent building material exists everywhere; the water is pure and abundant, and the climate is very healthy. On a little branch of the Big Blue there is an exposure of the rocks favorable for a section.

6. Dark-brown ferruginous sandstones of variable color and texture; used for buildings; contains many deciduous leaves; 50 to 60 feet.
5. Yellowish-gray sandstone; soft, easily crumbling, and wearing away exposed; on Blakely's Run, two miles west of Beatrice; 30 to 50 feet.
4. Slope in most places, but composed of variegated clays of doubtful age; potter's clay; 40 to 50 feet.
3. Loose layers of yellow limestone, full of geode cavities; porous, spongy. *Syntrilasma hemiplicata*, *Pinna peracuta*.
2. Yellow, rather compact limestone; good for building purposes; 2 to 2½ feet.
1. Dark-gray argillaceous limestone, becoming light gray on exposure, filled with geodes, with cavities full of crystals of carbonate of lime. This bed is at times massive, heavy-bedded limestone, of a beautiful cream-color; 10 feet.

Beds 1, 2, and 3 of the above section are undoubtedly of Permian or Permo-Carboniferous age, though they contain fossils common to both Permian and Carboniferous rocks.

Bed 4 is of doubtful age. Beds 5 and 6 are exceedingly interesting in a geological point of view, from the fact that they represent a new geological formation not before seen east of this point.

Bed 4 seems to form a sort of transition bed between the Permian* and Carboniferous formations. The Permian rocks pass beneath the water level at Beatrice, westward, and over a belt ten to fifteen miles wide, in a northeast and southwest direction, the brown sandstones prevail to the exclusion of all other rocks.

I should think that the Cretaceous sandstone in this region was 30 to 50 feet thick, though from the sliding down of the rocks it was impossible to tell with accuracy. The sandstone is very variable; sometimes coarse, friable, dark-rust color; sometimes in flat masses, with the appearance of pot metal; then a laminated sandstone. Underneath the rusty sandstone is a friable bed which will afford an abundance of good coarse sand. When unmixed with the drift gravel this sand is pure and beautiful. In the sandstone are many impressions of dicotyledonous leaves. I did not see the sandstone at all on the east side of the Platte, but on the west side it crops out of the high, thin-soiled hills. In the deep ravines the abrupt cliffs of limestone are seen. Between the rusty sandstone and the limestone there is a considerable thickness of fine sands and clays, which everywhere wear down in gentle grassy slopes, concealing the junction of the two formations. Therefore I was not able to find them in apposition.

The west side of the Big Blue is a little hilly, but the east side, between Indian Creek and the Blue, is very level and gently rolling, as if the underlying rock was of a soft material, and had been worn down over a large surface nearly alike. Gage County, so far as I have seen it, begins to reveal some of the indications that show we are on the borders of the dry western belt. The long intervals without living water, the peculiar flowering plants, and some of the prairie grasses, and the absence of trees, even the small shrubs, show that we are verging on the dry prairie. The *Schrankea uncinata*, sensitive plant, is seen here for the first time.

The soil of Gage County does not equal that of Pawnee County, or

---

* It is not certain that the true Permian beds, as recognized in Kansas, extend northward into Nebraska, though thin beds may occur in some of the southern counties.

the counties along the Missouri, as a whole. The bottom-lands are excellent, but the upland soil is thin. The grass is less luxuriant and the timber along the streams less abundant. For wheat, however, this soil, composed as it is largely of the eroded materials of the Cretaceous sandstones, contains a large amount of silica, and seems to be most favorable. A bushel weighs more than that of the river counties, but the corn and other kinds of grain are not quite so good. Yet too much cannot be said in favor of Gage County as an agricultural and grazing region. No coal will ever be found there, and the sooner the farmers commence planting trees the more prosperous they will be. Comparatively little peat will be found in the county, so that the question of fuel must be determined by the intelligence and industry of the people. If they plant trees now, they cannot suffer for fuel, for before that which they now have is gone, the planted forests will be ready for use. In regard to fruits, garden vegetables, &c., the same may be said of Gage County as of the other counties before described. Success will attend all well-directed efforts that way. The excellence of the water in springs and wells in this county is a most important feature, in a sanitary point of view.

There are no minerals that can be worked to advantage in this portion of the State. In the Cretaceous rocks there are large masses of limonite, (hydrated sesquioxide of iron,) but they are so full of siliceous matter that they can never be of much value. Even if there was an abundance of iron in this county there is no fuel to prepare it for use.

Jefferson County.—From Beatrice to Rock Creek, a distance of twenty-two miles, we passed over an open prairie without wood or water. Indeed, as we go westward there are whole townships of land without a tree or a particle of running water. The grass is good, and though the soil is thin on the high hills, the lower slopes and bottoms are as fertile as ever. Wheat, oats, corn, and all the cereals grow well. Water can be obtained at moderate depths, for the drainage is not good. There are miles of level surface without a ravine or channel to receive the water, and therefore it must quietly sink through the superficial beds to some reservoir of clay. It is plain that a little above Beatrice, on the Big Blue, No. 1 Cretaceous comes down to the water level by a rapid dip, and that the Permian and Carboniferous beds cease to appear in their westward extension.

On Rock Creek, a small branch of the Little Blue River, we find the rusty sandstones of the Dakota Group largely developed, and the readiness with which they yield to the erosive action of water has given to all the ravines of the Little Blue a very rugged appearance. The rocks also are so porous that the water readily permeates them, so that there is comparatively little in the ravines a large portion of the year. Still, in the valley of the Little Blue there are some as fine springs as are to be found in the State, but they are somewhat rare. A section in descending order along the Little Blue, below the Big Sandy, would be as follows:

5. Yellow and dark-brown rust-colored sandstones of the Cretaceous or Dakota Group, so well known in many other portions of the West. A few dicotyledonous leaves were found. This bed is of irregular thickness; from 50 to 100 feet.
4. Moderately-coarse, yellowish-white sand, with irregular laminæ of deposition; 50 feet.
3. Dark-colored, arenaceous, laminated clays, with particles and seams of carbonaceous matter. All through are beds of carbonaceous clay,

18 inches to 3 feet thick; much sulphuret of iron and silicified wood; 30 to 50 feet.

2. Variegated arenaceous clays; the slopes exposed are so great that I cannot give the exact thickness; probably 50 to 70 feet; some seams of excellent potter's clay.
1. Dark bluish shaly clay, upon which the foundation of Mr. Jenkins's mill rests. It is undoubtedly Permian or Permo-Carboniferous, but is not exposed to view by natural excavations until we reach a point south of the Nebraska line near Marysville, Kansas.

Beds 1 and 2 of the above section are seen at Mr. Jenkins's mill. Bed 2 is also shown, two miles above, along the bottom hills of the Little Blue. The variegated clays come under the white, incoherent sandstone. The dark band in bed No. 3 has been regarded by the settlers with a good deal of interest, as indicating the proximity of a workable bed of coal. I gave all the exposures a careful examination and found them of no possible value. At Jenkins's Mill there is a very fine spring of pure water flowing out just under the sandstone. The water, as it percolates readily through the porous sandstone, reveals the hard clay bed and remains upon this almost impermeable surface. I am convinced that the black hard shales upon which the mill rests belong to the Permian, and that the variegated sands and clays are intermediate between the sandstone and the true Permian. The excavations are going on now below the bed of the river, and Mr. Jenkins informs me that the shale becomes harder as he proceeds downward. The Kansas line is only about one and one-half miles below this mill, and about two miles below on the Blue, the yellow limestones appear. Not more than fifty or sixty miles south of this point, at Junction City, Kansas, the Permian rocks are well shown, and we would reasonably expect they would extend northward into Nebraska in the valley of the Little Blue. On Rock Creek there is not water enough in wells a portion of the year for culinary purposes, and wells have been dug in that neighborhood sixty feet in depth without reaching any water. In the intermediate variegated sand and clay beds, which I regard as belonging to the Dakota Group, just below the mill, is a locality where there is a band of laminated clay and sand, with seams of coaly matter, petrified wood, sulphuret of iron, selenite, &c. The whole bed looks much like an ash-heap. There are many fragments of wood, much of it as light as dry wood; others seem to be simply charred. This bed is also full of bits of arenaceous rock, thin layers of rust-colored, fine-grained argillaceous concretions, full of oxide of iron, all of which give to the bed a singularly sterile or barren appearance. On the west side of the Little Blue, on the high hills, there is an extensive quarry of the gray sandstone in the form of broad, flat masses, which will work into good building-stone. In the absence of better stone, this becomes very useful to the settlers. On Rose Creek, just above its entrance into the Little Blue, the variegated bed, wood and pyrites bed, and the incoherent sand bed, all come together. On Coal Creek, a little branch of Rose Creek, there is a perpendicular bluff showing the following succession of beds:

4. Fine loose sand, varying in color from white to a dark rusty red. In this bed are some thin layers of harder rock, 30 feet.
3. Black carbonaceous clay filled with bits of charcoal, clay, &c., 18 inches, and passing up into 2 to 4 inches of yellow clay mixed with bits of coal.
2. White fine clay, 2 feet.
1. Variegated clay, 20 feet.

Bed 4 is very irregular in its composition; sometimes there are 3 or 4 feet in thickness of very fine, whitish, slightly-arenaceous clay, with waved laminæ and sesquioxide of iron; then again the sand-bed is almost pure white sand. Just above bed 3, in the section, there is a thin layer, one-fourth to one inch thick, which has the appearance of having been baked from below, caused by the ignition of the lignite beneath. On the way up to the Big Sandy, for twelve miles or more, these clays and sands are seen occasionally with the red rocks. On the summits of the hills, but above the mouth of Rock Creek, the hills are not rugged, and the sandstone does not form any bold cliffs. About two miles below the mouth of the Big Sandy, No. 3 is seen on the summits of the hills, not in ledges, but in outcrops of loose slabs of limestone. These slabs or large flat masses are really only an aggregate of the shells of *Inoceramus problematicus.* South of this point there is no limestone to be found in the State along the valley of the Little Blue, and, therefore, it becomes a matter of some importance; it is much used for lime by the settlers. As we proceed up the valley of the Little Blue, the bottoms become more sandy and gravelly, and the vegetation of both lowland and highland exhibits a marked change. The basis of the alluvial material is the eroded materials of No. 1. In the hills are some naked patches of flesh-colored marl, and over the surface are scattered very thickly, water-worn pebbles and some quite large quartzite granitic boulders. The entire features of the country exhibit a distinct approach to the arid belt. That we are verging upon it there is not a doubt. Yet a finer country for stock cannot be found in the world. Horses, cattle, and sheep would thrive on the short nutritious grasses. The numerous broken masses of rock and the drift pebbles among the hills would be favorable for the sheep. For good farms a suitable quantity of good level land can be found, but the successful farmer must combine stock-raising with agriculture. Wheat grows here in great perfection, while oats, corn, and all kinds of garden vegetables do well. At the entrance of the Big Sandy there is a very wide bottom, at least three miles long and one mile wide, with soil from 6 to 10 inches deep. The Big Sandy is almost twelve yards wide where it enters the Little Blue, and is a stream of fine clear water. As we go westward from this point the limestones of No. 3 take possession of the country. On the high hills the limestones of No. 3 crop out in great abundance, and they everywhere seem to be mostly made up of *Inoceramus.* Lower is a bed of laminated shale which seems to pass gradually down into Cretaceous No. 2, though the line of demarcation between No. 3 and No. 2 cannot be seen. It is only the upper portion that so strongly resembles No. 3, as seen on the Upper Missouri. Below it is a gray shale, with some sand between the thin sheets of clay or limestone, but the *Inoceramus problematicus*, which everywhere testifies to the presence of No. 3, is seen either entire or in fragments on the bluffs around. At one locality about three miles above the mouth of Big Sandy, on the Little Blue, section 20, township 3, range 1, west, there is an exposure of the shale 40 to 50 feet high, showing the slaty character perfectly, with slabs containing the *Inoceramus.* There are also seams of gypsum, colored with the oxide of iron, and crystals of the carbonate of lime. The saw-mill near this point rests upon a conglomerate which precisely resembles portions of No. 2, as shown along the Missouri, and in these conglomerate masses here, there are traces of *Inoceramus* and fragments of wood in great abundance.

One important feature in the valley of the Little Blue, are the terraces, which are quite marked north of the Nebraska line. There is really but one, though sometimes there is a low one just above the immediate bed of the river, but the principal terrace is 25 to 50 feet above the water level. From that terrace the ascent is gradual to the summit of the high lands.

Leaving the valley of the Little Blue we took a course nearly northeast over the divide between Big Sandy and Swan Creek, a distance of about twenty miles, most of the way without water or wood. The soil is fertile and the surface is covered with a thick growth of grass. The valley of Swan Creek, a branch of Turkey Creek, which is a branch of the Big Blue, is a beautiful fertile one, with a large number of most productive farms. This region is settled with a large number of thrifty Germans, who are cultivating the land very successfully. But very little stone of any kind is seen in place in this valley. Yet the limestones of No. 3 furnish the materials for the foundations of buildings. The eroded materials of the limestones of No. 3 mixing with the sands of No. 1 make a most excellent soil. There is no rough land in this region; the bottoms are broad and the ascents on either side are very gradual, so that the hills bordering on the creek seem quite inconspicuous. The bottoms of Swan Creek will average three miles in length. All the basis rocks on the high lands are concealed from view by a large deposit of alluvial, which is composed of eroded materials from the basis rocks, with some water-worn pebbles; but strewed over the surface in many places are moderately-large blocks of quartzite, evidently deposited there at a comparatively modern period. Many of the hills are paved with pebbles as if they had formed the bottoms of small streams. The junction of Swan and Turkey Creeks with the Little Blue, being near each other, makes a broad valley which is now a fine farming region. Near this point there are the remains of an old Indian village. Fragments of pottery have been dug up three feet beneath the surface. I notice everywhere in the ploughed land that small pebbles are mixed with the soil, which is not the case in the two tiers of counties next to the Missouri River.

As I shall hereafter have occasion to refer often to the divisions of the Cretaceous group throughout this report, they will be better understood if defined more clearly in this connection. The Cretaceous rocks, as developed along the Missouri River, exhibit five quite well-marked divisions, which are characterized by peculiar fossils for the most part.

In a paper published in the proceedings of the Academy of Natural Sciences of Philadelphia, December, 1860, was given a general section of the Cretaceous rocks of the Northwest; long prior to that time Professor Hall and Mr. Meek had published the divisions by the numbers 1, 2, 3, 4, 5, but afterward Mr. Meek and the writer gave to them special geographical names. The sandstones which we have referred to in this report we denominated the Dakota Group or Formation No. 1, because these rocks were then supposed to reach their largest development along the Missouri River, near Dakota Territory. Formation No. 2 was called the Fort Benton Group, from the fact that it occurred in greatest thickness around Fort Benton, near the sources of the Missouri River. Formation No. 3 was named the Niobrara Division, from the conspicuous thickness of the bed near the mouth of the Niobrara River. These divisions constitute the lower series of Cretaceous rocks in the West, and are supposed to be the equivalent of the lower or gray chalk and upper greensand, by geologists. Formation No. 4 we called the Fort Pierre Group, because it reaches its greatest thickness near this point along the Missouri River. Formation No. 5 was called the Fox Hills Beds, from the fact that they form a conspicuous range of hills between the Big Cheyenne and Missouri Rivers. These two groups constitute the Upper Cretaceous series, and though they are quite distinct in their lithological characters, a large number of the species are common to both groups, so that the line of separation can only be determined by the mineral nature of the beds. What I have said above is sufficient to render all references to these divisions clear to the reader, and they will be discussed more in detail in another part of this report. We find that as we proceed westward from Pawnee County, and in portions of that county, more and more recent formations come to view. We find that in Jefferson County Formations 1, 2, and 3 are revealed, and if we were to continue our course still farther westward we should probably find the geological structure the same as along the Platte, the White River Tertiary beds overlapping and concealing the chalky limestones of No. 3. So far as I have been able to make examination, Nos. 4 and 5 are not seen, though we cannot doubt of their existence somewhere underneath the Tertiary beds. They are all well shown on the Missouri River and along the eastern base of the mountains far south to the Arkansas, but along the Platte and all the little branches No. 3 is seen to a greater or less extent, with no trace of Nos. 4 and 5. These two groups may have been removed by erosion along the margins of the great Tertiary lake, prior to the deposition of the Tertiary beds, or they may be concealed from view by the great superficial marl or drift which forms the edges of the true Tertiary deposits. There is evidence to believe that the county next west of Jefferson is covered with the Tertiary beds, and that they continue westward to the western limit of the State. We know that at Fort Kearney the Tertiary beds are shown along the Platte, in their full development, with high, bluff hills, 200 to 300 feet in height. The hostile Indians were roving all over the county bordering on the Republican Fork, committing all manner of depredations, effectively preventing me from making an exploration of this interesting region, but it is safe to state that the colors on the map are very nearly or quite correct. As to the resources of that portion of the State covered with the Tertiary beds, it is safe to say it must be the same as that bordering on the Platte west of Fort Kearney. It will never make very good farming land, yet it may be inhabited sparsely by a pastoral people, for the surface is thickly covered with short, nutritious grasses.

CHAPTER V.

JOHNSON AND LANCASTER COUNTIES.

Johnson County.—The north branch of the Great Nemaha River runs nearly diagonally through Johnson County, in a southeasterly direction. It is the only important water-course in the county, and its value to the inhabitants cannot be overestimated. The entire county is underlaid by rocks of the age of the upper coal-measures; hence the geology is comparatively simple.

There are very few exposures along the Nemaha and its branches, and the high divides on either side present only rolling prairies covered with a luxuriant growth of grass, exhibiting every evidence of remarkable fertility, but having no timber and comparatively little living water.

From Beatrice our course was nearly northeast, passing over the divide between the waters of the Big Blue and those of the Nemaha. This divide, as usual, was treeless and nearly waterless for eighteen miles; yet, either to the right or to the left of our road, water and small trees could have been found within five or six miles. The grass was excellent, showing a fertile soil, and the surface was monotonously beautiful to the eye, but not an exposure of the underlying rocks could be seen.

On Yankee Creek, a branch of the Nemaha, the first exhibition of the rocks was observed. A few limestone quarries were opened for obtaining building materials. The beds are thin, not more than from 6 to 12 inches in thickness, intercalated with beds of clay and sand. The surface is rather rugged, some abrupt hills, but usually clothed with grass down to the water's edge.

At Tecumseh a thin seam of coal has been opened, and is now worked with some success by Mr. Beatty. The drift is very similar to that before described in my report of Pawnee County, and extends into the bank about 100 yards. Mr. Beatty has taken out about 1,000 bushels of coal, which he sells readily at the mine for 25 cents per bushel. It is undoubtedly the same bed that is opened on Turner's Branch and at Frieze's Mill, in Pawnee County, but it is not quite as thick or as good; it contains large masses of the sulphuret of iron and other impurities The coal seam here varies much in thickness, from 10 to 15 inches. The cap-rock is a bed of limestone not more than 2 or 3 feet in thickness. A well was sunk in the village of Tecumseh 60 feet; a drill was driven down through rock and hard clay a few feet farther, and passed through what the workmen thought to be three feet of good coal. This discovery created much excitement at the time, and increased the demand for the public lands in Johnson County. It afterward turned out to be the same seam of coal worked by Mr. Beatty on the Nemaha, and was only 11 inches in thickness. The prospects, therefore, for workable beds of coal in Johnson County are no better than in the neighboring counties already examined. The succession of the beds at Beatty's coal-drift is nearly as follows:

9. Alluvial with pebbles. Small fragments of limestone. In this bed the tooth of a mastodon *(M. giganteus)* was found.
8. Hard layer of rusty limestone with seams of carbonate of lime all through, 4 to 6 inches.
7. Coal, varying in thickness and quality, 10 to 15 inches; contains much sulphuret of iron.
6. Blue potter's clay, 4 to 6 inches; rusty or light-gray sandstone, 20 inches, which is regarded as useful for whetstones.
5. Micaceous laminated grit, 16 feet.
4. Argillaceous limestone, full of fossils, *Myalina subquadrata*, *Hemipronites crassus*.
3. Reddish clay, 6 inches.
2. Reddish siliceous limestone, 15 inches.
1. Micaceous sandy grit, 20 inches.

In a bed of limestone, holding a high position in the hills, the following fossils were found: *Spirifer cameratus*, *Athyris subtilita*, *Syntrilasma hemiplicata*, *Productus semireticulatus*. These rocks prevail all over the county, so that the geology is very simple, the Upper Coal-measures only being revealed.

Tecumseh is the county seat of Johnson County, a small town located on the elevated prairie near Nemaha River.

From Tecumseh to the source of the Nemaha, about forty-five miles, I did not discover a single exposure of rock, and I could not ascertain that any had ever been observed by the settlers. We must conclude, therefore, that building materials in the shape of rock are not well distributed over the country; indeed, I do not know of any one in which I observed less.

From the sources of the Nemaha we passed over to those of Salt Creek. This county is very beautiful and productive. As a compensation for the absence of rocky layers the slopes are gentle and the creeks and ravines are grassed to the water's edge. I was told that a few building-stone were obtained on one of the little branches of the Nemaha. On Salt Creek, most outcroppings were seen down to the forks. The best quarries of limestone are on the farm of Mr. S. B. Mills. The west fork has very little stone on it except for about a mile above the junction. It is hardly necessary to give the succession of the beds. There are alternations of clay and limestone, and the limestones appear to belong to the transition or Permo-Carboniferous series. The fossils occur somewhat rarely, as *Productus semireticulatus*, *P. prattenianus*, *P. punctatus*, *P. costatus*, *Retzia punctulifera*, *Fusulina cylindrica*, *Fenestella Shumardi*. These limestones are exposed over an area of not more than five miles square. Their entire thickness cannot be more than from 15 to 20 feet, arranged in layers from six inches to two feet thick.

In abstracting the rocks from the quarry the fracture is so regular, breaking into massive square or oblong blocks, and the texture so fine, compact, and of light cream-color, that they are highly esteemed by builders, and make beautiful as well as durable houses. There are quite a number of large dwelling-houses made of this stone in the vicinity. It works quite easily. The finest springs of water in this country issue from this rock.

There are five or six of these quarries opened at this time, but the principal one occurs on the farm of Mr. S. B. Mills.

These fine quarries must become of great value to this country, for they yield the only good building material for thirty to fifty miles north, south, and west, and from ten to twenty miles east of the place.

The rusty, rather soft, friable sandstones of the Dakota Group are used, to some extent, for dwelling-houses. It presents an exceedingly somber and unpleasant appearance to the eye, and possesses no elements of durability. It can be relied on only in the absence of other building material. About twelve miles below these quarries, near the salt basins, Lincoln, the capital of the State, is located. Pretty good water is obtained here by digging, but there is a liability even then to strike brackish water, on account of the proximity to the salt lands.

From a point five miles above Lincoln to a point five miles above the mouth of Salt Creek, there is a scant supply of building material, of timber, and of fresh water; so that it can be seen at a glance that this valley is not as desirable as many other portions of the State.

Near Miss Warner's, about ten miles above Lincoln, a well was dug on the high hills, bordering the valley, to the depth of 60 feet, without striking rock. At Yankee Hill, two miles above Lincoln, a well was dug 66 feet, without reaching the basis rocks. These facts show the great thickness of the superficial alluvial deposits of this region, and also the skeleton form of the surface prior to the deposition of these deposits. I shall treat more fully on this subject at a future time.

The sandstones of the Dakota Group are quite largely developed in this region, and exhibit their usual variability of texture and color. The prevailing color is a deep drab rusty-brown, sometimes yellow, or nearly white. Some layers contain many impressions of dicotyledonous leaves. I was unable to find as large and perfect impressions as I have collected at many other localities.

So far as the surface of the country is concerned, in Lancaster County it may be regarded as remarkable for its beauty. It is always gently rolling, well drained, and from elevations the views are very fine, forming most excellent building sites. When the soil is not influenced by salt springs, it is equal to any in the State, but in an agricultural point of view there is no doubt that Salt Creek, with the numerous salt springs that issue forth near it, is a disadvantage to the valley. That portion about two miles above Lancaster does not seem affected by the salt. The farm of Mr. S. B. Mills, of over 1,000 acres, about ten miles above the county seat, is one of the most fertile and valuable in the State. Although the salt springs in this county may eventually be of some value to the State in the production of salt, yet I am convinced that if there was not a salt spring of any kind in the county, the difference in the value of the lands for agricultural and grazing purposes would much more than balance all income that will ever arise from the salt springs. In that case Salt Creek, instead of being almost useless, or rather an impediment, would be a fine fresh-water stream, making it one of the finest stock counties in the State.

The surface of the uplands lies very beautifully, is very attractive to the eye, but there is scarcely any timber in the county. The soil is excellent, and forest trees may be planted with success whenever settlers choose to do so, though very little has been done as yet.

The fact that the capital of the State is located in this county gives to it additional importance, and it also becomes a matter of no small interest to determine the true value of the salt springs that occur in this vicinity. The basins and scattering springs occupy a large area several miles in extent, but the main basin is located near the town of Lancaster. These basins are depressions in the surface nearly destitute of vegetation, and the white incrustations of salt give the surface the appearance in the distance of a sheet of water. The Great Basin, as it is called, is situated about one mile from Lancaster, township 10, range 6, section 22, and covers an area of about 400 acres. The brine issues from a large number of places all over the surface, but in small quantities. All the salt water that comes to the surface from this basin unites in one stream, and we estimated the entire amount of water that flowed from this basin at from six to eight gallons per minute. The second salt basin lies between Oak and Salt Creeks, and covers an area of 200 acres. Third basin is on Little Salt Creek, called Kenosha Basin, and covers 200 acres. Numerous small basins occur on Middle Creek, which occupy in all about 600 acres. Between Middle and Salt Creeks are several small basins, covering 40 or 50 acres. From the surface of all these basins more or less springs ooze out. In former years great quantities of salt have been taken from the surface and carried away. During the war as many as sixty families at a time have been located about these basins, employed in securing the salt.

Besides the numerous basins above mentioned, Salt Creek, Hayes's Branch, Middle Creek, Oak and Little Salt Creeks, have each a dozen springs coming out near the water's edge. One spring on Salt Creek issues from a sand rock, and gushes forth with a stream as large as a man's arm, at the rate of four gallons a minute. This is the largest spring known in the State. The geological formations in the vicinity are of the Upper Carboniferous and Lower Cretaceous age. The salt springs undoubtedly come up from a great depth, probably from the Upper Carboniferous rocks, and are the same in their history and character as those in Kansas. Two methods have been used to some extent in this region in preparing the salt—boiling and solar evaporation. The only method which can be employed profitably in this country, where fuel is so scarce, is solar evaporation, and this can be carried on more effectually than in any State east of Nebraska. The unusual dryness of the atmosphere, the comparatively few moist or cloudy days, the fine wind which is ever blowing, will render evaporation easy. The surface indications do not lead me to believe that Nebraska will ever be a noted salt region. It seems to me that if all the brine that issues from all the basins and isolated springs were united in one they would not furnish more than brine enough to keep one good company employed.

What the result of boring will be can be determined only by actual experiment. Some large springs may yet be found in that way, but I saw no brine that was much stronger than ocean water. The rains have been so frequent this spring that it is much diluted with rain-water. The Nebraska Salt Company made, from July to November, 1866, 60,000 pounds of salt. Another company, at work at the same time, made about the same amount. Good working days 6,000 pounds have been made in a day. The kettles used for boiling are very rude steam boilers split into two parts. In a vat 12 by 24 feet average evaporation was 125 to 130 pounds per day. An extra day was 250 pounds. I think it not improbable that a company with a large capital, and employing all the improved methods of manufacturing the salt, would succeed. The salt is said to be good, though not as strong as the common salt of commerce.

Passing down Salt Creek toward its mouth, we see on every side a gentle undulating region covered with a fine growth of grass, with all the indications of a soil fertile and productive. There is but little timber in the valley of Salt Creek and its vicinity; very little in Lancaster County; and this deficiency should be remedied at once by the planting of trees. There is no good reason why every quarter-section in the country should not have a proper amount of wood-land in a few years. There are no indications of the Carboniferous limestones until we reach a point near Dean's Mill, and then the river on the east side exposes a considerable thickness of them, with numerous fossils. On the west side of Salt Creek the limestones do not appear, and this point may be considered the western limit on the south side of the Platte. About a mile above Ashland there is a quarry of the rusty sandstone, which is much used for building purposes. Houses are not built of them generally, but walls or fortifications are sometimes made of them. These red sandstone quarries are quite abundant along the road from Ashland to Plattsmouth, but they thin out on the summits of the hills about ten miles before reaching the latter place.

We do not see the Cretaceous sandstones in actual contact with the Carboniferous limestones, even along Salt Creek or along the Platte on the south side, but this is due, undoubtedly, to the beds of fine sand and clay that intervene, forming slopes. About fifteen miles east of Ashland a bed of fine, white, indurated sandy clay, with red streaks all through it, upon which the red sandstone rest. This bed is 20 or 30 feet in thickness, and can be made very useful for economical purposes.

At Ashland, near Dean's Mill, there is a pretty good thickness of limestone with irregular cleavage, 20 to 30 feet in thickness, extending down to the water's edge, and forming a rock bottom or ford. Above this is a slope of clay, probably, and then 18 inches of hard, yellow limestone, which is also quarried out. From thence to the summit of the hills the slope is 150 to 200 feet, covered with a heavy deposit of yellow marl; and if the basis rocks crop out at all, they are red sandstone. From the lower bed at Dean's Mill I collected a large number of fossils,

Productus semireticulatus, P. prattenianus, P. punctatus, P. costatus, P. wabashensis, Athyris subtilita, Meekella striato-costata, Syntrilasma hemiplicata, Retzia punctulifera, Spirifer lineatus, S. planoconvexus, S. kentuckensis, Hemipronites crassus, Chonetes granulifera, Fusulina cylindrica, Allorisma, Pinna peracuta, Schizodus wheeleri, Myalina subquadrata, Scaphiocinis, hemisphæricus, Fenestella, Lophophyllum, &c.

The above list of fossils will show at once that the limestones here, even up to the apposition of the Cretaceous beds, are Upper Coal-measure, and that the transition and Permian, if they were ever deposited here, have been swept away.

About five miles east of Ashland the succession of the beds is thus shown:

7. Sandstone, cropping out from the summits of the hills of Cretaceous age.
6. Slope, probably fine sand and clay.
5. Yellowish magnesian limestone, gradually passing down into harder broken layers, 10 to 20 feet thick, in layers 6 inches to 4 feet. The upper part is mostly made up of comminuted organic remains. I detected *Meekella striato-costata, Athyris subtilita, Fusulina cylindrica.*
4. Slope, 20 feet.
3. Layers of whitish limestone, very hard and good for lime, and for building purposes, 4 to 6 feet; *Fusulina, Athyris, Archeocidaris, Meekella,* crinoid stems and corals.
2. Slope 150 feet above the Platte; about 100 feet thick at this point.
1. A heavy ledge of rather brittle limestone full of *Productus, Spirifer, Athyris, Chonetes, Fusulina,* crinoid stems, &c., 1 to 15 feet thick.

Bed 5 of the above section furnishes very excellent building-stone, which has been used in the fine dwelling of Mr. Dean and some others. It is very tenacious but soft, easily cut with a knife, can be smoothed with a common jack-plane, so that it makes excellent caps and sills. One special virtue in this rock for building is that it never sweats, that is, moisture never accumulates in the walls. So that a house built of it is dryer even than one of wood. Ashland is the county seat of Saunders County, and is very favorably located for a flourishing village. It is surrounded by a fertile and, productive country, and the water of Salt Creek has become so freshened by the time it reaches this point that cattle are not affected by it, but rather prefer it. All kinds of building material are abundant everywhere, and wood in moderate quantities. Farther westward the exposures of the basis rocks are very rare; occasionally a little stream will cut down into them. On Skull Creek the gray arenaceous shaly marls of No. 3 appear near the bed of the creek. The upper part of the bed is about one foot, of yellow or whitish-yellow limestone; over it is a considerable thickness of coarse, loose sand, much of it worn down from the Cretaceous sandstones of No. 1. In the manufacture of artificial building material this might be made very useful. A little higher up on this creek I found a man digging for coal. A layer of shale, which possibly represents No. 2, deceived him into the belief that coal must exist near. He passed through a bed of chalky marl with abundant specimens of *Inoceramus problematicus*, then a marl that disintegrates so rapidly on exposure. Along the Platte River, on the south side, about fifteen miles below the mouth of Loup Fork, there is a cut bank which exposes the following beds, which belong to the post-Tertiary, but which I regard as older than the yellow marl:

2. Three feet soil; 1 foot yellow marl with pebbles; 10 feet sand, gravel, and rounded granite boulders.
1. Black ash-colored plastic clay, 25 feet.

There is a line of separation between beds 1 and 2, that seems to indicate a break or interval of time; at any rate the materials change most abruptly. The clay bed below has much the appearance of the dark Cretaceous clays, but is full of small, water-worn pebbles, with now and then a rock of considerable size. I am inclined to the opinion that this clay will be made very useful for many purposes, but more especially for the manufacture of pottery. The geology of Nebraska, south of the Platte River, may be summed up as follows: The first two tiers of counties are underlaid by Carboniferous and Permian rocks; overlapping these in the third tier west, are the Cretaceous sandstones of No. 1, then No. 2 and No. 3, and possibly Nos. 4 and 5, but it is probable that the latter beds are not seen; but about thirty miles above the entrance of the Loup Fork the Cretaceous belt is overlapped in turn by Tertiary beds, and then continue uninterruptedly to the western limits of the State. We have, therefore, but these three portions that can be colored on the map. The simplicity of the geological structure of Nebraska south of the Platte can be seen at a glance, and I have perhaps devoted in this report all the attention to it which it deserves. The superficial deposits of that region are more difficult to study and require more attention than I was able to give them. They will be treated of more fully in the account of the geology of Nebraska north of the Platte.

CHAPTER VI.

GEOLOGY OF NEBRASKA NORTH OF THE PLATTE.

With the exception of a small portion of Douglas and Sarpy Counties, bordering on the Missouri and Platte Rivers, the whole State of Nebraska north of the Platte River is underlaid with rocks belonging to two geological eras, Cretaceous and Tertiary.

The Cretaceous rocks make their appearance in their eastward extension in rather thin beds, capping the summits of the hills, and only the more compact layers, resisting the eroding effects of water or atmospheric agencies, remain to indicate its boundaries and extent. I am inclined to the belief that the rusty sandstones of the Dakota Group once extended in full force directly across the Missouri into Iowa, and that the sandstones recently discovered by Dr. White on the Nishnabotna River form a portion of the series, disconnected only by the wearing away of the intervening rocks. There is no doubt that a great portion of northwestern Iowa is underlaid by rocks of the Dakota Group.

The green color on the geological map of Nebraska connected with this report will show the eastern boundaries of this group with accuracy. The Carboniferous limestones soon begin to disappear north of the Papillion River.

At Sarpy's old trading post, near Bellevue Landing, some thin layers of rock occur in the hills, and a thin seam of coal has been found, and at low water two or three layers of rock are revealed, which can be made useful for building purposes.

At Omaha five to ten feet of limestones are revealed near the water's edge. The rock is quarried to considerable extent; but from the fact that Omaha is almost entirely supplied with rocks and lime for building purposes from the Platte, we may infer that the quarries at Omaha are not extensive. The cost of stripping the vast thickness of superincumbent Drift and Loess at Omaha must render the working of this quarry very expensive.

The next exposure is at Florence, where the limestones are seen only at low water.

The last exhibition is at Rockport, near De Soto, where at very low water the limestones are seen at the edge of the river, but at neither of the localities above named are there quarries of any special value.

Along the Missouri bluffs there is no exposure of the underlying rocks again until we reach Tekama, Burt County. Here the nuclei of all the hills are sandstones and clays of the Dakota Group. From Florence to Tekama the bluffs or hills bordering on the Missouri are very rugged and high, but are composed entirely of Drift gravel at the bottom, and a great thickness of yellow marl at the top; indeed, this yellow marl or Loess is not unfrequently 50 to 100 feet in thickness. It is so soft and yielding in its nature that the little temporary streams flowing down the bluffs wear out immense gorges 100 to 150 feet in depth. The sides of these hills along the Missouri bottom, on the Iowa as well as Nebraska side, are often very deep, with angles of descent of 30° to 40°, and I have seen vegetation clinging quite thickly to their sides, when the descent was 50° to 55°, although the great geographer, Ritter, says that the grade at which it is possible for earth to cling is 45° At Tekama are some exposures of the sandstones of the Dakota Group, but mostly so soft and friable as to be of little value as building material.

In the absence of all other rocks the inhabitants quarry out the harder portions and use them. Underneath the sandstones are the usual variegated clays and sands, red, white, gray, and drab, with nodules of the sulphuret of iron. In the sandstones above, there is quite a variety in the texture of the rock. Sometimes there are thin intercalations of clay; then little pockets, as it were, of clay inclosed in a thin shell of iron; then the thin layers are oblique, as if the waters in which the sands were deposited were in currents, or in a disturbed condition. Indeed, it would hardly be possible to describe all the varied conditions which this rock presents. Between Tekama and Decatur, a distance of about sixteen miles, there are frequent exposures of the sandstones and clays, but none worthy of special notice until we reach the vicinity of the little town of Decatur, near the border of the Omaha reserve. Here some harder layers of rock are exposed, which are used for the foundations of buildings and other economical purposes. There is one layer of quartzite.

The range of hills, all the way from Tekama to Decatur, is well grassed over, and very abruptly steep. Numerous springs of the purest water flow from these hills at all elevations. One of them is quite remarkable, and is known as the Golden Spring, from which quite a stream flows through the coarse sand. The time must come, in the future, when this will become a popular watering-place for the West. Above Decatur the Missouri cuts the bluffs so as to show a mural escarpment as follows:

2. Massive yellow marl; no lines of stratification visible; immense masses slidden down, several yards having fallen in, 40 to 50 feet.
1. Variegated clays, with a soft, fine grit; there is a dark seam in it filled with bits of coaly matter.

These clays vary from a purplish hue to a dark drab; 100 to 200 feet exposed here. A little farther up the river the soft yellow sandstone appears above the clays in regular order. There are great quantities of iron ore, a kind of limonite, scattered all along the river in large flat or irregular masses, also some nodules or concretions. Another section, near the last, shows the following succession of beds:

4. Yellow marl, with small calcareous nodules; no stratification; 10 to 20 feet.
3. Yellow and gray indurated clays; 8 to 15 feet.
2. Soft, yellow, rusty sandstone, massive, with very little sign of stratification; 50 feet.
1. Variegated clays; 50 to 60 feet.

The rusty-yellow color prevails everywhere, arising from the oxydation of the iron. This gives complexion to everything in this formation. These different sandstones extend from a point near Florence, where they cap the hills above the mouth of Iowa Creek over one hundred miles. The fossil plants in the hills near Decatur, and at the Blackbird Mission, are very numerous; about fifty or sixty species have already been collected at these localities. The sandstone is quarried out and used for cellar walls, and for other purposes where stone is needed. It is useful, inasmuch as all other rocks are absent. Sand for making the patent concrete is exceedingly abundant. Near Tekama and Decatur there are thin seams of iron ore, which, when broken with a hammer, give forth a sound much like that from old pot-metal. It is really pretty good iron ore, but silicious and impure; and even if this ore were of the best quality, and in great abundance, there is no fuel in the country to render it of any value.

At the Blackbird Mission, on the Missouri, eight miles above Decatur, the bluffs of sandstone are quite conspicuous, and often present very high mural fronts, upon which the Indians have carved many rude pictures, doubtless portions of their hieroglyphical history. At this locality are quite numerous layers, from 1 to 4 feet thick, of a very compact, massive quartzite, the hardest and most durable rock in the State. It has the appearance of a metamorphic rock, so very hard and close-grained is it. The harder portions have been quarried out and used for the construction of a very large three-story building for the mission school.

As the construction of several railroad bridges across the Missouri are contemplated, no rock in the State would be so unyielding and durable for abutments as this, providing enough of it can be found. It seems to assume a concretionary form in the sandstone, and is of very uncertain thickness and extent. About two miles above the mission, the hills are cut by the river so as to reveal vertical bluffs, the rocks of which, in the distance, have a yellowish-white appearance, and from this fact are usually called chalk bluffs. The sandstone is massive, almost without stratification, and very friable and soft.

4. Yellow marl, recent, 10 to 50 feet.
3. Eight inches of earthy lignite, resting upon 12 inches of yellowish-drab arenaceous clay, underlaid by 8 inches impure lignite.
2. Massive yellow sandstone, with some thin intercalations of clay, soft and friable, readily yielding to the erosive effects of water, 60 to 80 feet thick.
1. Yellow, plastic, unctuous clay, toward the top becoming a grayish-blue; contains flat argillaceous concretions, 2 feet.

This is perhaps the finest and largest exposure of the rocks of this group along the river. The mural exposures of soft sandstone present good surfaces for the Indian to make use of to write his rude history, and on the chalk bluffs there are many of these hieroglyphics in positions totally inaccessible to the Indians at the present time. None of them now living know anything about them, and it is supposed that they must be very ancient, and that since they were made great changes must have been wrought in these bluffs by the waters of the Missouri. These markings are at least 50 feet above the water, and 50 feet or more below the summit of the bluff, so that they must have been made before the lower portion of the bluff was washed away by the Missouri. It seems strange that none of these hieroglyphical writings, which occur quite often on the chalk rocks of the Niobrara Group higher up the Missouri, are known to any Indians now living. Manuel's Creek is called in the Dakota language, the creek where the dead have worked, on account of the markings on the rocks.

The above illustration conveys an idea of the sandstones of the Dakota Group, as they front the Missouri, and shows the wearing away of the material of the rock underneath during high water. This erosion is continued for a series of years, until the superincumbent rocks fall down and are washed away by the river. Near the mouth of Omaha Creek are some very high vertical bluffs of sandstone, from which some rock has been taken for building purposes. It is useful, since no better can be found in the vicinity. For a considerable distance along the hills opposite Sioux City, beds of the gray quartzite are found, which are worked to considerable extent, and furnish a very good supply for the inhabitants. A few impressions of plants and a few fossil shells were found here. Near Sioux City, on the Iowa side of the Missouri, is a high cut bluff extending to the mouth of the Big Sioux River. Here was formerly a large exposure of the rocks of the Dakota Group, and these rocks exhibited well their variegated texture and composition. The color seems to differ, depending upon the amount of ferruginous matter in them. Only about 20 feet of the different layers are exposed, and only about 5 feet hard enough for building purposes. This quarry has been wrought for twelve years or more, and at this time seems to have given out, for very little suitable building stone can be found, mostly loose sandstone and clay. In former years I have obtained impressions of dicotyledonous leaves, as willow, laurel, &c., with some fossil shells of the genera *Pharella*, *Axinea*, *Mactra*, and *Cyrena*, which are in part estuary and in part marine in their habits.

Near the northern boundary of the Omaha reserve traces of a whitish, chalky limestone, almost entirely made up of the shells of a species of *Inoceramus*, make their appearance on the high hills. This rock indicates the first appearance of the Cretaceous division, No. 3, or the Niobrara group.

In passing northward, as we continue up the Missouri, we find this formation becoming more and more conspicuous, until opposite Sioux City it is 50 to 100 feet in thickness. It is of much value to this region of the country, on account of its qualities as a material for lime, and it supplies a large district with that valuable material. Omaha is largely supplied with lime from the region of the Platte. Between Omaha and the northern boundary of the Indian reservation, a distance of eighty miles or more, extending southward to the Platte, near Columbus, there are five or six counties entirely destitute of limestone. This limestone of the Niobrara Group becomes very valuable, therefore, and it will be from this upper district that the counties underlaid by sandstones of the Dakota Group must obtain their supply of lime. Number 2, or Fort Benton Group, seems to be wanting until we reach a point near the mouth of Iowa Creek. This is a thin bed, not over 40 feet in thickness at any one point, and is characterized by black plastic clay filled with beautiful crystallized sulphuret of iron. It is pretty well exposed below the mouth of Iowa Creek, where the Missouri cuts the bluffs, and here we see all the rocks in their order:

4. Yellow marl, a recent deposit.
3. Niobrara Group, layers of white and yellow chalky lime, passing down into gray marly rock.
2. Black plastic clay, with hard layers, containing *Inoceramus*, a species of *Ostrea*, like *O. congesta*, remains of fishes, many crystals of sulphuret of iron, selenite, &c.
1. Dakota Group, sulphuret of iron, fragments of wood, impressions of leaves, willow, laurel, &c.

Near the mouth of the Niobrara River the black shaly clays of the Fort Pierre Group begin to make their appearance on the hills over the Niobrara Division, so that within the limits of Nebraska proper we have four out of five of the important divisions of the Cretaceous rocks of the west.

Near the mouth of Iowa Creek there seems to be a bed of impure lignite in the Fort Benton Group, or in the transition between the Dakota and Fort Benton Groups. This bed, which has been worked to a considerable extent, and the coal used by blacksmiths in this vicinity with some success, does not seem to be the same as that seen along the Indian reserve, which is undoubtedly in the sandstone of the Dakota Group.

I am inclined to the opinion that this bed of lignite near Ponka City is a local bed, or at least restricted in its geographical extent, and is the result of an accumulation of drift-wood in an estuary of the Cretaceous sea. I am informed that it is seen over on the Elkhorn River, about thirty-five miles west of this point.

Mr. Clark tells me that he dug 12 or 15 feet below this bed, and struck another seam of coal much better than the one cropping out. The lower bed must be the one in the Dakota Group. Lithologically, it is impossible to draw a line of demarkation between these formations here. No. 1 passes so imperceptibly into No. 2, and No. 2 into No. 3, that there is no break, and yet their principal characteristics are very distinct. The first is a sandstone; second, a black, plastic clay; third, a chalky limestone; and yet I cannot tell the exact point where one commences and the other ends.

The impressions of leaves have ceased to appear before the close of the Dakota Group. The sandstones of the Dakota Group occupy the whole country along the Platte from the mouth of the Elkhorn to a point some twenty miles beyond the entrance of the Loup Fork.

The intermediate counties between the Missouri and Platte have very few exposures of rock of any kind, so that quarries in this region, even though the rock is of inferior quality, are much prized. We have, therefore, within the limits of the State of Nebraska north of the Platte, the Carboniferous rocks underlying two counties—Sarpy and Douglas; then come the Cretaceous rocks, especially No. 1, overlapping the Carboniferous. These beds continue along the Missouri to the northern limit of the State. They extend in a southeastern direction beyond the south line of the State into Kansas, underlying a belt of country from fifty to two hundred miles in width. About the sources of the Loup Fork, and about thirty miles west of Columbus, the Tertiary beds begin to make their appearance, and then extend to the extreme western limits of the State. Even along the Platte, before reaching Columbus, their influence is felt to considerable extent, for the soil of the bottoms is quite sandy, and in some places too much so to be productive. The area drained by the Elkhorn and its branches cannot be much less than from two thousand to four thousand square miles, and yet over this large surface there are but few outcroppings of rocks of any kind. We know that the Cretaceous beds underlie nearly or quite all of it, and yet they seem to have yielded so readily to atmospheric influences that the surface is very gently undulating, the slopes along the streams being grassed over down to the water's edge. Along Elkhorn or Logan's Creek there is a bottom of greater or less width on either side, so that the basis rocks are concealed along the banks by a moderate thickness of superficial marl, a great part of which is composed of the disintegrated materials of the Cretaceous or Tertiary rocks. Over all this region there is very little timber, but the clays for making artificial building materials are without limit. The soil is wonderfully fertile, the water very pure, climate healthy, so that we cannot but believe that all this district will eventually take the highest position as an agricultural and grazing region. West of latitude 99° the Tertiary beds prevail to the exclusion of all others; the soil is less fertile, water and wood is less abundant, and yet the surface is covered with a thick growth of grass and other herbaceous vegetation. Although this portion of Nebraska cannot be cultivated successfully at this time, yet the climate is so healthy and the water so pure, the grass so abundant and nutritious, that it seems to me it will become in the future an excellent grazing region. The character of the surface, and the shortness, as well as the nutritious character, of the grass, would seem to adapt it especially for the raising of sheep, and immense herds of them might be raised by a pastoral people. The Union Pacific Railroad will transport all the products of this vast region to market, and I am convinced that even the sand-hills will yet become a fine pasture ground for herds of sheep, cattle, and horses, and thus every part of the State may be settled and made productive.

CHAPTER VIII.

THE COAL-FIELDS OF COLORADO.

I have endeavored in this chapter to take as complete a view of the Lignite formations of the West as the facts will admit, first confining my observations to the country west of the Mississippi, and then endeavoring to trace them beyond those limits. If the lignites of the West prove of economical importance, as we believe they will, they will be of inestimable value to all parts of the West, but to no portion more than to the State of Nebraska, and the time is now fast approaching when their true value will be determined. It is on this account that I give them so much attention in this report.

Although our knowledge of the Lignite formations of the West is not very extensive at this time, yet the discovery of them is by no means a new one, for they have been referred to in most books of travel since 1800. Lewis and Clark saw beds of "stone-coal" all along the Missouri. Wyeth says that the banks of the Yellowstone, below the mouth of the Big Horn, are in many places precipitous, with strata of "bituminous coal," and Captain Bonneville saw "anthracite coal" on Powder River. We now know that all these coal-beds are of Tertiary age, and cover a vast area bordering on the Missouri River. In 1842 Colonel Frémont observed the lignite beds on some of the tributaries of the Columbia, and Green River, a branch of the Colorado. This is undoubtedly a western extension of a portion of the formations described on the Medicine Bow River. Near longitude 122° and latitude 45° 30′, Colonel Frémont discovered a stratum. (Frémont, page 192.) On Muddy River, in longitude 111°, and latitude 41° 30′, he found a most interesting locality, with several beds of coal and clay, and in one bed of indurated clay was a great abundance of vegetable remains. The stratum containing the plants was about 20 feet thick, and above it were beds of coal, each 15 inches thick; below are three beds of coal, separated by layers of clay. Captain Stansbury, in his report of an expedition to Great Salt Lake, frequently notices the existence of coal in numerous localities. On a fork of Bitter Creek he saw a stratum 10 feet thick, exposed for 100 yards. During the whole day's travel coal was exposed in every favorable locality. The dip of the beds was northeasterly 5° to 10°. All along the Muddy it also outcrops, even to the summit, where the waters flow each side into the Pacific and Atlantic. On Rattlesnake Creek were indications of coal, also on Sulphur Creek, of so good quality that he calls it stone-coal. In his report he says: "Specimens of it, although much weathered, burned in a camp-fire with a clear, bright flame. It is bright black; but when cut with a knife, appears dark brown; and when weathered, light brown. It is superior brown coal."

Mr. Henry Engelmann, in General Simpson's report, 1859, observed the Utah and Western Colorado coal beds in numerous localities. He regards the coal strata at Fort Bridger and beyond of Cretaceous age. On Sulphur Creek he found extensive coal beds with fossils, identical with Professor Hall's *Turbo paludinæ formis*, from Muddy Creek. Captain Gunnison noticed coal-beds near the junction of Grand and Blue Rivers, in the Green River Valley; also on the eastern slope of the Wasatch Mountains, near the head-waters of Sevier River; also in the upper portion of the San Pete Valley, a stream which has its source in the eastern slope of the Wasatch range. This coal was used by the blacksmiths at Camp Floyd, and regarded by them as a superior "bituminous coal." Engelmann found the lignite all along Weber Creek, from Echo Cañon to Kamas Prairie. As this last locality is apparently in the midst of the mountains, Mr. E. speaks of the great tilting up of the beds. He also found fossils on Sulphur Creek that appear to be identical with those discovered by Colonel Frémont on Snake River, longitude 115°, latitude 43°, overlying a sandstone which contains in the greatest abundance an ostrea (*O. congesta*, probably) and *Inoceramus*, (*I. Crispii.*) The sandstone was white, rather soft and fine-grained, and undoubtedly represents No. 3, or the Niobrara Group of Cretaceous rocks. At the forks of Sulphur Creek the whitish sandstone occurs again, with *Ostrea*. Near the crossing of Sulphur Creek is a spring of petroleum, which has been referred to by several former travelers. Shallow depressions in the ground near the spring are filled with water, and rising on the surface is the oil or tar. The Mormons and other emigrants employ it as a liniment for wounds, and as a lubricator for their wagons. On exposure to the air it is changed from a green color to a dark brown, and it has an aromatic taste. This tar hardens as it flows from the ground, and becomes mixed with the soil around the sides of the spring. In 1860 I noticed the same phenomena in several places along the eastern side of the Sweetwater Mountains, and I was informed of the occurrence of these springs in many other localities. On Weber River, above the mouth of White Clay Creek, the *Ostrea* was found in several places in the white sandstones. Some specimens of coal have been found in Round Prairie, on the Timpanagos, and the little stream near this point has been called on that account Coal Creek.

Colonel Emory, as far back as 1848, speaks of the occurrence of coal between Bent's Fort, on the Arkansas River, and Santa Fé, to the north and south of Raton Pass. A few specimens of dicotyledonous leaves were obtained, which indicate that the age of the rocks in which the lignite is found is Tertiary. I have thus far noticed the existence of this coal over districts which are about to be rendered more valuable by the construction of the two Pacific railroads, covering an area between the 48th and 38th degrees of latitude. I shall refer again in another portion of this chapter to the geographical extension of these beds beyond these limits.

After having examined with some degree of care most of the settled portions of the State, I regarded it as my duty to pass beyond its limits into the neighboring Territories in the vicinity of the Rocky Mountains to ascertain the quantity, as well as quality, of the lignite which was reported to exist there. The construction of the Union Pacific Railroads has created a new demand for mineral fuel all over this portion of the West, and if these great deposits can be made available, their value in the future development of the West can not be overestimated. By the kindness of the managers of the railroads I received free passes for myself and party, and thus was enabled, with comparatively little expense, to go beyond the first range of mountains, into the Laramie Plains, and returning by way of the overland stage route to South Boulder Creek, thence northward to Cheyenne City, along the base of the mountains. Along the line of the Union Pacific Railroad the limestones of the Upper Coal-Measures extend from Omaha to the north of the Elkhorn River, a distance of about thirty miles. Before reaching this point westward, however, the rusty, reddish sandstones of the Dakota Group are seen on the summits of the hills, gradually increasing in thickness and importance, until at the mouth of the Elkhorn, and for some distance up that stream, it forms moderately high, bluff banks. The railroad then descends into the Platte bottom, which is very wide, and the yielding nature of this formation, as well as the other divisions of the Cretaceous, causes the gentle slopes and the almost entire concealment of all the underlying rocks. From the mouth of the Elkhorn River to a point at least one hundred miles west on the Platte and Loup Fork, indications of the white chalky limestones of the Niobrara Group are seen, with the ever present *Inoceramus*, but there are no prominent exposures, and this rock, although excellent for limestone, is not found in sufficient quantities to be of much economical value. In 1857 I observed an exposure of soft, rusty sandstone at the mouth of Loup Fork, which belonged undoubtedly to the Dakota Group, but this year I was unable to find the locality again. It is safe to say that the Cretaceous beds extend up as far as the mouth of Calamus Creek, but all the underlying rocks in this region seem to be concealed from view by a thick deposit of yellow, sandy marl. The Tertiary beds commence overlying the Cretaceous rocks near the south portion of the Pawnee reservation. The exact position and age of these beds, as they first appear, I cannot tell, but they must be Upper Pliocene, or perhaps, indeed, of the age of the yellow marl or Loess. Near latitude 99° some small bones were found, which indicate the true Pliocene, and from there to the foot of the mountains the unmistakable Pliocene Tertiary beds prevail. So far as any evidence that I have yet been able to obtain, the Upper Pliocene beds shade off into the Loess in such a manner that I have seen no locality where I could detect any break in the continuity.

At Cedar Bluffs the yellow arenaceous marl is 100 to 150 feet thick, filled with shells of recent species, and this marl diminishes in quantity as we proceed, until the true Pliocene beds are visible. At Fort Kearney the Tertiary beds are quite conspicuous, forming high, rugged hills on either side of the Platte. The bottom of the Platte varies somewhat in width from five to fifteen miles, but, for the most part, is very level, while the hills which border it are rugged or smooth, depending on the character of the underlying rocks. Sometimes, for miles, they will present softened, smooth outlines, clothed with grass, without a rock in view; then, again, they will be cut up into ravines, and from the sides will project thick walls of sandstone or marl. At Julesburg the sand-hills form a somewhat conspicuous feature. The soil of the bottoms is so mixed with this floating sand that it seems hardly possible ever to make any use of this region for agricultural purposes. From Columbus to Fort Kearney there is little or no hard rock for building purposes, but from Kearney to the North Platte two or three moderately-thick beds of a fine, gray, calcareous sandstone, which, though soft and porous, might be made useful in the absence of any other materials for building purposes. The bottoms below the junction of the North Platte are quite fertile, yielding from one and a half to two tons of hay to the acre. After passing beyond the junction of the North Platte, the sterility and aridity increases until the sand monopolizes most of the country. This sand is mixed with small, water-worn pebbles, grains of quartz, flesh-colored feldspar, &c. Near Sidney Station the hills on either side of Pole Creek are quite high and rocky, and the rock is used by the Union Pacific Railroad Company for the foundations of the buildings at the station. There are several beds of this rock, varying in texture from a fine-grained sandstone, with some water-worn pebbles, to a flesh-colored, marly limestone. A bed of reddish, flesh-colored, indurated, arenaceous clay is also conspicuous. The soil is everywhere from 1 to 2½ feet thick, and sometimes from 6 to 8 feet thick. The summit is formed of a bed of marly sandstone 10 to 20 feet in thickness, and large masses of the rock have been detached and fallen down the sides of the hill. The lower part is composed of a light yellow, indurated marl. A few bones have been found in the rocky, as well as the more yielding beds of this locality, but they were not sufficiently characteristic to be determined specifically. But there is one interesting feature of this region which is worthy of notice. Through all these rocks there is more or less silex in various forms. It sometimes assumes the purer forms of chalcedony, and is found in great abundance, forming specimens of considerable beauty. The Indians, in ancient times, would gather this flinty rock into certain localities, and there manufacture their flint arrow-heads, knives, chisels, &c. Around Pine Bluffs is an abundance of these chipped flints, with broken arrow-heads, &c. These more recent Tertiary beds are a portion of the great basin to which the "Bad Lands" properly belong, but they do not always present those rugged features which are characteristic between White and Niobrara Rivers. Now and then there will be a small area which will remind one of those wonderful "lands," and the strata exposed give evidence of the age and their relation to the White River beds. On the North Platte are several localities that form landmarks, as it were, which are of very great geological interest. Nowhere, except close to the base of the mountains, are there any dislocations of strata or any evidences of upheaval, and where there are any hills, or "buttes," as they are called in the West, shooting their summits above the surrounding country, we know that they are monuments of the past which show that the surface of all the country was once on a level with the summits of those high hills at least, perhaps much higher, for we cannot tell how much the summits have suffered erosion; we only know that the beds are horizontal up to the summit, and that there are fragments of beds which we never see elsewhere. The Court-house and Scott's Bluffs and other well-known localities are examples. The soft, marly nature of the rock causes it to yield readily to atmospheric agencies, and it is thus worn into the most fantastic forms, often giving to the scenery a wonderfully picturesque beauty. Near these localities have been found numerous specimens of fossil turtles and teeth, and bones of extinct animals. Some of them were of enormous size, and probably belonged to the huge species of Elephas (*E. imperator*), whose remains are so abundant in the Pliocene beds of Niobrara and Loup Fork. It would seem, from the organic remains found at various localities along the North Platte, that the Miocene beds of White River, and the Pliocene of the Niobrara, were represented there.

The formations along the North and South Platte will need a more careful study than I have been able to give them yet, before the details of the geology can be written with certainty. We know, however, in a general way, that all the rocks are of Tertiary age. The excavations made for the Union Pacific Railroad, along the valley of Pole Creek to the foot of the first range of the mountains, always reveal the marly beds of the Tertiary, varying in color from a chalky white to a deep reddish flesh color; but the superficial deposits attract the most attention, because more easily seen. Low down, almost to the mouth of Pole Creek, the small water-worn pebbles are seen mixed to a considerable depth with loose sands and marls. These pebbles increase in size as we approach the mountains, and all along the eastern slope are well-worn boulders, 3 or 4 feet in diameter, and the cuttings for the railroad show that they extend down to a considerable depth. What relation this superficial deposit of boulders, sands, and marls bears to the Tertiary beds beneath it is difficult to determine without a more thorough examination; but that it is connected with the latest period of upheaval of the Rocky Mountain range there can be scarcely room to doubt. I shall discuss this subject in another place.

From Omaha to Cheyenne City, five hundred and twenty-five miles, the ascent is quite gradual but persistent. According to the observations of the engineers of the Union Pacific Railroad, which were made with great care, Omaha is 925 feet above the sea level, while Cheyenne City is 5,085 feet, averaging a little over 9 feet ascent to the mile. This point is regarded by the railroad company as the base of the mountains, and here some of their most expensive buildings will be erected. The city was laid out into lots in July, 1867, and by December of the same year there were from 2,000 to 3,000 inhabitants, with some very costly and substantial buildings. Such cities, springing up all along the line of the road, at once called into demand all the resources of the country, and the search for building materials, fuel, &c., has commenced in earnest. Although the resources immediately around Cheyenne are not extensive, yet at a distance of not over twenty miles into the Laramie Mountains timber is moderately abundant, while the best of building materials, in the shape of stone, are found on every side. Even around Cheyenne some of the marly Tertiary sandstones are found to be sufficiently hard and durable for houses, and General Stevenson is erecting a large warehouse of this rock. It is used to a great extent, and will probably be found to answer all practical purposes, though entirely devoid of beauty. Some of the beds of Tertiary marl are found to contain so large a percentage of lime that it is burned, and is thought by some to be of excellent quality; but it has a dirty appearance and does not make a clean wall. Near the margins of the mountains there are some of the finest beds of limestone in the West. The rock itself is nearly white, and when burned in a kiln produces a lime that is as white as snow, and the walls of houses prepared with it present a most cheerful, light appearance. There is the greatest abundance of these rocks; these carboniferous limestones sometimes attain a thickness of 1,000 to 1,500 feet. The roads up to the margins of the Laramie Mountains are excellent, and the ascent is gradual, so that transportation is comparatively easy. From Cheyenne to the summit of Laramie Mountains the distance is about twenty to thirty miles, and the elevation is 7,297—an increase in the ascent in this distance of nearly 90 feet to the mile. From Cheyenne to the summit of the mountains the railroad passes up a ridge which is quite remarkable, and seems especially adapted to the location of this great national road. On either side for many miles no such range can be found, rising gradually, as it were, to the very summits, without any of the abrupt, rugged places which are so common everywhere along the margins of the mountains. Usually there is a deep valley scooped out by glacial action all along the immediate base of the mountains, which would prove a serious obstacle in the building of a road, but here, for a belt of five to fifteen miles in width, this erosive action seems to have been checked for the time, leaving one of the most remarkable inclined planes I have ever seen along the mountain slopes. The basis rocks are unquestionably Tertiary, probably later Pliocene, but scattered all over the surface are many water-worn boulders, and the cuttings of the road reveal a considerable thickness of superficial material filled with worn rocks of every size and variety, most of them evidently having their origin in the nucleus of the mountains. The rocks which compose the true Tertiary beds seem changed here. All the hard strata are formed of an aggregation of crystals of quartz, feldspar, and small, water-worn pebbles, making a kind of puddingstone, and interstratified are these layers of whitish and yellowish clay, loose sand, or marl, the whole indicating a deposition in disturbed waters. This deposition must have taken place here also after the Rocky Mountains had reached their present elevation, or nearly so, for these beds jut up against the inclined older formations, reposing unconformably on them. Indeed, in all the little valleys of the mountains, patches of these Tertiary beds may be found, sometimes resting high upon the margins, but in all cases nearly or quite horizontal. Farther north, about fifty miles above Fort Laramie, and near the Red Buttes, I have seen these beds dipping at an angle of 50°, showing that they partook of the last period of elevation. At another locality, near the margin of the mountains, at least 7,000 feet above the sea, a bed of Tertiary sandstone inclined eastward from the mountains 10°. These beds here are composed of materials more or less coarse; some layers made up of crystals of quartz and feldspar, with now and then a rounded pebble; sometimes loosely aggregated, then cemented into a firm rock; then coarse beds like conglomerate. These beds attain here a great thickness, evidently not less than from 500 to 600 feet. By examining the border lines between the Pliocene Tertiary and the Eocene coal beds, a little farther to the southward, it will be seen at once that they do not conform. This same want of exact conformability is seen all along the slope of the mountains to the South Pass. Near Carmichael's Camp, or Evans Pass, the older rocks appear from beneath the Tertiary, inclining in regular sequence from the mountain nucleus. The Carboniferous and the red or Triassic beds are all that are exposed to view. Just at this locality all the more recent formations, as Jurassic, Cretaceous, and Eocene Tertiary, are concealed by the immense deposit of recent Tertiary, yet all these formations are exposed in full force within a few miles on either side north or south of this point. Nowhere on either side of this range south of Laramie Peak was I able to find any rocks below the Carboniferous limestones and next to the metamorphic rocks that along the Wind River, Big Horn Mountains, and the Black Hills, I have hitherto called the Potsdam sandstone. A section along the upturned edges of the beds from the syenite nucleus would be as follows:

1. Light-gray, compact, siliceous limestone, 2 feet.
2. Reddish laminated sandstone, 10 feet.
3. Light-gray, limestone-like bed, 50 feet.
4. Fine reddish sandstone, 40 feet.
5. Light-gray, bluish limestone, 10 feet.
6. A reddish calcareous sandstone, 6 feet.
7. Light-gray limestone, 30 feet.
8. Reddish laminated sandstone, some parts quite soft, 20 feet.
9. Bluish limestone, 10 feet.
10. Reddish laminated sandstone, 10 feet.
11. Bluish, hard limestone, 15 feet.
12. Reddish laminated sandstone, 4 feet.
13. Excellent white limestone, with *Productus Prattenianus*, *Athyris subtilita*, &c., 150 feet.

The above section is given simply to show the lithological character of these Carboniferous beds, that the brick-red character is not confined to the supposed Triassic. The thickness stated is merely approximative. The whitish limestones are quite persistent in their characters, but the other beds are variable, sometimes massive or laminated, siliceous or arenaceous, bluish or reddish here, &c. The dip is also variable from 20° to 45° southeast. Looking from the top of the hills into the valleys below, we can see the white beds of the Tertiary jutting up against the older rocks, or in isolated patches high on the sides, easily detected by the eye far distant, by the numerous eroded surfaces. When the eye has become familiar with the lithological characters of the different geological formations it is easy to detect them even in the distance, and as far as the eye can reach the rocks belonging to the great Tertiary basins, the Miocene and Pliocene White River and Niobrara deposits, and the Eocene and Miocene lignite beds. The former always has a light color, and in almost all cases presents many eroded bare spots, which seem of a much lighter color in the distance, while the rocks of the latter basin always exhibit a brown, somber hue, each one possessing numerous characters which to the practiced eye are unmistakable. In 1859 I detected, through a glass, from the base of the Black Hills, some isolated ridges, thirty miles distant, which appeared to be the White River Tertiary beds superimposed on the lignite beds. Anxious, at that time, to show the relations of the two great basins to each other by actual contact, I rode to them and found them to be as I had conjectured, and thus I was enabled to solve the desired problem. Indeed, all the geological formations of the West possess certain lithological lineaments caused by the varied effects of atmospheric influences which are unmistakable to the practiced eye. As I have before remarked, no unchanged sedimentary rocks are seen in this region between the true Carboniferous and the metamorphic, and, consequently, we pass directly to the nucleus of the Laramie range. The principal rocks are syenite, and it presents almost every variety, from an exceedingly compact, coarse-grained mass, almost destitute of hornblende, to a very coarse aggregate of quartz, feldspar, and hornblende, the large, reddish crystals of feldspar being the most conspicuous. Most of these metamorphic rocks contain considerable iron, probably originally in the state of sulphuret, which has caused the separation of the crystals of feldspar, and there seems to be all over the surface of the country on the summits of this range a heavy superficial deposit which is mostly made up of small masses of feldspar. When this material is abundant it forms the finest of roads, which are very smooth. This tendency to disintegration seems to characterize all the metamorphic rocks of this region, and thus they are often worn into some wonderfully fantastic forms. These massive granite piles are not uncommon all along the summits of the mountain ranges, and, standing so clear above the general level of the country, seem more like immense boulders transported from some more northern region. Yet these massive piles, while they do not seem to reveal any distinct signs of stratification, extend along the mountains in definite, rather narrow belts, with a course about northwest and southeast. The illustration given above shows the forms of these massive piles of rocks as they are seen in hundreds of localities. The evidence is quite clear that all the rocks which are exposed in this Laramie range are really stratified metamorphic rocks in nearly or quite a vertical position. Sometimes the seams are of trap or greenstone, then quartz or clay slate, extending in definite lines, and all parallel. The corners are still very sharp, as if time, which rounds off the hardest granite, could not affect this. It forms exceedingly beautiful and picturesque scenery, and must prove an elegant and durable rock for building purposes. Some of the massive piles of syenite near the sources of Dale Creek are very close-grained, and would, if polished, be almost as handsome for building purposes or monuments as the Scotch syenite. Mr. S. B. Reed, engineer of construction, Union Pacific Railroad, first called my attention to this fine syenite, and it is now proposed to transport this rock from the summit of the Rocky Mountains to Omaha to construct the piers of the bridge across the Missouri River. Again, these granite masses assume forms sometimes looking like a figure with its broad cap, then apparently just poised ready to tumble down into the ravines below, where thousands of similar masses may be seen which have fallen at different times.

Flowing among these hills are some of the purest streams of water that the mountains afford, well stocked with trout. Sometimes in the level, prairie-like portions of the mountains, these streams are so deep and the borders so boggy that it is difficult to cross them, and the dams of the beaver add to the difficulty; all through the mountains are thousands of beautiful valleys filled with springs and small streams, with a fine growth of grass; quite large, prairie-like areas, sometimes nearly level, or slightly undulating, dotted over with monumental piles of granite, syenite, gneiss, or slate. Continuous ridges, rising up in the prairie, are not uncommon, while the eroded or disintegrated materials, clothed with vegetation, so conceal the rocks sometimes for miles around that the country looks like a plain. Occurring quite frequently, but not as common as the syenite rocks, are stratified gneiss ridges which extend also across the summit of the mountains in definite lines. In most cases these rocks seem to be vertical, but sometimes they are inclined; but taken in the aggregate no dip in any particular direction could be obtained.

The timber in these mountains is mostly pine and spruce, but it is by no means abundant. There are no dense forests just along the immediate line of railroad. The soil in these mountains is excellent and produces a good growth of vegetation, but it is hardly probable that any attempt will be made to cultivate land so much elevated above the sea level. As we descend the western slope of the Laramie range into the plains, the brick-red beds which are seen on the eastern margin again appear, inclining westward. It is then evident at a glance that these beds are the counterparts of those on the opposite side, and that prior to the elevation of the mountains they extended without a break, and in a horizontal position over the area occupied by the metamorphic rocks. Even the moderately-thick covering of vegetation on the western margin, and extending far into the plains, cannot conceal the brick-red color of the earth beneath. This fact gives a singular appearance to the scenery, and as these red beds are gypsiferous, a white efflorescence covers the surface in low wet places in the dry season of autumn. This white material, which is most abundant everywhere throughout the Plains, is called alkali by travelers. These red beds seem here to rest directly upon a dull-red granitoid rock, which on exposure to the atmosphere readily disintegrates, so that there is a great thickness of débris composed for the most part of feldspar crystals. The Carboniferous beds extend down the western slope in ridges, as if the erosive power had acted in grooves or channels, leaving alternate valley and ridge. A section of one of these will include most of the beds visible here, and convey a pretty clear idea of their composition.

6. A compact, thin layer of arenaceous limestone, containing a species of *Orthis*, forms the summit, 4 feet.
5. Fine sandstone, easily disintegrated, with very irregular laminæ; a rather light-red rock, 100 to 150 feet.
4. Yellowish-white limestone, excellent quality, 2 feet.
3. Brick-red fine sand, with spots of white sand-rock; irregular laminæ of deposition.
2. Very hard, bluish limestone, filled with comminuted crinoidal remains, 6 to 8 feet.
1. Loose, brick-red material, gradually passing up into a brick-red, fine, compact sandstone, 100 feet.

The brick-red beds I have been accustomed to regard as Triassic, but it is evident that this color is not confined to them, but is also common to the Carboniferous rocks. The evidence seems to point to the conclusion that in this immediate vicinity the beds are of Carboniferous age, and that there are none of the true gypsiferous beds occurring here, on either side of the range. The disintegration of beds 2 and 3 produces the superficial red earth which covers so much of the Plains. The lower part of bed 1 is composed of loose, red material, with the seams of harder laminated rock, with one or two seams of gray, rather coarse-grained sandstone. The rock seems to be an aggregate of particles of quartz. This is to some extent a gypsiferous deposit, and it is not strange that the soil composed of the disintegrated materials of these beds should be alkaline. Bed 2 is a solid, massive limestone, composed almost entirely of fragments of crinoids, one of which was nearly perfect. This bed will be very useful for building-stone or lime. The inclination of the beds, as shown by these ridges, varies from 3° to 25°. But what adds much to the character of the scenery is the fantastic shapes these sandstones present through atmospheric action. Sometimes the ridges taper down to a point, the sides presenting the most rugged appearance, while the summits project in domes, columns, &c. Then again there are rounded masses, looking in the distance like boulders, and on near approach the well-furrowed sides can be seen.

This sandstone is much in use at Fort Sanders for the construction of houses. It works easily, and seems to be durable, but some of it must disintegrate too readily. There are portions of it, when quarried out, of a light-brown color, which do not seem to have been penetrated with the sesquioxide of iron. There are also, in the different layers, fine examples of waved surfaces.

The limestone has been used at the fort, and found to be of the best quality.

The Laramie Plains are surrounded with mountains as with a wall, and the scenery on a clear day is extremely fine. Along the west margin of the first or Laramie range the hills slope gently down into the plains, and are apparently lost in the level beyond. It is an important question to determine whether these plains can be cultivated with any success. That towns and cities will be springing up all along the line of this great national highway is already rendered certain, and much of the land has been taken by preëmption. General Gibbons made an attempt to raise garden vegetables in 1867, but he began too late in the season. In October, beets, lettuce, turnips, cabbage, &c., were still growing, and from the appearance of the vines I thought there were some good potatoes. Irrigation is required, and another season General Gibbon is confident of success. Fort Sanders is the most comfortable military post I have seen in the West. Fine streams of water are caused to flow all through the grounds, so that the best of spring water is at every one's door. About eight miles from Fort Sanders is a sulphur spring, which may eventually become of some importance. There are many other large springs in the plains. On both sides of the mountains, and about them, places of resort for health and pleasure will be erected at no distant day. East of the Big Laramie River the Plains seem to be covered with the débris of the red beds, and I looked in vain for some evidence of the existence of the Cretaceous rocks which should be represented. Near the Big Laramie Stage Station about 10 feet of black slate is exposed, which is evidently of the age of the Fort Benton Group. There were an abundance of fish scales and *Ostrea congesta*. This bed is exposed only for a short distance, and the strata are apparently perfectly horizontal. On the east side of the river, farther up about five miles, is a long ridge, which is formed of yellow arenaceous marl, with a layer 2 feet thick of rusty concretionary sandstone, with fish scales, sharks' teeth, and finely-comminuted shells, 30 to 50 feet in thickness. These Cretaceous beds seem to be lithologically like those that occur along the slope of the Big Horn Mountains, and they do not preserve their distinctive characters anywhere in the vicinity of the mountain ranges as well as on the Missouri River. The subdivisions are not always as distinct, yet it is hardly possible to mistake them. The ubiquitous *Ostrea congesta* always marks divisions 2 and 3 at localities however widely separated geographically. About four miles above the station, immediately along the river, a bed of yellow chalky shale was observed, in which well-defined specimens of *Ostrea congesta* were found, and its calcareous character showed it to belong to the Niobrara Group. The rusty layers on the east side, which hold a higher position, might belong to the Fox Hills Group. The layer of harder rock was a rusty, rotten sandstone, with some clay intermixed. Many of the high ridges that come down from the slopes of the mountains on each side of the river seem to be composed of Cretaceous beds, so that there is ample room for all the divisions to be represented. So far as I could determine, the Cretaceous beds extend about fifteen miles above on the south, and ten miles below on the north of Fort Sanders, and on the east side of the Laramie

River about five miles, so that the red beds slope down from the mountains and pass from sight directly under them.

About eight miles west of Fort Sanders, near the valley of the Little Laramie River, there is an exposure fifty feet or more in thickness of thinly-laminated shaly, light, rust-colored beds of the Niobrara Division, containing a great abundance of well-defined specimens of *Ostrea congesta*, in most cases attached to another large shell with a somewhat remarkable fibrous structure, observed long since on the Missouri River and in other places. This shell is here broken into fragments for the most part, yet the outline is seen complete, though it is quite impossible to procure perfect specimens of it. There is no doubt, however, that it is an undescribed species of *Inoceramus* of large size, sometimes a foot or more in diameter, to the smooth sides of which this little oyster seemed to have a strong tendency to attach itself. At the base of this exposure there is a dark-bluish, marly shale, which has the peculiar jointed structure of the Fort Pierre Group. These vertical joints run nearly northeast and southwest. Everywhere this bed is cut through in a remarkable manner by these vertical lines. All through these beds are seams varying from 1 to 6 inches or more in thickness, of calc-spar, often cutting the beds at right angles. There are also some apparently fibrous seams that are horizontal. Over the surface are abundant fragments of a rusty, argillaceous limestone, like that scattered over the cretaceous hills bordering on the great bend of the Missouri River, which would indicate the former existence of the Fort Pierre Group. On the summit of this exposure are layers of whitish, chalky rock, which is unmistakably No. 3, so that we may conclude that we find in the Laramie Plains certainly Nos. 2 and 3, and probably No. 4, of the divisions of the Cretaceous as developed along the Missouri River. There is another remarkable feature about this valley, as I have called it—that is, that it has no outlet. It is about two miles wide and ten or fifteen long, and has a drainage like the valley of some stream, and yet it has no connection with any permanent stream and ends abruptly at either extremity. Through the kindness of General Gibbons, commanding Fort Sanders, I was enabled to visit the coal fields on Rock Creek, a branch of the Medicine Bow, under the most favorable auspices. A few inches of snow had fallen the day before we left for Rock Creek which obstructed our examinations somewhat, but we obtained facts enough to show that there is here the eastern limit of a most valuable coal-basin, which will yet have a marked influence on the development of the West. The beds of lignite are exposed in hundreds of localities over an area at least forty miles in width and over two hundred miles in length. The report of Mr. Van Lennep, geologist to the Union Pacific Railroad, indicates more than fifty localities where good beds of lignite are exposed along the Medicine Bow River alone. And other facts show that we have here from 5,000 to 8,000 square miles of coal at least. It is only necessary in this connection to state the fact that this railroad is now being cut directly through the length of this great coal basin.

The first exposures that I saw of the coal were about eight miles west of Cooper's Creek. Here we found several beds, which seemed to have been brought to the surface by the upheaval of the mountains. The inclinations of the strata are from 5° to 25° at the opening, but gradually becoming horizontal as they recede from the mountain. The coal beds vary from 1 to 10 feet in thickness, and the coal resembles in its appearance the best quality of coals in Pennsylvania. In some of the beds are solid seams 2 to 6 inches in thickness, of a material very like cannel coal, as if the vegetable matter of which it was formed was originally in a pulpy state. Even when taken from the outcroppings, where it had been more or less exposed to the atmosphere, this lignite burned quite well, with some draught. The interest in these fields, which the supposed future demand for fuel in this region has brought about, is very great, and large numbers of claims have been taken up. Iron ore is found also in the vicinity, but to what extent I could not determine. Stray fragments of the carbonate of iron have been washed down the valleys of the little streams from the mountains, and it no doubt exists in large quantities somewhere toward the sources of these streams. All these beds repose on well-marked Cretaceous rocks, and among some plants found in concretions in a bed beneath the first bed of coal seen, were a species of *Populus* and a *Platanus* apparently identical with those occurring on the Missouri river. As we proceed westward the coal formations become much more extensive, and a number of beds of coal are exposed. From the best information I could obtain exposures have been observed in a hundred places or more all the way to Great Salt Lake. A remarkable feature in this vicinity was the evidences of Drift or glacial action along the foot of the mountains. The Medicine Bow range extends nearly northeast and southwest, and all along the northeastern base broad deep valleys are scooped out, leaving lofty ridges sometimes broken and sharp at the top, 500 to 1000 feet in height, formed to a great depth of materials of various kinds, with different degrees of fineness, mingled with water-worn rocks, while the southwest sides are completely paved with rocks, some worn and others angular, as if the transporting power had been checked at the summit of the hill, and melting away, had dropped its burden on the summit and side of the hill. The opposite side is usually almost free from rocks, and not unfrequently reveals the basis formations as near Cooper's Creek, the rusty arenaceous concretions projecting from the sides of the hills were filled with a species of *Inoceramus* that indicate several hundred feet of the Fox Hills Group. After having made some brief examinations as far as Rock Creek we returned to Fort Sanders, and prepared to start for Denver, October, 1867, along the overland stage route over the mountains. Our course across the mountains was nearly southeast over a most excellent road macadamized with feldspar crystals. As we passed up the western slope the same red beds and layers of limestone as before described could be seen inclining at various angles.

Before reaching the summit also piles of rotten syenite occur. Far to the southwest the sharp snow-clad peaks of the main range could be seen, much of the time invested with clouds. A light snow had fallen the day before, adding much to the beauty of the scenery, but concealing the surface in most instances. The rugged piles of syenite, however, could be seen on every side, like monuments left after erosion. I am inclined to think that they are due to atmospheric agencies, and not to any local outbursts or upheavals, and that all these beautiful valleys which, like that of Dale Creek, have been admired by all travelers, are entirely due to the wearing away of the rocks by water. The soil is good, and the superficial materials or débris, as we may call the more recent deposit which forms the immediate surface, is quite thick. The vegetation which clothes the ground is quite abundant—grass, reeds, and smaller shrubs, but the larger trees are scarce over large areas. Along the northeastern margins of the Medicine Bow Mountains there are some dense forests of the Douglass spruce (*Abies Douglassi*), which is used extensively by the Union Pacific Railroad for ties, and it surpasses all other timber for that purpose. These trees are often as straight as an arrow for 80 feet. They seldom attain a larger size, so that the logs are hardly large enough for lumber. Houses are built of it by sawing the logs into two portions, and laying the sawed side out, and thus some of the handsomest buildings at Fort Sanders' are constructed. Pine is not uncommon, yet there are no dense forests of it, as we might expect in so mountainous a region. On the summits of the mountains it seems to grow low and scraggy, as if it drew but a scanty nourishment from the soil. In other localities it rises to the dignity of a forest tree 80 to 100 feet high, and has a trunk large enough for mill logs. The little cottonwood, quaking asp, alder, and some other small trees and bushes occur, but they are very thinly represented, and would furnish a very scanty supply of fuel. After passing the summit we descend into the most beautiful and picturesque valley along the overland route, Virginia Dale. Dale creek is a branch of the Cache la Poudre, and is composed of a number of branches ramifying through the mountains, filled with trout. The water is most pure and delightful to the taste. It is a curious fact that while some of the branches of the North Platte rise in the divide with those of the South Platte, yet not a single trout has ever been seen in any of the waters of the North Platte; while in every mountain stream emptying into the South Platte the trout are very abundant. I have regarded the cause of the absence of trout in the North Platte due to the existence of the great amount of alkaline matter all along its valley, which, mingling with the water, is destructive to the trout, but not to the ordinary kinds of fish. On either side of the road high mountain peaks rose like a wall, but in a southwesterly direction the beautiful mountain called Long's Peak, with the extensive and lofty range to which it belongs, could be seen. I found along the road in passing through this valley even to the Cache la Poudre, comparatively little rock, but the different varieties of syenite. In one instance I found some mica schist, but as a general rule mica is wanting. In Virginia Dale I found some high masses of the red feldspathic granite with small, thin plates of mica. This is the first true granite I have noticed. There are also thin seams of large crystals of feldspar. The granite here is elevated in huge piles with vertical and horizontal partings, so that it is separated into massive blocks 10 to 15 feet on a side in dimensions. This granite would be very durable for buildings—is not very coarse, although it is a mere aggregate of the three constituents. There are also dikes or wedges of hornblendic slate and greenstone, and seams of white quartz are not uncommon. Near Virginia Dale Station there is a huge massive pile of rock forming the side of the cañon of the stream 600 feet high. It is a coarse feldspathic granite, the feldspar predominating in large crystals of a reddish hue. Near Stonewall Ranch, on the east side of the range, there is an immense pyramidal pile composed of the red beds, which are nearly horizontal. At the base there is a whitish, gypsiferous sandstone, and from there, in ascending order are 350 to 400 feet of alternate layers of fine brick, red grit and harder red sandstone, the harder layers projecting out from the nearly vertical sides. These rocky layers vary from 1 to 10 feet, and the loose material 4 to 30 feet in thickness. These reddish rocks form a sort of wall along the road for miles. Near the summits are some beds of brown limestone like the limestone layers near Fort Sanders, and although I found no fossils, I do not hesitate to regard them as of Carboniferous age. This range of hills, after extending southeastward along the road for a mile or more, in a nearly horizontal position, commence, inclining 1°, then a little farther 5° to 7°. The direction of the dip is southeast again; the dip is 14°, and still farther the whole series inclines 10°, 11° to 17°, and in some instances 30°. On the summit there is a bed of yellowish limestone resting upon the reddish sandstone, 10 to 15 feet, then above this a whitish limestone, then a bluish, hard limestone, 6 to 10 feet, all of which would make excellent lime. Farther down the margins of the mountains we pass over a great thickness, 1,000 feet or more, of brick-red grit beds, which are plainly gypsiferous, entirely destitute of organic remains, inclining at various angles varying from 15° to 24°. There seems to be no break in the continuity of the layers, at least none is perceptible to the eye; and, therefore, in the absence of fossils it would be unsafe to pronounce upon its age, but they are either a prolongation upward of the Carboniferous or possibly Triassic. Continuing our course eastward, we pass over a bed of light-gray, massive sandstone, 50 feet thick, nearly horizontal; above this there is a thin bed of greenish-gray, arenaceous marl, 15 to 20 feet in thickness, capped with a fine, rusty-gray siliceous rock, which occupies the position of the Jurassic beds farther north, and resembles them somewhat, lithologically. After searching with considerable care over a large area, I was unable to find even a trace of any organic body. On the east base, along the roads, and extending to Laporte, on the Cache la Poudre, we found a full development of the Cretaceous rocks. They seem to be 300 to 500 feet in thickness. No. 2, or Fort Benton Group, is well shown as a dark shaly clay, with seams of marly clay, containing *Ostra congesta* and an *Inoceramus* undistinguishable from *I. problematicus.* Below No. 2, which we will regard as a fixed horizon, is a bed of gray silicious grit, which may represent No. 1 or Dakota Group, or it may belong to the Jurassic. Although this bed is quite persistent all along the margins of the mountains, from the Wind River range to Pike's Peak, I have never yet been able to find a trace of any organic object, not even silicified wood. I have usually regarded this bed as No. 1, with a query. Above No. 2 we find a good thickness of No. 3, but becoming more arenaceous in its southern extension. Yet portions of it retain the same light color and chalky character as heretofore. Separating No. 2 from No. 3, there is a bed of light-gray sandstone, which gradually passes down into the plastic clay of No. 2. No. 3 contains the *O. congesta* and *I. problematicus* in considerable numbers. As we look eastward from the summit of the range toward the level prairie, the upturned edges of the different beds recede like waves of the sea until they die out in the plains. These ridges of upheaval, grow smaller and lower; the inclination is usually at a smaller angle until the beds that are exposed above the surface show no perceptible dip. It seems quite strange that no rock is to be found south of the Red Buttes that can be referred to the Potsdam sandstone. Along the Wind River and Big Horn range it is seen resting on the metamorphic rocks and charged with its peculiar fossils; but as we proceed southward it seems to disappear, so that, after leaving the central metamorphic portion of the Laramie range we pass over Carboniferous, Triassic? Jurassic? Cretaceous, and finally, in the plains, the Lignite-Tertiary. Before leaving this nucleus, however, we see an abundance of the hornblendic slates and other metamorphic rocks which occur in the principle ranges, showing the close proximity to the main crest of the mountains of which Pike's Peak, Long's Peak, &c., form a part. Somewhat extensive valleys occur between all the ridges of upheaval, varying from a fourth to half a mile in width, through which some little stream makes its way, usually. Sometimes, however, the rivers cut their way directly through these ridges, as is the case with Cache la Poudre. Nowhere south of Fort Laramie do we see those massive beds of whitish limestone that occur farther north. Near the head of Deer Creek, about one hundred miles north of Fort Laramie, the Carboniferous rocks attain a thickness of 1,000 feet or more, composed of layers of massive limestone with hardly any tinge of red; but as we proceed southward the lime gives place to sand, and the color becomes more reddish, until even the thin layers of limestone are tinged with the red color. It is quite possible that this change is due to the decomposition of the sienite rocks, of which the nucleus of the Laramie range is mostly composed, and that these brick-red beds are formed of this débris. In these red beds are considerable quantities of silex, and in many localities the fragments cover the ground. This was the favorite material of which the Indians in former times fashioned their arrow-heads and knives. The distance from the plains across these ridges of upheaval to the nucleus rocks varies much at different localities.

At Laporte, on the Cache la Poudre, the distance is not more than a mile, and there are not more than three of these ridges. At a point about ten miles north, along the stage road, the distance is five to ten miles, with a dozen or more of the ridges of upheaval. At Cache la Poudre, Cretaceous beds form the low ridges, inclining 16° to 17°, the Jurassic, Triassic, and Carboniferous, two to three ridges, from 500 to 800 feet in height, and inclining 15° to 17°, with a strike about northwest and southeast. As we descend the Cache la Poudre River, we pass over the black shales of No. 4, and the rusty-yellow arenaceous rocks of No. 5 for about four or five miles, when they are overlapped by the Lignite-tertiary beds in nearly or quite a horizontal position. The soil along this valley is quite good, and by irrigation the farmers raise good crops. Two and a half miles south of Cache la Poudre the first conspicuous ridge is composed of Carboniferous rocks and inclines 25°, again a little farther, 21°. Near Thompson's Creek the road runs along the immediate base of the first hills of the range. Between the upturned edges of the Cretaceous and the slope of the red beds and Carboniferous rocks is a beautiful valley, about half a mile wide, worn out in a concave manner, with little side valleys coming into it from the western side. This valley was finely grassed over, and seemed more like a park walled in on either side. The Cretaceous beds formed two or three ridges inclining 5° to 8°, and finally sloping into the prairie, and passed under the Tertiary beds. Here also I found, in No.'s 2 and 3, *O. congesta* and *I. problematicus*, with numerous fragments of the fibrous shell of another species of *Inoceramus*. In the bed of a little creek No. 4 is well shown, dipping 15°. In a number of localities between Cache la Poudre and Thompson's Creek, No. 4 is revealed with a thickness of from 100 to 200 feet. These dark shales have been thoroughly prospected for coal, as is shown by the numerous excavations. Near this point is a fine exposure of No. 3, dipping in such a manner as to pass directly under No. 4. Without doubt No. 5 would be seen by proceeding eastward into the prairie a short distance before we come to the point of overlapping of the Tertiary beds. The truth is, all the divisions of the Cretaceous rocks of the west, with the exception of No. 1, are well developed all along the eastern flank of the mountains from the Red Buttes to Pike's Peak at least, a distance of more than four hundred miles. Sometimes some one of the divisions is concealed, for a short distance, by superficial material, but it reappears again under favorable circumstances. Between Cretaceous divisions No. 2 and the supposed Jurassic, or the Trassic, if the former is wanting, is a massive bed of irregular, fine grit, in many instances so silicious as to form a fine quartzite. The upper part shows irregular laminæ of deposition. It is yellowish-white on exposure, with some masses of rusty rock. The lower part is a massive fine pudding-stone, composed of an aggregation of small, water-worn pebbles, apparently cemented together with iron. This bed, opposite Long's Peak, near Thompson's Creek, forms one main ridge and one sub-ridge. The sub-ridge comes about half-way up to the main ridge, forming a sort of break or notch; dip 17°. The main ridge has a dip of 19°, as observed from the summit. Then a very beautiful concave valley intervenes, about half a mile in width, to the next ridge, which is composed of gray limestone with brick-red beds. This is also composed of the ridges, forming really one main ridge. The first ridge has its southeastern slope covered with dark-blue limestone like a slate roof, with dip 24°. Beyond this dividing ridge we find the metamorphic rocks, so that everywhere the Carboniferous beds seem to rest directly on the metamorphic rocks, though not conforming. In almost all instances between the main ridges there is the bed of a stream sometimes dry, sometimes with water. Taken in the aggregate, these ridges, although inclining at so many different angles, at different localities, when not badly eroded or concealed by a superficial deposit, exhibit great system, convincing one at once that the upheaval was not a sudden paroxysm, but a long-continued, slow movement. These ridges are divided by little streams, which have cut their way through them, into separate parts, each portion 100 yards to half a mile in length. The harder layers usually protect the sides and summits of these ridges from too much erosion. Thence in a direct line from the metamorphic rocks to the level prairie southeastward cannot be over four miles. The lower or southern end of this ridge (for it runs out at Thompson's Creek), shows the red beds, and on the opposite side the rocks dipping in a contrary direction, forming one side of an anticlinal. The side next to the mountains dips 58°, and the pudding-stone stands up like a high wall for a mile or more in extent in its massiveness. This anticlinal curves around westward, so that the ends of the two sides meet at Thompson's Creek. At this creek there is a jog in the upheaval of at least ten miles; that is, north of Thompson's Creek the ridges die out, while immediately south, in the open prairie, the ridges commencing again far to the westward, and continually running out in the prairie southward, the line of fracture seeming to be northwest and southeast, while the trend of the aggregate range is nearly north and south, so that, passing along the foot-hills of the mountains from the south, northward, we see the ends of the ridges of upheaval, as it were, *en echelon*. These jogs are not uncommon all along the base of the mountain ranges. At Cache la Poudre there is one that is quite remarkable. Passing over the Cherokee trail from Thompson's Creek southward about six miles, we come to a ridge of sandstone near the road. Toward the mountains a terrace-like ridge is visible, which is composed of the yellow, chalky layers of No. 3. This sandstone ridge is No. 5, dip 13°, strike very nearly northwest and southeast. The black shale of No. 4 gradually passes up into the yellow and gray arenaceous sediments, in which are huge round concretions with foliated layers, and many shells of *Inoceramus*. This passes up into regular stratified sandstone, the layers varying in thickness from half an inch to a foot. This sandstone is mostly a rusty yellow, and moderately coarse-grained, and possesses nearly the same characters as No. 5, seen at the head of Teton River and Fox Ridge, on the Missouri. *Baculites ovatus* also occurs here in considerable numbers. The concretions are somewhat calcareous, as the shells indicate. Just before reaching St. Vrain's Fork the wall-like character of these beds is again seen. In the upheaval No. 1, sandstone is broken off, leaving a portion in a nearly horizontal position on the summit of a ridge 300 feet high, while the fallen part is near the base, inclining 65°. This wall extends for nearly a mile without interruption. I was informed that about three miles above this point, on St. Vrain's Fork, a man discovered a bed of coal while digging post holes. Further examination showed the seam of coal to be three feet thick, and that it had been used by a blacksmith in the vicinity, and pronounced by him as good as the bituminous coals of Iowa. This is quite possible, from the fact that isolated portions of the lignite formation may have been left after the erosive action which followed the elevation of the mountains. We know that the great thickness of Tertiary which we find eastward of the foot of the range, once extended in unbroken continuity across the area now occupied by the Laramie Mountains; and while it is impossible for a bed of coal to be discovered of very great extent, indications of it may be found even in the valleys, or on the sides of the mountains. Tertiary beds, with lignite, are found very near the summits of the Wind River Mountains, 12,000 feet above the sea. From Little Thompson's Creek, No. 5 approaches the foot of the mountains, and at St. Vrain's Fork the coal beds are near the mountains, or perhaps form a part of the foot-hills. It would seem as though the erosive power of water had not acted as strongly on the sedimentary rocks in this region, and, in consequence, Cretaceous beds form ridge-like walls against the sides of the mountains, which may have partially protected the Tertiary beds from being washed away. This is especially the case from St. Vrain's Fork to Boulder Creek, and doubtless beyond. Some of the tipped-up ridges of Carboniferous, Jurassic, and Cretaceous rocks are nearly 2,000 feet high, and incline 40° to 60°. Long's Peak is seen most distinctly from this point, though we are still forty miles distant. Near Left-Hand Creek there is a high conical hill, at least 800 feet above the bed of the creek, finely rounded, which is composed of lignite beds, which, as they were elevated and tipped away from the mountains, escaped erosion. The surface of this region for miles from the foot-hills of the mountains is covered with water-worn rocks of every variety, some of which are evidently of eruptive origin. The valley of Boulder Creek, with its border, is about six miles wide, and derives its name from the vast numbers of worn rocks which literally pave the whole surface. It is on South Boulder Creek, where the Marshall mines are located; and at this locality is the best exposure of all the rocks in this region. Just north of the cañon of the South Boulder is a nearly vertical wall of what appears to be basaltic or eruptive rocks rising at least 3,000 feet above the bed of the stream. On examining this closely, it would seem to be the Carboniferous beds, standing nearly vertical, presenting the most remarkable mural front I have ever seen along the mountains. The rocks are almost metamorphosed by heat, so that they are consolidated and hardened. Their dip is from 40° to 60°. Deep furrows are worn down the sides of this wall by atmospheric agencies, so that narrow vertical columns seem to stand out, adding to the wonderfully picturesque beauty of the view. At the foot of the highest ridge, myriads of massive, nearly square blocks have fallen down, gray with lichens. These blocks are a fine compact sandstone, siliceous, light gray, yellowish; no loose red material; stratification perfect. The red rock is hard and massive. Most of the rocks are brittle, when broken showing a vitreous fracture. The outside rocks seem to be less affected by heat, and would make the best kind of building material. Between the main ridge and the next succeeding, there is an interval of one-fourth of a mile. This valley-like interval is covered with grass and a few pine trees growing from a rich black soil. The next ridge is perhaps

1,500 feet above the bed of Boulder Creek, but only a few feet above the débris at the base, while the main ridge is at least 1,500 feet above this. This lower second ridge is a fine gray and yellowish-gray sandstone, with irregular laminæ, siliceous, of unknown age. One hundred yards farther is a third ridge, inclining at about the same angle, composed of pudding-stone, the same bed seen all along the margins of the mountains, from the Red Buttes to Pike's Peak, and supposed to be occupying the position of No. 1 Cretaceous. It certainly lies between well-marked Cretaceous and Jurassic beds on the North Platte. There is another interval grass-covered, and then the fourth low ridge, composed of shale, fine sand, and clay intermixed, inclining at an angle of 56°. Upon the surface of the thicker layer are numerous mud markings. The yellow arenaceous clay terminates quite abruptly, and a bed of gray, quite-hard limestone comes in. In this limestone I found in considerable numbers *Ostrea congesta*, and a large, rounded species of *Inoceramus*, referred to by Professor Hall in Frémont's report, but not named. Professor H. compares it with *I. involutus*, Sow. (Min. Con.,) but it is doubtless an undescribed species.* At any rate we can fix the position of this bed of limestone as No. 3. Between this point and the first opening for coal the distance is about two miles, and the intermediate rocks are concealed by a large deposit of drift material, but it is easy to see by the inclination of No. 3 that in this interval there is ample room for the existence of Nos. 4 and 5. I shall speak of the lignite beds, therefore, as Lower Tertiary until some more definite evidence is given to the contrary. I should have remarked that all these beds, even including the lower lignite beds, exhibit evidence of having been subjected to moderate but long-continued heat. These marks of heat decrease as we proceed outward from the nucleus of the range, but some of the layers of Carboniferous sandstone seemed to have been changed to a mica schist, the plates of mica being very distinct. No. 3 was quite changed, exhibiting a compactness, a hardness, and a fracture never before seen in any part of the West. The value of the coal in the lower bed, now worked by Mr. Marshall, has undoubtedly been greatly enhanced by the heat to which it has been subjected. Dr. Leconte observed similar phenomena, but on a much larger scale, in New Mexico and the Raton Mountains.

By the close proximity to the foot of the mountains, and the inclination of the strata, probably the most remarkable and complete section can be obtained that can be found anywhere. It seems hardly possible that so great a thickness of the Tertiary beds can occur, but the section taken at this point is the result of a pretty careful examination of one day under the immediate direction of Mr. Marshall, the owner of the mine, and the information he gave me is the result of over four years' experience at the mine. It is hardly possible that beds 6 to 13,† inclusive, have been broken down from some higher beds in the series. A more thorough examination at some future period will determine this. Beds 1 to 10, inclusive, incline 8° to the east; the remainder, 35° to 40°. Lignite beds 39 and 43 of the section have not been fairly opened yet, but were discovered in searching for iron ore. In clay bed 21 there are some features which cannot be well represented in the section. In addition to the main coal-bed are a couple of smaller seams. As the drift passes the upper seam becomes very hard, resembling anthracite, but it is so thin that no notice is taken of it by the miners. The following are the divisions of clay bed 22:

---

* Mr. Meek has since named it *I. deformis*, in Mr. King's report.
† See Report of Colorado and New Mexico, page 29.

7. Drab clay.
6. Lignite, 12 inches.
5. Drab clay, 6 to 12 inches.
4. Sandstone, 2½ feet.
3. Drab clay, 6 inches
2. Lignite, 1 foot.
1. Drab clay, 1 foot.

It is probable that the lowest bed of coal mentioned in the section will furnish the most desirable fuel. The day of my visit to these mines 73 tons were taken away, and Mr. Marshall informed me that an average of 50 tons a day were wrought. This coal brings readily $4 a ton at the mine, and from $12 to $16 at Denver, twenty-two miles distant. The coal has very nearly the hardness of anthracite, which it very much resembles, but it falls in pieces more readily. It can be exposed in a dry atmosphere without much injury, but water causes it to crumble in pieces at once. I spent two evenings at Mr. Marshall's, burning this coal in an open grate, and I found that with a moderate draught it burned with a clear, bright-red flame, produced a good amount of heat, gave off no offensive odor, leaving scarcely any ash, and no clinkers. It contains very little sulphur, and, indeed, no erosive elements. For all domestic purposes it will undoubtedly prove equal or superior to any bituminous coals, and, in a sanitary point of view, there is no comparison. I will here quote a few paragraphs from the report of Dr. John Torrey, United States assayer at New York, who analyzed specimens of coal for the Union Pacific Railroad from this place and Coal Creek, three miles south:

The mineral has nearly the hardness of ordinary anthracite, but is much more brittle. The fragments are often cuboidal or rhomboidal, and in some of them a little amber was detected. The luster was bright and shining. The coal does not stain the fingers. The powder is black when viewed in a heap, but when a thin film of it is spread upon a white surface it has a slight tint of brown. Specific gravity, 1.29. When heated in a glass tube, the temperature of which is gradually raised to 400° or 500° F., it gives off water, the last portions of which a little empyreumatic oil or tar. At a dull-red heat it takes fire, burning with a bright-yellow and smoking flame, emitting an odor between that of heated bituminous coal and that of imperfectly-burning wood. Some of the fragments gave out a slight odor of sulphur, which was traced to minute scales and spangles of iron pyrites, scattered here and there among the lumps. Compared, however, with most bituminous coals, this mineral fuel is remarkably free from sulphur. When submitted to analysis it yielded the following results:

Water in a state of combination, or probably its elements as in dry wood	20.00
Volatile matter, expelled at a red heat in the form of inflammable gases and vapors	19.30
Fixed carbon	58.70
Ash, consisting chiefly of oxide of iron, alumina, and a little silica	2.00
	100.00

The ash is mostly reddish, but sometimes light gray. Another specimen contained only 16 per cent. of water.

The coal from Boulder Creek, which occurs in a bed 4 feet thick, and in another 10 feet, has a general resemblance to that from the other locality. It is, however, more dense, having a specific gravity of 1.4, and is less brittle, and the fracture is not so glossy. It contains, also, flakes of mineral charcoal, scattered through the mass, and the proportions of its constituents differ considerably from those of the Coal Creek bed, it being a stronger fuel. It contains a little sulphur like the other; the composition is as follows, viz:

Water in a state of combination, or its elements	12.00
Volatile matter expelled at a red heat in the form of inflammable gases and vapors	26.00
Fixed carbon	59.20
Ash of a reddish color, sometimes gray	2.80
	100.00

From the characters and analyses of the specimens here described, it will be seen that the Rocky Mountain belongs to the class of lignites, and that it is not technically a bituminous coal, neither cannel nor an anthracite. Still, in common parlance, it will be regarded as coal. In calorific power the Rocky Mountain coal may be placed between dry wood and bituminous coal, and therefore it is a most valuable fuel, especially where bituminous coal and anthracite are not likely ever to be found, and firewood is difficult to procure. I see no other reason why it may not be used for the smelting of iron and other ores. For locomotives it could be employed to advantage, with some modification of the fireplace. The ash is so small in quantity, and so light, that most of it would be carried off by the blast of the furnace. From my own trials, I find that the coal burns freely in a small stove, making a hot and clear fire, and leaving no clinkers. The specimens that I have examined show a tendency to break up and crumble after they have been soaked with water and allowed to dry; so that it would be well to preserve the coal as much as possible from being wet by rain. The lumps that you brought home from your journey show no disposition to crumble in a dry place. In conclusion, I remark that the discovery of such extensive beds of a good mineral fuel is of the highest importance to the section of country in which they occur. The iron ore is limonite, commonly known by the name of brown hematite or brown iron ore. It is a compact variety, and is certainly derived from carbonate of iron, some of which, in an unaltered state, is evident in one of the specimens. The carbonate will probably be found in larger proportion as the beds are worked further in beyond the reach of atmospheric influence. There is reason to believe that the iron obtained from this ore will be of good quality.

The bed of coal opened on Coal Creek corresponded in every particular with the sixth bed, or bed 23, of the section on South Boulder, at Marshall's Mine. The drift was carried in about 150 feet. It was first opened in the summer of 1860, while Marshall's Mine was opened in the fall of 1858. The sandstone that lies above the coal seems to be in many instances nodular or concretionary, like that which occurs in a similar position on the Missouri River. This main bed of coal is 7 feet in thickness, and, with the beds above and below, inclines at an angle of 43° to the east. It is the great dip of the beds which renders these coal-beds far less desirable than those on South Boulder. Seven beds of coal have been opened on Coal Creek, and underneath the third bed is a layer of excellent fire clay, 6 or 7 feet thick. In this clay are found nodules of iron ore, containing impressions of leaves of deciduous trees. This nodular iron ore is abundant everywhere, however, in all the beds. Some of the nodules are filled with clay or sand. The coal has not been worked here for over two years, and, although only seven beds have been examined, there is no reason to suppose that the whole series will not be found on further examination. The distance from the Marshall's Mine, or South Boulder, to the Coal Creek opening, is just three miles; yet intervening, is a high plateau, 1,200 to 1,500 feet above the bed of the Boulder, extending close up to the foot of the mountains. Although this plateau is mostly composed of coal strata, yet it is covered with a vast thickness of drift material washed down from the mountains, so that the intervening formations are for the most part concealed. As in the South Boulder Valley, no Jurassic beds, and only traces of one or two Cretaceous can be seen in Coal Creek Valley. Even where it emerges from the mountains the drift material juts up against the almost vertical sandstones of the Carboniferous period, yet we cannot doubt from what we have already seen that the complete series of rocks exist underneath. All along the foot of the mountains in this region, in the valleys and on the hills, there is a wonderful accumulation of particles of worn, and even angular rocks, so abundant as greatly to impede traveling. They present also every variety of texture and composition; but what most strikes the attention is the partially metamorphic condition of many of these immense masses of the pudding-stone, which are found so abundant in the Carboniferous rocks, lying scattered about; the inclosed pebbles appearing not to have been affected by heat, while the matrix is almost changed.

There are also worn masses of gneiss, quartz, and most of the varieties of rocks that probably exist in the heart of the mountains. In the lignite beds on South Boulder are indications of spontaneous combustion, by which the rocks also have been baked. In all the coal-beds are quantities of wood, and in sinking shafts stumps are brought out which show the woody structure very perfectly. In the beds are masses of wood which are partially changed, the surface of which is composed of beautiful, jet-like coal, while the inner portions show the layers of growth. One of the beds is called the "peacock vein," on account of the iridescent hue of the coal, and this peculiarity seems to be confined to this bed. Before leaving this locality I will say a few words in regard to the iron ore which is found quite abundantly in the sand and clay beds inclosing the coal. The iron ore is, as Dr. Torrey has determined, a limonite, commonly known as brown hematite, or brown iron ore. It occurs in nodular masses, varying from an ounce to a ton or more in weight, and is never found in layers or strata, but is distributed through the beds. On breaking these nodular masses, we find regular concentric layers, looking much like the bands of an agate, oftentimes varying in color from brown to yellow. From the immediate vicinity of South Boulder Creek Mr. Marshall has already taken out more than 500 tons of the ore, and from the examinations which have been made it must occur in great abundance over an area of at least fifty square miles. I am convinced that for all practical purposes there is no limit to the supply of this ore. The experiments of Mr. Marshall, in the furnaces at this locality, show that 4,400 pounds of ore will produce one ton of pig-iron in about the following proportions: ore, 200 pounds; limestone, 20; charcoal, 13 to 15.

This produces a very excellent quality of gray iron. Although on Boulder Creek the coal-beds seem to be most largely developed and to present the greatest facilities for their study, yet there are openings in many other points of the Territory, at Golden City, and various localities farther down on Coal Creek and Boulder Creek. I was unable to visit any of these points, but I saw some of the coal which had been obtained from there. It is used more or less for fuel at all the little towns in this region, but I do not think it is as good as that from the Marshall's mine. That which I saw came from a mine opened along the stage route, low down on Coal Creek, and it had a dull-brown color and crumbled readily on exposure. Mr. Collier, editor of the Mining Register, gave me a list of exposures of coal in various parts of the Territory, and there is no doubt this list might be greatly extended. Twenty miles east of Cañon City coal occurs in bars or seams, also seven miles south of the same place. At Colorado City small seams are seen and it is taken out in bars. This is a very curious form and must have existed originally in the form of asphaltum, for it is now found in bars half an inch thick, two inches wide, and sometimes several feet in length. On Cherry Creek divide, twelve miles east of Denver and two miles above the stage road on Coal Creek, there is a 7-foot bed which Mr. Collier opened in 1860; two miles below this point another bed has been opened. There are here five distinct beds. Again, on Running Creek, twenty miles southeast of Denver, and on Bijoux Creek, sixty miles southeast of Denver, also on White River, western end of the Colorado. There is an excellent bed of coal, 7 feet thick, at Golden City, and three miles north of the city. Seven miles south of Golden City, in Bald Mountain, the jet is found. Seventeen miles up the Platte River from Denver, a bed 2 feet thick exists. It is said also to be found in the Middle Park. Leaving Boulder Creek I prepared to return to Cheyenne City, along the main stage road, which is usually distant four to fifteen miles from the foot-hills of the mountains. Near a little town called Valmont, on North Boulder Creek, there is the most remarkable trap dike I have seen in the Territory. It rises about 200 feet above the little village at its base, and trends nearly east and west. It is about a mile in length and seems to have been thrust up through the Tertiary beds with vertical sides, so that it looks like a wall. The base of the dike is so covered with débris that it is not possible now to see what effect it had on the rocks through which it protruded itself. The rock might be called a hornblendic trachyte of a dark, greenish-brown color, the crystals of hornblende appearing quite distinctly. In the valleys of Boulder and St. Vrain's Creeks are a number of little lakes, many of them alkaline. The settlements are quite numerous and farming and stock-raising seem to be carried on extensively. It is only in the valleys of the streams, where irrigation can be employed, that any crops can be raised. The grass is so nutritious here that sheep and cattle thrive well and seem to be healthy. About four miles east of the base of the mountains the ridges of yellowish-gray sandstone are seen, dipping at a slight angle. They may be traced to a point within twenty miles of Pole Creek, and are plainly Tertiary. Near Thompson's Creek the most conspicuous feature in the scene is a high wall of sandstone with a hole through it, which has received from the old trappers of the country the name of the Bears' Church. Mr. Carbutt took a most excellent photographic view of it. The Cretaceous beds here form a belt about five miles wide. Eighteen miles south of Cheyenne City. No. 5 is well developed, inclining at an angle varying from 10° to 15°, and containing in some hard concretionary masses, a species of *Inoceramus*. Reposing on No. 5, without any apparent break in the continuity, are the Lignite-Tertiary beds. The following section will show the order of the beds here, as they are exposed over a considerable area:

7. Arenaceous clay with an abundance of *Ostrea subtrigonalis?*
6. Coal, 5 feet.
5. Clay.
4. Reddish rusty sandstone in thin layers.
3. Drab, arenaceous, indurated clay.
2. Massive yellow sandstone.
1. No. 5, Cretaceous, and arenaceous clay, a yellow sandstone.

The bed of coal is 5 feet 4 inches thick, 200 feet from the entrance of the drift. In the bed above the lignite there seems to have been a layer which must have been 3 or 4 feet in thickness. The oyster-shells are scattered about in the greatest abundance, reminding one in their abundance, size, and form, of the small oyster along the shores of South Carolina and Florida. It is said that among the foot-hills of the mountains outcroppings of the coal have been found. These irregular seams of jet, before referred to, occur abundantly in this region. I shall dwell more fully on these western lignites in my chapter devoted to that subject.

CHAPTER IX.

SKETCH OF THE GEOLOGICAL FORMATIONS ALONG THE ROUTE OF THE UNION PACIFIC RAILWAY, EASTERN DIVISION.

The facilities for traveling afforded me by the kindness of Colonel Lamborn, secretary of the Kansas Pacific Railway, induced me to make a hasty examination of the geology along this route so far as the road had been completed. Accordingly I left Leavenworth City November 13, 1867, and proceeded directly to the valley of Kansas River by way of Lawrence. This valley is one of the most fertile in the West, quite broad, the banks sloping gently down, presenting, however, a good many exposures. All the rocks from the Missouri River to a point near Fort Riley undoubtedly belong to the age of the Upper Coal-Measures, and, as these rocks have been described very fully in a paper by Mr. Meek and the writer, published in 1859, I shall not dwell on them in this connection. Even at Fort Riley the upper beds, over the summit of the hills, indicate their Permian character, and must be included in the series called Permo-Carboniferous. From Manhattan to Fort Riley the outcropping beds of limestone are quite conspicuous on the sides of the high hills, showing also a considerable thickness of the intervening clay or sand beds which form the grassy slopes. The bed that furnishes the beautiful building-stone at Manhattan, Fort Riley, and Ogden, and which has been used for the construction of some of the finest houses in Kansas, shows a marked dip to the westward at least 10 feet to the mile. The rock can be wrought into any shape with great ease, and is most beautiful and durable. At Junction City the Permian magnesian limestone is so soft that it is cut into any desirable form with a circular saw, and is transported to all points along the line of the railroad. Masses, in the form of bricks for chimneys, caps, sills, &c., are made in the greatest abundance. Junction City, which is a city of 1,200 to 1,500 inhabitants, is almost entirely constructed of this Permian limestone. At the one hundred and forty-fifth mile-post west of the Missouri the bed of limestone from which so much building material is taken crops out quite conspicuously from the sides of the hills, about 30 feet above the river, but dipping so rapidly that it soon passes beneath the bed of the river. After passing the one hundred and fifty-fifth mile-post, the soft beds, or those intermediate between the well-known Cretaceous No. 1 and the Permian, begin to show their influence on the surface of the country, giving very broad level bottoms to the river, and the low, gently-sloping hills on either side. Very few exposures of the basis rocks are to be seen.

Near Salina the grass-covered, sloping hills, the entire absence of any rock exposures, and the wide, level bottoms, show that the underlying rocks are composed of soft, yielding sands and clays. We know these belong to the series intermediate between the Permian and the rusty sandstones of the Dakota Group, and, so far as we have yet determined, are of doubtful age. At Fort Ellsworth the cuts along the road show that the Dakota sandstones have appeared, and then continue on as far as Fort Harker and beyond. Everywhere we find a thick, superficial covering of arenaceous material which was evidently derived from the disintegration of No. 1. Our road was along the valley of the Smoky Hill Fork, and the bluffs in the distance look as though they were composed of loose sand. It is plain that as we pass southward the sandy beds below the red sandstone, as seen on the Little Blue River, increase in thickness and retain their variegated color. The soil is still good all along the road. Where the surface crust is cut through by the road we observe from one to three feet of vegetable soil. It is somewhat arenaceous and would be very favorable to the production of cereals and all kinds of garden vegetables, but there must be more or less irrigation. There is no timber here except a little fringe along the Smoky Hill Fork and some of its branches, and this is being fast removed as fuel for the use of the road. The surface is very rolling, showing the loosely aggregated character of the underlying basis rocks. The prairie grasses grow thickly over the ground, especially the short buffalo grass, which forms a mat in most places. No finer grazing country can be found in the world. At Wilson's Station I saw the chalky limestone of the Niobrara Group filled with *Inoceramus problematicus.* A part of the bed is in slabs or thinnish layers, as it usually appears wherever it occurs south of the Missouri River, but a part also is more massive, arenaceous, and rust-colored. Between the two hundred and forty-fifth and two hundred and fiftieth mile-stone west, the road cuts through No. 3 very distinctly, the whole country appearing to be underlaid by this rock. The superficial deposits are quite heavy in places, composed of a deep yellow or flesh-colored marl, with small white concretions of lime or chalk. This disintegrated material seems to take the place of the newer pliocene deposits on the Loup Fork and other places. Between the two hundred and fiftieth and two hundred and fifty-fifth mile-post we saw a small herd of buffalo within a short distance of the track, also a herd of forty or fifty antelope. As the locomotive moved along they seemed more like native cattle grazing over these limitless pasture fields. We saw the mirage which is not uncommon in this country. Even small ridges rose up into lofty mountains and every valley reflected the light, so that it appeared like a pond of water. At the two hundred and eightieth mile-post there seemed to be indications of Tertiary beds, and in the cuttings are thousands of the small white limestone nodules so common in the marls in other localities. I made inquiry at the station, but no rock has yet been found in that vicinity. The bluffs along the creek have a whitish appearance. At Hays City the massive rocky layers of No. 3 are sawed into blocks with a common cross-cut saw, and employed in the construction of buildings. It is a rather hard chalk, but is very easily worked, the shells, as *Inoceramus*, *Ostrea congesta*, &c., obstructing the saw to some extent. The builders informed me that this rock hardens after it is laid in the wall. The quarry from which this chalky limestone is obtained is about four miles from the village, and the bed is 40 feet thick. Fort Hays is located on Big Creek, a branch of Smoky Hill Fork. It is surrounded by a broad, rolling prairie country, and no timber in sight except the very narrow, irregular fringe of cottonwood along the streams. The soil appears to be fertile; indeed, it could not be otherwise with the basis rock composed of the limestones of No. 3, but the difficulty of irrigating will prevent the country from being used otherwise than as a grazing region, for which purpose it seems eminently suitable. The short nutritious grasses which are peculiar to this dry prairie region cover the surface like a mat, and it is not strange that the buffalo are so fond of this portion of the West, and still visit it every summer. About eight miles west of Hays City there are about 60 feet exposed, of the dark clays of No. 2, of the Fort Benton Group. The road cuts through it so as to show its character well. It is a bright bluish-black slaty clay, covered with a thin coating of iron rust whenever the water or air can have access to it. It is full of arenaceous concretions of every size, which are lined inside with crystals of calc-spar. On the summit of the hills, resting direct on No. 2, are the massive layers of the yellow chalk No. 3, containing a few oysters and a huge *Inoceramus*. On the Smoky Fork are exposures of the bluish clays, with thin layers of arenaceous limestone containing fossils. At Fort Wallace the buildings are all made of this whitish chalk. This rock is also used in the construction of bridges along the road in many places, and, although not a durable stone, seems to serve a good purpose. We have, therefore, between Salina and Fort Wallace, exposures of three divisions of the Cretaceous, Nos. 1, 2, and 3. In No. 1 are found in great abundance well-preserved impressions of dicotyledonous leaves, as magnolia, oak, sassafras, poplar, sycamore, willow, fig, walnut, &c. There are also various casts of shells of unknown species. In No. 2, not far from Fort Wallace, have been found, the remains of gigantic reptiles 40 or 50 feet in length. No. 3 contains an abundance of the usual species, *I. problematicus* and *O. congesta*, with fragments of fishes. The interval between Wallace and Denver has not yet been examined by a competent geologist, and, therefore, nothing definite is known in regard to its geology. This chapter contains such observations only as I was able to make during a brief trip along the line of the railroad.

In regard to the resources of Kansas, I cannot do better than to close this chapter with an extract from a memoir prepared by Mr. Meek and the writer from observations made in the summer of 1858, and published in the Proceedings of the Academy of Natural Sciences at Philadelphia:

"Although this paper is merely designed to give a brief sketch of the leading geological features of those portions of Northeastern Kansas visited by us, we cannot close it without alluding to the truly great agricultural and other natural resources of this new and interesting territory. We mean no disparagement to other portions of the Mississippi Valley when we state that after having traveled extensively in the great West, and after having seen many of its most favored spots, we have met with no country combining more attractive features than Kansas Territory. Her geographical position gives her a comparatively mild and genial climate, intermediate between the extremes of heat and cold, while the rich virgin soil of her beautiful prairies is admirably adapted to the growth of all the great staple grain and root crops of the West. It is true that in some districts there is rather a deficiency of timber, but as a general thing there is along the streams sufficient for the immediate wants of the country. In addition to this, the wonderful rapidity with which forests are known to have sprung up on similar prairie lands in Missouri, as the country became settled so as to keep out the annual fires, shows that the present scarcity of timber should not be regarded as presenting any serious obstacle to the settlement of the most extensive prairie districts in Kansas. Before going out into the interior of the Territory we had expected to find the whole country immediately west of Fort Riley comparatively sterile; on the contrary, however, we were agreeably disappointed at meeting with scarcely any indications of decreasing fertility as far as our travels extended, which was about 60 miles west of Fort Riley. Here we found the prairies clothed with a luxuriant growth of grass, and literally alive with vast herds of buffalo that were seen quietly grazing as far as the eye could reach in every direction.

"Even on the high divide between the Smoky Hill and Arkansas Rivers, south of this, we found the soil rich and supporting a dense growth of grass; and from all we could learn from persons who have gone further out, the same kind of country extends for a long distance beyond this toward the west. Hence we infer that the belt of unproductive lands between the rich country on the east and the eastern base of the Rocky

Mountains on the west is much narrower than is generally supposed; and even this so-called desert country is known to possess a good soil, which may be rendered fruitful by artificial irrigation.

"In regard to the mineral resources of Kansas we have at present only time and space to say a few words: As already stated, coal is known to exist, though its extent is not yet fully determined, at several localities in the region of Leavenworth City, while the geological structure of the country, as well as discoveries already made, warrant the conclusion that this important and useful mineral abounds at many localities south of there. Limestone suitable for building purposes and the production of quicklime exists throughout large areas, while inexhaustible beds of gypsum are known to occur at several places not far west of the mouth of Solomon's River. Near this place we likewise saw in the lower cretaceous rocks crowning the summits of the Smoky Hills deposits of iron ore, but were unable to determine, in the limited time at our command, whether or not it exists in large quantities. Of the discoveries of gold in the mountains on the western border of Kansas much has been said; nothing, however, but a thorough geological survey, by authority of the State government, can lay before the public such full, accurate, and reliable information on these subjects as will bring from the older States the capital, skill, and enterprise necessary to develop the great natural resources of the country."

CHAPTER X.

THE EXISTENCE OF BEDS OF PEAT IN NEBRASKA.

The subject of peat is also of the highest importance to the West, and one which deserves much more attention than it has yet received. While the dry climate of Nebraska would necessarily prevent the accumulation of vast deposits of this useful fuel, such as we know to occur in many portions of Europe and on the Atlantic coast, yet even if beds of two feet or more in thickness can be found of peat that can be used as a fuel, it will prove a discovery of inestimable value to the State. Peat is usually regarded as holding a position next to coal in its value as a fuel, but of course its real value will depend upon the amount of carbon it contains. Low, marshy, swampy places occur all over the State, and, so far as I could determine, in a somewhat hasty examination, the conditions indicated were favorable to the existence of valuable beds of this fuel in almost every county. It is my opinion that at no distant day peat will become an article of great pecuniary value in the West, and that large fortunes will be made in its preparation from lands which are now regarded of little value. The first step to be taken is to ascertain the elements of success; and the proper way to answer the question whether there are actually any peat-beds in the State, and whether they can be wrought with profit, is to proceed at once to some neighboring State where the working of peat-beds is in successful operation, and determine the character of the peat and the mode of its preparation. Much discredit is often thrown upon new enterprises on account of the want of knowledge and experience on the part of the persons concerned, and the subsequent almost certain failure is attributed to other causes. It has already been shown by Dr. White, the State geologist of Iowa, that peat exists in great quantities in that State, and it is even now wrought with profit by companies which have been organized for that purpose. This subject is regarded of so much importance to the people of Iowa that it is receiving a large share of the attention of the State geologist. In Minnesota, also, there are said to be vast deposits of peat, so that the supply is almost unlimited. Now if these statements are true, if we had no other knowledge, we should at once infer that peat beds of greater or less size and value must occur in Nebraska and the neighboring Territories. The conditions, as well as the character of the vegetation, must be nearly the same in all these States. Peat, so far as the West is concerned, may be defined as the slow decomposition of flags, rushes, and the common sedges and grasses under water a portion or all of the year. But very little accurate information in regard to this matter is known in the West. Many people who have read much about the peat-bogs of Ireland suppose that in order that there should be peat deposits there must be a vast growth of mosses. The great dryness of the climate prevents the growth of mosses to any extent in the West, and very few species occur there. Still in many of the Eastern States this subject is exciting much attention, and enormous prices are sometimes paid for peat-bogs. Some expensive machinery has also been invented for its manufacture into fuel. At Pittsfield, Massachusetts, there is a company that manufactures 100 tons of crude peat per day, which, when dried, is reduced to 30 tons. In New York State a peat bog was purchased at the rate of $400 per acre, and a gentleman in New Jersey refused $25,000 for 28 acres. Now, if the reader will refer back to the special reports on the different counties he will observe that there is scarcely a county in the State which has not more or less of those low bogs, in which there is a rank growth of reed grass and bulrushes, which are the resort of myriads of muskrats. In most of the true peat-bogs the conical huts of the muskrat may be seen. Now, where these masses of vegetable matter are covered with water so that they do not have access to the air for the greater part or all of the year, a process of slow decay or combustion goes on by which this vegetable matter is converted into peat. There are no large beds of mosses as we so often see east of the Mississippi, and even in the region of the Rocky Mountains this family of plants is very meagerly represented.. In Ireland, Scotland, and some other portions of Europe, there are beds of peat or turf 20 to 40 feet in depth. The heather grows in the greatest luxuriance, and especially the sphagnum, which has a wonderful avidity for moisture and remains fresh all summer. These mosses decay at the roots while the tops keep fresh and green, so that these peat-bogs increase in depth from year to year. In our Western as well as Eastern States there are certain kinds of aquatic plants that contribute largely to the vegetable matter in a bog. In all the marshy bottoms or low-lands of the West we find, oftentimes in great abundance, the water *Polygonums*, the pond lilies (*Nymphæa* and *Nuphar*), duck-weeds (*Lemna*), pickerel-weed, (*Pontederia*), arrow-weed (*Sagittaria*), pond-weed (*Potamogeton*). Many of these plants grow in deep water, with stems several feet in length, and with such luxuriance as to contribute yearly a large amount of vegetable matter to the bog. There are many kinds of peat which may be readily detected by the experienced eye. These differences are sometimes in color, some kinds being red, others gray, others black; sometimes they are almost entirely destitute of fiber or any traces of vegetation; again they are light and porous, and do not seem to be far advanced in the stage of decomposition. Some kinds of peat are so pure that, burning, only a small per cent. of ash remains. Others contain much soil, lime, iron, and other mineral substances.

The geological formations of the country seem also to determine the character of the vegetation which forms the peat. The surface rocks of Nebraska are mostly calcareous, and the vegetation of the low places is composed of the coarse grasses and sedges. In granitic and silicious districts, according to Professor Johnson, the peat-bogs are likely to be filled with mosses.

Peat as an article of fuel.—Perhaps it is hardly necessary, at this time, to enumerate, in detail, the multiplied uses of peat as an article of fuel, and yet the want of information on the subject has been felt by the writer, even in his more recent investigations. He has made use of all the writings within his reach, but for the facts contained in this chapter he is mostly indebted to a thoroughly-scientific work by Professor S. W. Johnson, "Peat and its Uses," and an interesting volume by T. H. Leavitt, "Facts about Peat." To these books, which are readily accessible to any one, the reader is referred for any further information. Mr. Leavitt has invented a machine which will work off 100 tons of crude peat, making 25 tons or more of condensed fuel, per day. The crude peat is dumped into a hopper, wet, as it comes from the ground, completely pulverized, so as to destroy its porousness, and worked into molds like bricks. When we consider that there is, in all probability, hundreds of acres of peat ground in Nebraska that would, with this machine, yield from 100 to 500 tons of condensed peat per acre, worth at least $10 per ton for fuel, we can at once conclude that the subject is well worthy of our earnest attention. This prepared peat seems to be a favorite fuel for all domestic purposes. It is commonly cut or molded into blocks or sods like bricks, with a length of 8 to 18 inches, a breadth of 4 to 6 inches, and a thickness of 1½ to 3 inches. There are other forms made, sometimes in circular masses or in balls. Mr. Rogers, of England, found by experience that peat was prepared by the drying process, in the month of March, by the strong winds, rather than in the hotter months of June, July, and August. So when there was little or no wind he created an artificial wind for that purpose by placing the sods in wicker frames and swinging them through the air, thus drying the sods in less than half the time. When thus prepared it is used in the heating of dwellings, by means of a furnace, stove, or open grate, and is said to give a more steadily-intense and mellow heat than any other kind of fuel. It is also used in manufacturing establishments in the production of iron, and the fact that it is free from sulphur makes it a favorite fuel for that purpose. It has been used also very successfully for generating steam. Experiments are said to have been made on the locomotives of the New York Central Railroad, where a half ton of peat performed the work of one ton of coal. Another experiment was tried on a steamboat on the Hudson River, when 1,200 pounds of peat lasted two hours and twenty minutes, while 1,200 pounds of coal were used in one hour. In France turf-peat is a favorite fuel for the manufacture of iron. At one of the shops 4,500 pounds of turf and 2,300 pounds of pig-iron produced one ton of puddled iron; 26 cubic feet of turf and 2,500 pounds of pig-iron yielded 2,000 pounds of bar-iron of superior quality. In Bavaria, also, 3,000 pounds of dense turf to 2,450 pounds of puddled iron produces a ton of small bars of fine quality.

Peat charcoal and turf charcoal are much used, and their freedom from sulphur and acids renders them very useful. In Holland the manufacture of brick, alum-works, breweries, bakeries, are all carried on with turf as fuel. Mr. Rogers, at the request of the Emperor of France, made experiments on the Paris and Orleans Railway, and he found that the quantity of steam produced by peat was three times greater than that of coal coke, while the steam was got up in one-half the time. The blaze was so great as to pass out of the furnace of the engine. The use of peat as fuel is almost universal in European countries; and east of the Mississippi the attention has been turned in that direction within a few years. It is found on examination to be widely distributed, and it is probable that it will be found to a greater or less extent in every State and Territory of the United States. The area covered with peat will, of course, depend much on the geological formation and the climate. We cannot look for such unlimited deposits in the West as in the East. The climate of the West is too dry, and there is comparatively little wet or boggy land, and an extensive swamp is hardly known. It is estimated that there are 120,000,000 tons in Massachusetts, which at the low price of $1 per ton would yield the enormous revenue of $120,000,000. In the State of Virginia several companies have been organized to work the peat-beds of that State; and it has been ascertained that the great Dismal Swamp, which had heretofore been regarded as waste land, is one of the most valuable deposits of peat in America. The Dismal Swamp Peat Company has already commenced operations with marked success. In a report from the latter company the following paragraph occurs, which at once reveals the interest connected with these great deposits:

> The peat of the Dismal Swamp has been the growth of uncounted ages. Recent geological investigations have established the fact that 'The Bare Garden,' which is the richest and best portion of this enormous peat field, was once covered by a gigantic forest of resinous woods, principally the gum-tree, cypress, juniper, and pitch-pine, which flourished in primeval luxuriance. When these forests were prostrated by convulsions, or the slow process of decay, they were decomposed and covered with mosses and grasses which accumulated for many centuries until, by gradual chemical changes, peat, extending to undiscovered depths and richer in caloric ingredients than almost any other peat hitherto discovered. Analyses of specimens by the ablest chemists in this country have fully attested its value. And its future influence on this country will be surpassed only by the great coal deposits.

If the few words written on this subject aid in awakening an interest in this matter throughout the West, the object will be attained.

CHAPTER XI.

ARTIFICIAL BUILDING MATERIALS.

The importance of ascertaining what kind of building material can be used, not only in the State of Nebraska, but in a large portion of the West, which can be manufactured without the use of fuel, cannot be overestimated. In this chapter I have gathered such facts bearing on this point as seemed to me to be of value to the people of the West, and to them I would call their earnest attention. I am convinced that if a company could be formed which would enter into the business of making pressed brick, pisè, patent concrete, or any excellent building material which could be afforded cheap to the settlers of these almost treeless plains, it would confer a very great benefit on the inhabitants, and secure a large pecuniary return. The absence of fuel either above or beneath the surface of so large a portion of the West will be my excuse for the suggestions which I shall make in this chapter, whether any of them are ever adopted or not.

1. In regard to pisè, (pronounced peè-za,) a French term given to a method of building which has been long in use on account of its cheap-

ness and economy in many of the departments of France, and has always commended itself on account of its neatness, strength, and durability, and as the earthy materials for such buildings are very abundant all over the State, and indeed in every part of the West, I will here give a brief statement of the method of preparation, referring the reader to Rees' Cyclopedia, under the article Pisè, for the details and illustrations. The walls of the building to be erected are formed by means of a mold made of boards, which are placed on their sides so as to inclose a space equal to the desired thickness, and then the earthy materials are placed in this mold in a damp condition, and rammed into a compact mass by a rammer of a peculiar kind, when it is left to dry. This rammer should be wedge-shaped, and made of the hardest kind of wood. Any small lumps of earth are thus crushed and the materials are well mixed, while the surplus water in the earth is pressed out, and the particles of earth are more closely united. Hence these houses or walls are constructed with great rapidity, and are remarkable for their healthiness, cheapness, strength, and durability. It is said that these walls may be made at the rate of three courses of three feet each in length in a day, and that as soon as the walls are high enough to receive the roof the heaviest rafters can be placed upon them without danger. The next question that arises is, where shall the farmer look for the earthy materials he needs for his purposes, and are they abundant in the State? The materials are all around him in unlimited quantities. In Rees' Cyclopedia we have the kinds of earth given which are suitable for this purpose: 1st, all earths in general are fit for such use when they have not the lightness of poor lands, nor the stiffness of clay; 2d, all earths fit for vegetation; 3d, brick earths; but these, if they are used alone, are apt to crack, owing to the quantity of moisture which they contain. This, however, does not hinder persons who understand the business from using them to good advantage. 4th, strong earths, with a mixture of small gravel, which for that reason cannot serve for making either bricks, tiles, or pottery. These gravelly earths are very useful, producing the best work of this sort. Any one who is familiar with those vast superficial deposits of sand and gravel which cover all of Western Nebraska and the greater portion of the treeless surface of Colorado, Montana, and Dakota, know how abundant the best of materials are for this kind of work. There are also certain marks by which the commonest observer may determine whether the right kind of materials are in his vicinity. When a spade or plow brings up lumps of earth; when arable land lies in lumps; when field mice are able to make subterranean passages; when roads or streams form ruts or channels in the ground and the sides remain firm, but especially are these materials found in great abundance at the foot of the hills where vines are planted, and all sloping cultivated lands. The presence of lime greatly improves the quality. It therefore follows that the yellow marl which attains such a great thickness all over the eastern portion of Nebraska, must be especially valuable for this purpose. In case the exact kind of earth is not found at the place where it is necessary to build, it is important to learn how to mix the earth so as to render it suitable. Strong earths must be tempered with light; when clay predominates there must be a mixture of chalk or lime and sand, and the proportions in which this mixture is made must be determined by the judgment of the builder. It is better to mix some small pebbles, gravel, rubbish, indeed any small mineral substances, but nothing of an animal or vegetable nature. These harder substances bind the earth firmly, and being pressed, and pressing in all directions, renders the whole more firm and solid, so that at the end of two years the wall becomes as hard as freestone, and a chisel must be used to break it, like any solid rock. This material can also be used for the building of walls, as the inclosures for parks and gardens, for fences, for fields, &c. The ancients used it for making rough walls; the Italians now use it for the construction of terraces that cover their houses; the Moors for all their walls; the Spaniards, French, and others, for some of the floors of their apartments. The strength and durability of these houses, built in this way, is shown from the fact that some of them have been pulled down, the title-deeds of which, in the possession of the proprietors, showed them to be one hundred and sixty-five years old, and they had been kept in poor repair. The rich traders of Lyons have no other way of building their country houses. These buildings, by painting in fresco, may be made to look as beautiful as the owner may desire by the use of red or yellow ocher, or any other mineral paint. Along the Rhine many of the beautiful mansions which gladden the eye of the traveler as he glides along its banks, are made of nothing but these earths. In the dry climate of the West these houses could be inhabited almost as soon as finished. Besides the great strength and cheapness of this method of building, it has the advantage of speed in its execution. One may calculate to a day the time when a certain work can be completed. It is known that two men can build in one day six feet square of the Pisè; then six men, which is the number required (three in the mold, and three to dig and prepare the earth), will build in sixteen days, or at the end of three weeks, at least 228 square feet of wall, or a solid and lasting habitation sufficient for the necessities of a family. Six men can complete a wall 540 feet long and 6 feet high in one month, and twelve men can finish the same work in fifteen days. Thus, by multiplying the number of molds and workmen, the rapidity of construction will be proportionately increased. The building is floored, roofed, and finished within like a stone or brick house, except that lathing on the walls is not necessary. The above facts are given merely to show the simplicity of this process of building, and its admirable adaptability to the wants of the western people.

Patent concrete.—Although Ransom's patent concrete, which is so popular in England, may never be brought into general use in this country under that name, yet the principle involved must be adopted to a greater or less extent; and as it seems to be so valuable, even to the production of a better and stronger building material than can usually be found in a natural state, I will here notice a few of its prominent features. It seems to me very proper that some of the methods of making artificial stones be briefly described in this connection, in order that the attention of builders may be called to it and some advantage be taken of the vast supply of materials that cover the country. The materials which are used in forming the "patent concrete" are mostly sand, which may be molded into any form. Even rubbish, fragments of rock, and almost any kind of loose sandy material, may be employed, and there is scarcely a county in the State that has not an abundant supply. The rock is made with great rapidity, and is ready for use without drying or burning. It hardly requires a temporary shed in this dry climate, for its manipulation is not affected by the weather. The rock becomes, when made with sand and cemented with the silicate of lime, a true sandstone. It is made by saturating the blocks in a solution of the silicate of soda and then applying a solution of the chloride of calcium. This produces a rapid double decomposition, leaving an insoluble silicate of lime within the stone, and a soluble chloride of sodium (or common salt,) which could afterward be removed by washing. A coating of hard silicate of lime has actually formed and deposited on the surface. Mr. Ransom made small blocks of various forms in the presence of a committee appointed for the purpose of examining the merits of this invention, mixed coarse sand with the fluid silicate of soda, and then dipping the mold into the chloride of calcium, there came out almost instantly a compact, solid, and evidently durable material. In such solids there appears to be no element of destruction. Many experiments were tried to ascertain the strength of this material, and it was found that very few of the natural rocks could compete with it. It was compared with the celebrated Portland and Caen stone, and found that a bar of the concrete stone 4 inches square and 8 inches between the supports sustained 2,122 pounds, while the Portland and Caen broke respectively at 750 and 780 pounds. It was also shown that the adhesion of the concrete by suspension was 1,980 pounds, while the others separated at 1,104 and 768 pounds. The lectures of Professor Ansted, of England, on the various kinds of economical rocks are of great interest, and he particularly commends this concrete as one of the cheapest, most beautiful, and durable of our artificial building materials.

In connection with this subject I would call the attention of the western people to the American Building-Block Company, the office of which is at No. 24 Vesey street, New York. The principle involved is very similar to the one already described. The merits of this invention may be best explained by the following extracts from a pamphlet published by the company:

A good, durable material for the construction of buildings, that combines facility of manufacture, cheapness, beauty of color and shape, convenience in handling, resistance against changes of atmosphere, and complete fitness for the uses for which it is intended, has been a want so well understood, that much time, capital, and labor have been devoted to its attainment.

The building-blocks manufactured by this company are believed to fully and completely meet all the above requirements.

These blocks are easily and rapidly made, each machine being capable of turning out about five thousand per day, and three or four machines may be run by one steam engine of thirty-horse power.

They are composed of the cheapest known materials—mainly sand and lime—and are made in such form and size that walls can be construted from them as cheaply as with good common brick.

In their external appearance these blocks make a building of unsurpassed beauty. The shape is entirely uniform, with sharp, well-defined lines, and they can be made of every variety of shade, from a pure white to a dark-brown stone color.

These blocks, as now manufactured, are ten inches long, five inches wide, and four inches thick, containing two hundred cubic inches, and weighing about eleven pounds each; they have an air-chamber running through the center, which facilitates the handling of them, both in transportation and in their construction into buildings.

The fact that air is one of the best non-conductors, and a wall built with these blocks having air-chambers running vertically through the entire wall, shows conclusively that changes of temperature will be less felt in houses constructed from them than in those built of any other material.

The blocks, from the nature of the material used, and the severe pressure to which they are subjected in process of manufacture, are very durable in their character, as it is a well-known and established fact that mortar composition, properly made, is the most enduring of all substances, withstanding exposure for centuries, and constantly growing harder by atmospheric changes, until it becomes a perfect stone.

These blocks have been subjected to every conceivable test—have been immersed in water until they have absorbed all the moisture which they could hold, and in that condition have been exposed to severe frosts and then thawed, and the same process repeated again and again. After being subjected to all the alterations of the atmosphere, the result in all cases has but proved the *indestructibility of the block*. And it is believed to possess all the enduring qualities of the old Roman mortars, which have existed unchanged for over three thousand years, except to become harder and harder.

Blocks identical in *shape* and *size*, but not in hardness, known as the Foster block, and composed of much the same materials as those manufactured by this company, have, fortunately for the actual test of their durability, been made, and houses built from them ten or eleven years ago. These houses are stronger and better to-day than when first erected. The material would long since have been in general use, but for the difficulties and obstacles which have now, happily, been overcome by means of devices and improvements which are covered by several letters-patent owned by this company. The block, as before stated, is mainly composed of lime and sand, and is molded under immense pressure; but in addition to the pressure, the blocks, as formerly made, required time to become sufficiently hard to be fit for use; this was an objection. The machines used to mold them were presses made in various ways, not one of which proved strong enough to do the work for any profitable length of time. This was a very serious obstacle. And the fact of the blocks being insufficiently solidified by pressure, and the want of experience and knowledge in the manipulation and after-treatment of the materials forming the blocks, caused some of them to be made of an inferior quality, which did the composition injustice, and has retarded their general introduction,

The first of these difficulties has been overcome by improvements in the composition and mode of treating the mixture before and after pressure.

The second has been overcome by the invention of a machine so constructed as to produce all the density attainable.

The theory of this composition, relative to its strength and durability, is simple and easily understood, and has been fully demonstrated to be practically and chemically true.

Clean, sharp sand, when brought in contact with caustic lime in correct proportions, and under right conditions, with proper manipulation and after-treatment, produces a partial decomposition of the sand, by which silicate of lime is generated—*an indispensable condition and an invariable result.* In this condition the composition is ready for the mold and the press. The block thus formed, when exposed to the air, gradually parts with a portion of its moisture, and begins to absorb carbonic acid from the air, for which the lime has an affinity. This process goes on slowly; a portion of the water present, and that afterward absorbed, being chemically taken up and combined in the form of crystallization. This process goes on until the interstices of the stone are filled up, and the block becomes non-absorbant, growing harder for years, by the action and effect of the very agents that destroy freestone, marble, brick, &c., viz, water, frost, and carbonic acid from the atmosphere.

The novel form of this block, having an air-chamber running vertically through the center of each block, six inches long by one inch wide, is so arranged as to secure a circulation of air through the entire wall, operating as a perfect non-conductor of moisture, cold and heat, making a perfectly dry wall inside, a cool house in summer, and a warm one in winter.

This perforation likewise facilitates the seasoning or hardening of the block evenly and thoroughly, presenting more surface to the action of the elements that perfect the block. The size of the block above alluded to is considered, for general purposes, the most convenient, but the manufacturer is by no means confined to it; for it is contemplated to make all sizes, from that of the common brick to the ordinary-sized brown stone and marble fronts, plain or fancy, as it may be molded in any desired form, and is, therefore, adapted to all kinds of buildings, as churches, halls, factories, houses, stores, cottages, stables, barns, or any other class of buildings, including bridges, sewers, docks, water and drain pipe, lawn and garden ornaments, tiles, statuary, vaults, ornamental work, &c., combining, it is believed, more positive advantages than any other building material now in use.

While this block has heretofore been rather imperfectly made, it has stood the test of time and the elements. Ample evidence is at hand to prove to the entire satisfaction of the most skeptical the durability of the block. Buildings constructed ten years since from it have clearly demonstrated its indestructibility. The intelligent, practical mason, familiar with its component parts, admits it to be reliable; and those learned in the science of chemistry, pronounce it imperishable; and the only reason why this material has not been more generally introduced to the public for building purposes was for the want of proper machinery to manufacture the blocks.

Years of time and thousand of dollars have been expended in efforts to reach the point where we now stand. We are now prepared to offer to the public the results—a building material which possesses the following merits, many of which are peculiar to itself:

1. Its material composition is such that, so long as the laws of chemistry hold good, time will but make it more durable.

2. It is made in such form, and with such exactness of outline and dimensions, that walls built with it give complete protection against atmospheric changes, and are fire-proof. Buildings can be finished cheaper and better than if of brick or wood. Interior walls may be left unplastered when desired, or, if plastering be preferred, it can be had at one-third of the cost, and of a better and more durable quality than by ordi-

nary modes, as a coat of hard-finish may be applied directly upon the face of the wall, for which it has a strong affinity, attaching itself so firmly that it cannot be scaled off, thus saving the expense of studding, furring, lathing, and two coats of plaster, and leaving no hiding place for vermin that so frequently find shelter in a lath-and-plastered wall. The saving of the item of plaster is estimated to be about seven-eighths over the quantity required in constructing a brick lath-and-plastered house. There is another advantage gained, which is, the certainty of having walls that will never crack in the plaster.

3. The affinity of mortar for the material—both being of the same nature—is such that the bond is much stronger than in the case with brick or stone. In this respect the superiority is so great that it must be seen to be fully realized, as a wall built from this material becomes, in fact, one compact mass of stone.

4. The sharpness of outline, the shape and size, the exactness of dimensions, (which are not warped or impaired by burning, after being molded,) and the color of the blocks, which can be modified or varied to suit the taste, all furnish the skillful architect with means for producing the most beautiful effects.

For the interior of churches and public buildings, designed to be finished in imitation of stone-work, it is most admirably adapted, either for columns, walls, or ceilings, especially if arched and groined, as it is not liable to discoloration or cracking, as is the case with lath and plaster. You get all the effects and durability of real stone at a small part of the cost, and the subdued color of the blocks gives a tone and quality most grateful to the eye.

5. The economy attainable by their use is not the least advantage. These blocks may be made wherever good sharp sand can be found, (the places where sand is not obtainable are rare,) in sections where there is neither stone nor clay for bricks, and in such districts the value of these inventions can hardly be overestimated.

The above statement being true, have we not that which will satisfy the great want that has been felt throughout the ages, from the time when men became wise enough to build houses rather than huts? Combining the great elements of durability, availability, the greatest possible protection from atmospheric changes, strength of walls as a whole and in all their parts, beauty in color and texture, in form and proportion, cheapness without a sacrifice of any requirement—what more can be desired?

Professor E. N. Horsford, of Cambridge, Massachusetts, one of the most eminent chemists and scientific experts in this country, made a most careful examination of the merits of the building block manufactured by this company.

Foster's patent is *new*. It differs from common sand and lime mortar in essential particulars. It is a STONE, while mortar is a material for binding stones together. The latter cannot be substituted for the former. Foster's stone owes its distinguishing characteristic to its having been produced under pressure. By means of this pressure and the small proportion of water employed, a product is obtained from common materials, suited to new and distinct uses, where uniform size and determined form are important—as in the erection of buildings—and in which great density and strength are required. By means of this pressure the sand grains are settled into positions of less mobility, and more surfaces have been brought into contact or close proximity. Within the stone there is developed silicate of lime, a cement, at the surface of each sand grain and binding it to its fellow. This cement is more effective because of the sand grains. The less the thickness of the layer of cement required the more effective it is. The hardness of Foster's stone is due in some degree to the formation of hydrate of lime, and double hydrate and carbonate of lime, as in common mortar, and to the formation of silicate of lime at the surface of the sand grains, but more to the pressure, which, by reducing the thickness, has increased the effectiveness of the cements. Common mortar, in setting or drying, becomes porous from the escape of surplus water. Through the pores thus left, the carbonic acid of the air enters and forms the double hydrate and carbonate; but, at the same time, prevents the formation of the more tenacious silicate of lime.

I have analyzed a series of samples of artificial building stone, made by Foster and others—one of them that had served eleven years as part of a foundation wall, and others of various lesser ages, some of them only a few days old. In the oldest of them only was the lime all combined and rendered substantially insoluble. This sample was nearly as hard as Connecticut River sandstone. Others of only a few months' age were equally hard at the surface, but less firm in the interior. The surface of a freshly-made block, exposed to the air, rapidly hardens, while the changes within are slower. A fresh fracture of a block several years old shows a zone of peculiar shade, extending from the outer surface toward the heart of the stone. This zone marks the progress of certain chemical changes which attend the hardening process, and illustrates the fact of the improvement of the stone with age. These changes are accompanied by increase in weight, due to carbonic acid absorbed from the air, and moisture absorbed and combined with the steadily forming silicate of lime. I have found, by experiment and analysis, that a block of Foster's building stone, eleven years old, contains from ten to fourteen times as much silica combined with the lime in the process of manufacture, and after it was molded and pressed, as was contained in a fresh block made by the same process. This increased amount of silicate of lime, as well as of double hydrate and carbonate of lime, is in keeping, as already suggested, with the observed increased hardness of the stone as it grows older.

The newness of Foster's patent is in the process and in the proportions of the ingredients he employs, rather than in the character or kind of the ingredients themselves. Of these ingredients he makes a portable stone of definite form, having the same kind of advantage for building purposes that bricks have. The originality and validity of the patent seem to me beyond doubt.

Van Derburgh's invention is the application of heat to the mixture of sand and moist hydrate of lime, to increase the amount and effectiveness of the cement before the blocks are molded. The *amount* of cement is increased, inasmuch as the production of silicate of lime from a mixture of sand, lime, and moisture is facilitated by heat, as I have demonstrated by experiment, and this has been brought to bear for a length of time before the materials are molded and pressed. The effectiveness of the cement is increased, since the mode of manufacture spreads the cement more uniformly over the surface of the sand grains.

It has been found in practice that blocks made by Foster's process are subject to the condition of the atmosphere, whether moist or otherwise—that is to say, under the influence of a dry atmosphere, immediately after molding and pressing, they do not harden; whereas blocks made by Van Derburgh's process are not influenced by the hygrometric condition of the atmosphere, but the process of hardening goes steadily and uniformly forward. It is also found, in practice, that by first slaking the lime and then mixing it, as a powder, with moist sand, as in the process of Foster, the slaked lime does not so effectually and uniformly coat the sand as when the mixture is made with moist sand and finely-ground unslaked lime, and the whole subjected to heat during the process of slaking, as in the method of Van Derburgh. By slaking the lime in confined space, with, or by means of, the moisture at the surface of the sand grains, the nearness of the hydrate of lime formed to the silica, on which it is to act, is greater, and its chemical affinity is greater, both from the nearness and from the heat due to slaking, and to the admitted steam, than it can be where the hydrate is cold and dry, and encounters moisture at the surface of the sand grains. In the latter case the condition approximates to that of milk of lime and sand, in which, according to Fuchs and Petzholdt, little or no action takes place. In the former, chemical affinity between the silica and lime is aided, not only by the nearness and heat, but also by the nascent condition of the caustic lime, at the instant of its slaking, in contact with the sand. In Van Derburgh's process the ingredients enter the block in a more advanced stage of the chemical action that is to result in solidification; more freshly-formed silicate of lime exists in the block, when molded and pressed, under Van Derburgh's improved process than under the process of Foster. Microscopic examination shows the individual sand grains in a Van Derburgh block to be coated by a more transparent crystalline cement than is the case in Foster's block. This transparency and crystalline character are evidences of greater tenacity in the cement, because of the greater extent of surface and thickness through which the cement exerts its binding force. They are due in part to the hydrated silicate of lime, in part to the double hydrate and carbonate, and, doubtless also, in a fresh fracture, to the crystallized hydrate. This hydrate, on exposure to the atmosphere, absorbs carbonic acid, forming additional double hydrate and carbonate, which imparts greater tenacity and hardness to the extent of the action, and accounts, for the most part, for the rapid hardening which a fresh surface experiences on exposure to the atmosphere.

I have prepared samples of building blocks from Berkshire sand and chemically pure lime, as nearly as might be in the laboratory, according to the practical working of Van Derburgh's, and according to the patent of Foster. These blocks, upon analysis, gave the following results:

Silicic acid combined with lime, for every hundred parts of quicklime employed—

In Van Derburgh's	5.02
In Foster's	2.76

I have also analyzed a block of Van Derburgh's stone, some twenty months old, and find the proportion of silica derived from the sand of the block, by the action of the lime at the time the block was made, and during the period that has since elapsed, to be very large. Assuming the composition of the block to be nine of sand to one of quicklime, I find the proportion of silica, produced by the action of the lime, to be 49.67 for every hundred of lime. This percentage, compared with the silica of a block of Van Derburgh's stone freshly prepared, using chemically pure lime, gives a ratio of 0.7430, 0.0751, or nearly 10 to 1. Compared with the silica in a block of commercial quicklime, it gives a ratio of about 42 to 1; while Foster's block, eleven years old, gives a ratio in the best specimen of 13.5 to 1.

I have found, by experiment, that a fresh block, made by Foster's process, placed in an atmosphere made artificially dry, does not harden so rapidly, or so thoroughly, as in ordinary atmospheric air, and especially as in moist atmospheric air. I have also found that, in an atmosphere highly charged with carbonic acid, the hardening of the block does not seem to be promoted, if the atmosphere be dry. These results might have been inferred from the general principle that the chemical action of the lime on silica, as well as the absorption of carbonic acid, and its combination with lime, demand moisture. They confirm the results of experience with the Foster building blocks, before mentioned, and enable one to see why a block in which more silicate of lime is formed at the outset, as in Van Derburgh's, should be less dependent, for its hardness, on the hygrometric condition of the atmosphere.

I regard the practical working under the Van Derburgh patents, as now carried out, as based upon a new and distinct invention, producing results decidedly superior to those formerly obtained by the process of Foster.

I have given an opinion to Mr. Strout, of Portland, on the claims of Foster, Van Derburgh, Ransome, Ruschhaupt, and other patentees, in which my attention had been drawn to only three out of seven of Van Derburgh's patents bearing on the subject. From the language of the patents submitted to me, I did not gain a just conception of Van Derburgh's processes, as practically carried out, and objected to them on account of the apparently great time required, and the moderate production to be expected. I have become satisfied from actual inspection of the process of manufacture that this great time is, in practical working, unnecessary, and that the whole operation is very simple, and permits continuous and rapid mixing, molding, and pressing.

The claims of Van Derburgh, Ransome, and Ruschhaupt, to some extent, cover common ground.

Van Derburgh may justly claim:

1st. The slaking of the lime in contact with the moist sand, which is to be made into building blocks, and, after subsequent dampening, subjecting the mixture to pressure to form building blocks.

2d. The application of heat to a mixture of lime, in the process of slaking, and moist sand, the heat being derived from the slaking lime, or from this and other sources combined, preparatory to pressing into blocks.

3d. The slaking of the lime in contact with the moist sand in confined space, by which very great heat is brought to the aid of the chemical action of the caustic lime on the silica of the sand.

4th. The subjection to pressure of a mixture of moist sand and lime, slaked in contact with the sand of the mixture, after having been subjected to heat.

5th. The use of a mixture of lime, slaked or unslaked, with sand, with the addition of a liquid silicate immediately before molding and pressing.

6th. The saturation, with a solution of an alkaline silicate, of blocks made by pressing a mixture of moist sand and caustic lime.

7th. The use of saccharate of lime and soluble silicates in the formation of silicate of lime in building blocks.

8th. The use of freshly-broken sand grains, to fill interstices between coarser grains, and thus produce a smoother and harder artificial stone.

9th. The application of pressure, by percussion, to a mixture of sand and lime, in whatever form his patents cover.

That the manufacture of artificial building materials without the use of fuel will be an object of zealous pursuit is an absolute necessity. The various earths for that purpose are distributed all over the country in the greatest abundance, and the peculiar dryness of the atmosphere renders the climate especially adapted to success in that direction. The above extracts have been made in this chapter for the purposes of conveying information to the people of the West on this most important subject, and we propose to pursue it much farther in future reports.

PART II.

PALEONTOLOGY.

REPORT ON THE PALEONTOLOGY OF EASTERN NEBRASKA,

WITH

SOME REMARKS ON THE CARBONIFEROUS ROCKS OF THAT DISTRICT.

BY F. B. MEEK.*

INTRODUCTORY REMARKS.

There are probably few well-informed geologists who will, at the present time, maintain that the occurrence of a very similar, or even the same group of fossils, at widely separated localities, necessarily proves the rocks in which they are found to be of exactly contemporaneous origin. The most that is now generally maintained in this regard is, that such identity or correspondence of types at very distantly separated parts of the world, indicates that the strata in which they are embedded were formed during the prevalence of identical or similar physical conditions at some time during the same great geological epoch, and that they hold the same, or nearly the same, *relative* position in the geological column of their respective districts. For instance, although a stratum in the Rocky Mountains, containing the remains of very nearly the same fauna as some particular subdivision of the Devonian system of Europe, might, for aught we know, be hundreds of years older or newer than that particular division, we would have little or no room for doubting that it belonged to the great Devonian series, or possibly even to some definite known horizon in that series. We could moreover very positively assert, in such a case, that it would be, according to all past experience, useless to seek there at any lower geological horizon for workable beds of coal, or to expect to find Silurian rocks or any of their peculiar products above, supposing there had been no overturning of the strata at the particular localities.

Hence, although paleontology does not enable us to ascertain the exact *actual* ages of rocks—when applied with due caution and skill in connection with a careful observance of their stratigraphical arrangement and lithological and other physical characters—it does afford the means of fixing their *relative* ages, as well as of identifying the same beds at different localities, within given fields of observation, with very considerable precision. It is therefore not merely *one* of the more important aids to the geologist in his investigations, but in the present state of geological science, it is the only sure guide in classifying and determining the order of succession of rocks, where this cannot be done by their actual continuity or obvious superposition. For these reasons, it is now the universal practice, in all geological surveys conducted upon sound scientific principles, to devote especial attention to this department.

In the present instance, this becomes the more particularly desirable

* I am under obligations to Professor Henry for all desired facilities while preparing this report, at the Smithsonian Institution.

in order to throw as much light as possible upon questions of both scientific and economical bearing, since rather widely different views are entertained by geologists in regard to the age and position in the geological column, of some of the rocks in Eastern Nebraska, with relation to the Coal-Measures and other Carboniferous deposits of this country and Europe.

This discordance of opinion has caused the writer to enter into more minuteness of detail in giving local sections, and repeated lists of fossils found in each particular bed and seam of rock seen at different localities, than would otherwise have been done. To all geologists familiar with our western Coal-Measures and their characteristic organic remains, it may also seem superfluous to bring forward arguments to prove that no part of the rocks under consideration can either belong to the Lower Carboniferous, or to the Permian. My apology for discussing at length a question already so clear to all in this country acquainted with the subject, is the fact that an eminent foreign geologist (Professor Geinitz, of Dresden) for whom I have great respect, from an examination of an incomplete series of specimens from these rocks (and doubtless partly from not being acquainted with the range of species and genera in the Carboniferous system of this country) has referred the outcrops here described along the Missouri in part to the Lower Carboniferous, and in part to the Permian. As his views on these points have probably been accepted, at least by some geologists in the Old World, it has been thought desirable to place before the reader all the facts now known that have a bearing on the questions at issue.

In order, therefore, to afford a fixed standard of comparison between the organic remains of the Nebraska deposits under consideration, and those of well-established horizons in the Coal-Measures of Iowa, Missouri, Illinois, Kentucky, &c., I have considered it proper to give figures and descriptions of as many of the species of these Nebraska fossils as possible. With this view, eleven plates of closely arranged figures have been prepared, illustrating a few more than one hundred species, belonging to the several groups, *Foraminifera*, Corals, *Polyzoa*, *Brachiopoda*, *Lamellibranchiata*, Trilobites, &c., together with a few fishes, all of which are more or less fully described in the following text.*

Although most of the species are known to range through the entire series, the figures of those from the several outcrops that have been referred by others to widely different horizons are arranged on different plates, in order to illustrate, as far as it can be conveniently done, the vertical range of the various types. To do this completely, however, would require the repetition of most of the figures on each of the plates from specimens obtained at Bellevue, Plattsmouth, Rock Bluff, and Nebraska City. But as the Omaha, Bellevue, Plattsmouth and Rock Bluff outcrops so obviously all belong to one unbroken series of Upper Coal-Measures,† and are even all acknowledged by Professor Geinitz, after a study of their organic remains, to be Carboniferous, the figures of their fossils given are here arranged together on Plates 1, 2, and 3; though the particular locality and position of each is mentioned in the text in

* These fishes have been described in a separate section by Professor O. St. John, of the Iowa and Illinois geological surveys, who has devoted several years to the study of that branch of paleontology with Professor Agassiz, at Cambridge.

† It may be proper to explain here, that by the use of the terms Upper, Middle, and Lower Coal-Measures, we simply mean the upper, middle, and lower parts of the true Coal-Measures, and not any divisions that are separated by constant paleontological or physical breaks in the series. We have no Coal-Measures below the horizon of the Mountain limestone of the Mississippi Valley, such as are sometimes, in other countries, called Lower Coal-Measures.

connection with the descriptions, and further illustrated in the accompanying sections. For convenience, this division of the series might be provisionally designated as the Platte division, from its development in the vicinity of the mouth of Platte River, where the various outcrops seem to exhibit altogether a thickness of between two and three hundred feet. It should be borne in mind, however, that there is no paleontological or constant lithological break between these beds and beds A and B, at Nebraska City, Wyoming, &c.; while bed B is closely related by most of its organic remains with the bed C above, at the first-mentioned locality.

The fossils figured on Plate IV, with the exception of a few noted in the explanations as being from outcrops on the Missouri below Nebraska City, at Brownville, Aspinwall and Rulo, believed to hold entirely, or in part, a position above the horizon of division C at Nebraska City, were all found (at various depths) in a shaft sunk nearly one hundred feet, on Hon. J. Sterling Morton's place, one mile and three-quarters west of the Nebraska City landing, commencing at an elevation of 73 feet above low-water mark of the Missouri. In regard to the exact relations of the beds penetrated by this shaft, to the outcrop at the Nebraska City landing, there is some room for doubts. If Professor Marcou was right in referring the beds exposed at Mr. Morton's place, at the top of the shaft, to the lower part of the section seen at the landing, then the beds passed through by the shaft must hold a position immediately below the base of the section at the landing, and between the latter and the horizon of Rock Bluff section; but if the outcrop at the top of shaft really holds a higher stratigraphical position than any part of the section at the landing, as there are some reasons (to be mentioned in another place) for believing, then the fossils figured on Plate IV, as the species would indicate, may belong in part to the horizon of division C, and in part to division B, of the section at the landing. This question, however, being involved in some doubt, it has been thought the better plan to figure the species from the shaft on a plate by themselves, distinct from those obtained at the Nebraska City outcrop; and this plate is placed provisionally between those containing the forms from bed B, and the plates including the fossils from the still lower Platte division.

The species found in division B, at Nebraska City, Bennett's Mill, and Wyoming, are arranged together on Plates V and VI; while all the known forms from bed C are arranged on Plates VII, VIII, IX, X, and XI. As we did not succeed in finding good specimens of all the numerous species known to occur in division C at Nebraska City, and it seemed highly desirable that they should all be here fully illustrated together, a few of Professor Geinitz's figures of species from this horizon have been copied, and in each instance duly acknowledged in the accompanying explanations.

Of all the species figured and described, comparatively few are entirely new to science, though many of them have not before been illustrated by figures; while the excellent illustrations of many of the others published by Professor Geinitz, in his recently issued work on the fossils of these rocks (Carbonformation und Dyas in Nebraska, 1866), are accessible to comparatively few in this country.

As Professor Geinitz has referred a number of these Nebraska fossils to European species, from which they seem to me to be entirely distinct, figures of the European species to which he referred them are also given on the same plates, for comparison. These have been copied with great care, from the most reliable figures accessible, and in most cases from those originally published by the authors of the species, in order to give the student the means of forming his own conclusions in regard to the relations of the American forms to the foreign species.

Having enjoyed an opportunity, during the spring of 1867, in connection with the Nebraska geological survey, to examine personally, not only in Nebraska, but through rather wide areas in Iowa, and portions of Kansas, numerous natural sections, quarries, shafts, drifts, borings, &c., in the rocks from which the fossils described in this report were collected, some of the facts observed may be appropriately stated here, before entering upon the paleontological descriptions.

Although thin beds of these Upper Coal-Measure limestones are known to exist under the bases of the hills bounding the Missouri Valley, a little above the surface of the river, for eight or ten miles above Omaha City, the quarry just below that place was the highest northern point at which these rocks were examined during the late survey. The base of these quarries is almost on a level with high-water mark of the Missouri, and the whole thickness exposed is only about eight feet, consisting of hard grayish and bluish gray, rather impure limestones, in four to ten inch layers, with more or less flinty concretions, and arenaceous clay partings, all showing a slight local east or northeast dip. The fossils found in these layers were *Fusulina cylindrica*, *Scaphiocrinus (?) hemisphæricus*, *Archæocidaris triserratus*, *Productus costatus*, *P. semireticulatus*, *P. longispinus*, *P. Prattenianus*, *Spiriferina Kentuckensis*, *Athyris subtilita*, *Spirifer cameratus*, and teeth of *Peripristis sermicircularis*.*

We were informed by one of the workmen at the quarry that a thin bed of black shale occurs beneath the limestone of the quarry, but it was not exposed at any place examined. The limestone is said to make good lime, and is well adapted to most kinds of masonry, while in hardness and strength, and the thickness of the layers, it is found to be an excellent material for door and window caps and sills. Unfortunately, however, the heavy deposits of Drift and Loess overlying these rocks, add materially to the expense of working the quarries. The superincumbent Drift is at some places from 10 to 40 feet in thickness, resting directly upon the hard, upper layer of limestone. It consists of arenaceous clays and sand, with at places oblique planes of deposition, and contains some pebbles and a few boulders. Above the Drift the Loess rises back with the slope from a few feet to perhaps 150 feet.

As noticed by Professor Egleston in his report of geological examinations made along the first hundred miles of the Pacific Railroad, as well as by Dr. White, in a paper published in the American Journal of Science, the upper bed of limestone at the above-mentioned quarries shows very distinct glacial scratches. The upper layer was not sufficiently uncovered at the time of my visit to show very clearly the direction of the striæ, but Professor Egleston mentions seeing them at some places with a direction south 8° east. Dr. White also saw them at other places ranging south 41° west by the magnetic needle, the variation of the compass being here about 11° east of north. From all the facts observed in this region, it is quite probable that a considerable thickness of Upper Coal-Measure strata, consisting of some beds of hard limestone, has been here ground off and swept away by glacial agencies, leaving the upper surface of the remaining limestones planed off, and sometimes beautifully striated or even polished.

As an indication of the nature of the rocks below the horizon of the

* The only tooth of this fish found at this locality was discovered by Dr. White, the State geologist of Iowa, who accompanied the writer while examining these quarries.

quarry at Omaha City, the following statement of the beds perforated in a boring made at that place to a depth of four hundred feet, by the Union Pacific Railroad Company, is added:

*Statement of a boring made in the Missouri Valley at Omaha City, by the Union Pacific Railroad Company, starting 22 feet above low-water mark of the Missouri.**

No.	Nature of strata penetrated.	Depth.
		Feet.
1	Black soil (alluvial)	20
2	Sand and lime rock. [Doubtless the rock quarried below the city]	10
3	Slate mixed with coal. [The black shale under the quarry rock]	5
4	Fire-clay	8
5	Lime-rock	16
6	Fire-clay	3
7	Shale and lime rock	5
8	Green-sandy slate	7
9	Sand-rock with soft seams	11
10	Lime-rock	29
11	Magnesian [?] limestone	5
12	Coal slate. [Bituminous shale]	6
13	Sand and lime rock	10
14	Slate	4
15	Lime-rock with soft seams	17
16	Red slate with hard layers. [Red shale and indurated clays]	31
17	Sand-rock	4
18	Lime-rock	3
19	Blue slate	40
20	Lime-rock	24
21	Soap-stone with hard layers lime-rock 15 inches thick. [Alternations of limestone and indurated clay]	28
22	Coal slate. [Bituminous shale]	7
23	Fire-clay	4
24	Lime-rock	35

* I am under obligations to General Dodge, the chief engineer of the Union Pacific Railroad, for the foregoing memorandum of the beds passed through by this boring

No.	Nature of strata penetrated.	Depth.
		Feet.
25	Soap-stone. [Indurated clay]	5
26	Lime and sand rock with seams of flint	8
27	Black slate	3
28	Gray soap-stone. [Indurated clay]	13
29	Yellow soap-stone. [Indurated clay]	20
30	Lime-rock	2
31		18
	Total	401

In this section the names by which the several beds were designated by the parties who kept the memorandum of the same, are in all cases retained, as I had no opportunity to examine the borings. I have, however, added in brackets a few suggestions, as it is evident some of them, such for instance as "soap-stone," "gray soap-stone," "yellow soap-stone," &c., are not applied as generally understood in geology. It is also quite probable that there may be some errors in the details of the different seams and beds passed through, as it would be very difficult for any person not versed in lithology to be always strictly right in regard to the true nature and composition of rocks perforated in this way, by merely examining the borings brought up by the auger. The nature of the rock penetrated by the lowest 18 feet was not given in the section copied, and I failed to get any information in regard to it.

As we know nothing respecting the organic remains contained in these rocks, of course we cannot positively demonstrate that they belong to the same general series of Upper Coal-Measures as the beds exposed at the quarry just below the city, but it must be evident enough to any one familiar with the lithological characters of our western Carboniferous rocks, that they certainly correspond to nò part of that system holding a position below the horizon of the Millstone grit, while they do agree quite well, in their general features, with portions of the Upper Coal-Measures of Iowa, Kansas, and Nebraska, as well as of Northern Missouri.

The most important fact revealed by this boring is that no workable beds of coal exist beneath the horizon of the Missouri Valley there within 400 feet of the surface. This, however, is not surprising, as all the investigations of these rocks in Western Iowa, Northern Missouri, Kansas and Nebraska, tend to the conclusion, as will be seen further on, that there is a considerable thickness of Upper Coal-Measure strata in that region, nearly or quite barren of workable beds of coal. As this is a question, however, in which not only the stockholders of the Pacific Railroad, but the whole community of this region are interested, it is to be hoped that this boring will be continued on deep enough to decide, positively forever, whether or not deep mining here can be expected to supply in quantity a good quality of fuel.

The first locality south of Omaha City, at which I had an opportunity to examine an exposure of these Upper Coal-Measure strata satisfactorily, was at Bellevue, 10 miles south of Omaha, where the following succession of beds was observed.

Section of beds exposed at Bellevue, with an enumeration of the fossils found in each.

No.	Nature of the beds, with the fossils found in each.	Thickness.
		Ft. In.
11	Heavy deposits of Loess, underlaid by Drift	200 0
10	Hard, dark-gray silicious limestone	2 0
9	Bluish and drab clays	6 0
8	Hard, dark silicious limestone. *Productus longispinus* (?), *P. Nebrascensis*, *Aviculopecten occidentalis*, *Myalina subquadrata*, *Fusulina cylindrica*, *Spirifer planoconvexus*, *Phillipsia major*, and *Peripristis semicircularis*.	2 0
7	Dark shale, with a 2-inch seam of good coal, showing together about 15 inches. *Spirifer planoconvexus*, *&c.* Unexposed space, about 5 feet.	6 3
6	Gray and yellowish argillo-calcareous bed, splitting into thin pieces where exposed.	2 6
5	Light clay above and dark laminated shale below	2 6
4	Light-gray and greenish clays	2 5
3	Soft yellowish, slightly gritty layer, passing into harder impure limestone.	2 6
2	Very hard massive gray and bluish-gray impure limestone, with *Fusulina* and traces of embedded oolitic particles.	3 0
1	Yellowish, light-gray, and hard brittle whitish limestones, with some argillaceous layers, all below high-water mark of the Missouri. *Fusulina cylindrica*, *Fistulipora sp.*, *Archæocidaris*, *Erisocrinus typus*, *Fenestella sp.*, *Hemipronites crassus*, *Chonetes granulifera*, *Productus punctatus*, *P. Nebrascensis*, *P. symmetricus*, *P. costatus*, *P. semireticulatus*, *P. longispinus* (?) *P. Prattenanus*, *Rhynchonella Osagensis*, *Athyris subtilita*, *Spirifer lineatus*, (=*S. perplexus*, McC.,) *Sp. cameratus*, *Spiriferina Kentuckensis*, *Terebratula bovidens*, *Myalina subquadrata*, *Pleurotomaria sphærulata*, *Macrocheilus inhabilis*, *Allorisma subcuneata*, &c.	18 0
	Total thickness of Carboniferous beds	47 +

At the time of our visit to this locality the Missouri was so high as to cover all of the lowest member of the section, but in 1853 Dr. Hayden and I had a fine opportunity to examine it when the water was low, while waiting there for a steamboat, on our return from an expedition to the Bad-lands. These lower beds were at that time best exposed directly above the old landing. The beds above No. 1 of the section are seen at various places along the shore and base of the hill for about a quarter of a mile above the landing.

The whitish, hard, brittle layers of the lower member of the above section, make beautiful white lime, while some of the other beds afford good building-stones. Excavations have been made at places under the bed No. 8 in search of coal, in the dark shale No. 7, in which there is a thin seam of coal from one to two inches in thickness. This coal is of good quality, but is of course useless, in consequence of the thinness of the seam. At another place a little farther up the river, an excavation or short drift was made in search of coal, under No. 10 of the foregoing section, in the clays No. 9, but no indications of coal were observed, though a very fine spring of clear, cool water was developed.

It is worthy of note in this connection that the rocks forming the foregoing section, as well as those mentioned at Omaha, were referred by Professor Marcou (see Bull. Geol. Soc. France, 2d ser., vol. xxi, p. 132) to the Mountain limestone series. Any one, however, having the slightest knowledge of the fossils of our Carboniferous series will at once see from those mentioned in the section, as well as from those found at Omaha, that these beds belong to the true Coal-Measures. Some of the species, it is true—such as *Productus semireticulatus, P. costatus, P. punctatus, Hemipronites crassus,* &c.*—are known to be common to our Coal-Measures and the Lower Carboniferous rocks, but all the others mentioned are peculiar to the Coal-Measures, not one of them, so far as known to the writer, having ever been found in the West below the horizon of the Millstone grit. It will also be observed further on, that all of these same species, with perhaps some two or three exceptions, occur together, along with other Coal-Measure forms, in the very beds referred by Professor Marcou to the Permian or Dyas south of the Platte, and the few exceptions alluded to, such as *Macrocheilus inhabilis*, and *Pleurotomaria sphærulata*, are well known Coal-Measure forms.

On the north side of Platte River outcrops of Upper Coal Measure rocks, that had been previously noticed by Dr. Hayden and Professor Egleston, were examined at several places along the bluffs from three to four miles up from the mouth of that stream. The particular localities examined are about five miles in a southwest direction from the exposures mentioned at Bellevue. These beds are said to rise considerably higher somewhat farther up the Platte than at the localities here alluded to, where the northeastward dip of a few degrees brings them lower. The exposures examined present the following succession of beds:

Section of the beds exposed on the north side of Platte River, between three and four miles from its mouth, with an enumeration of the fossils found in each bed.

No.	Composition of the beds.	Thickness. Ft.	In.
12	Drift and Loess	150	0
11	Hard, massive, light-grayish limestone, composed mainly of oolitic particles, and *Fusulina* embedded in a compact calcareous base. *Syringopora multattenuata, Productus costatus, P. Nebrascensis, Myalina subquadrata.*	3	0
10	Light-grayish clay	6	0

*Professor Marcou also mentions among his collections from these rocks, *Productus pustulosus, P. scabriculus, Athyris plano-sulcata and A. Royissii*, but the first two are almost beyond doubt, only *P. symmetricus*, and *P. Nebrascensis*, as they appear in the condition of casts, when broken from hard limestone.. At any rate, Professor Geinitz identifies them also in the so-called Dyas at Plattsmouth. The *A. plano-sulcata* is the same Coal-Measure form described by McChesney under the name *A. orbicularis*, and the other sp. (*A. Royissii*) was not identified by Professor Geinitz among the Bellevue collections.

No.	Composition of the beds.	Thickness.
		Ft. In.
9	Hard, yellowish limestone, with some flinty concretions, and toward the upper part, soft decomposing isolated masses, from a few inches to 1 or 2 feet in diameter. *Productus Prattenianus, Athyrus subtilita*, &c.	2 6
8	Whitish, hard, brittle limestone, in layers 5 to 8 inches thick, with light-colored clay partings. *Fusulina cylindrica, Axophyllum sp., Scaphiocrinus (?) hemisphæricus, Archæocidaris triserratus, Synocladia biserialis, Productus longispinus, P. costatus, P. Nebrascensis, Chonetes Verneuiliana, Spirifer cameratus, Spiriferina Kentuckensis, Athyris subtilita*, &c.	6 0
7	Light drab or grayish indurated marly clay. *Rhombopora lepidodendroides, Axophyllim sp., Fusulina cylindrica, Scaphiocrinus (?) hemisphæricus, Lophophyllum proliferum, Synocladia biserialis, Hemipronites crassus, Orthis carbonaria, Chonetes granulifera, C. Verneuiliana, Productus semireticulatus, P. costatus, P. longispinus, P. punctatus, Rhynchonella Osagensis, Spirifer cameratus, S. lineatus, Spiriferina Kentuckensis, Athyris subtilita, Terebratula bovidens, Chænomya, Myalina subquadrata*, &c.	4 0
6	*a.* Hard, dark-gray compact limestone .... 8 inches.. *b.* Light-gray indurated marlite .... 24 do.... *c.* Black shale .... 4 do.... *d.* Light-gray indurated marlite .... 14 do.... *e.* Hard, dark impure limestone .... 6 do....	4 8
5	Light-grayish, slightly gritty, soft marlite	2 6
4	Bluish indurated crumbling clay or marlite, stained reddish purple......	3 0
3	Bluish ash-colored (sometimes stained purple above) soft argillo-calcareous rock, becoming harder and in thicker layers below, with partings and irregular layers of soft yellowish marly clay, containing numerous *Chonetes granulifera, Rhombopora sp., Synocladia biserialis, Productus symmetricus, P. Nebrascensis, Hemipronites crassus, Athyris subtilita, Spiriferina Kentuckensis, Fusulina*, &c.	7 6
2	Black laminated shale	6
1	Very hard, bluish, compact argillaceous limestone	1 0
	Thickness of Carboniferous beds	40 8

The base of this section was estimated to be about 20 feet above the valley of Platte River, and about 23 feet above high-water mark of the Missouri. All the beds show a dip of 2° or 3°, in a direction to the east of north. They were observed to rise quite perceptibly in tracing the outcrops a little south of westward, up the Platte; though there is probably a reverse of dip farther up.

In these rocks, here as elsewhere, the different subordinate beds are sometimes liable to change materially their lithological characters and thickness within comparatively short distances, as can be occasionally clearly seen in following an uninterrupted outcrop a mile or so. For instance, the bed of clay, No. 10 of the above section, which is 6 feet in thickness at the particular place where the upper beds of this section were observed best developed, is reduced to a mere seam of an inch or two in thickness, or in places entirely wanting, so as to cause the bed 11 to rest directly down upon No. 9, at other localities within less than a mile further eastward.

It will be observed that nearly all of the fossils mentioned in the Bellevue section also occur here, along with a number of other well-known Coal-Measure species. It is also evident that the lower member of the Bellevue section is represented by Nos. 8 and 9 of this section, and No. 2 of the former by No. 11 of the latter, the clays No. 10 of this section not being represented at Bellevue. This elevation of the lowest member of the Bellevue section (which, at the time of my visit during this survey, was entirely below the level of high-water mark of the Missouri) to something more than 40 feet above the same horizon, on the Platte, in a distance of about five miles, is a tolerably accurate indication of the northeastward inclination of the strata here, though farther south the inclination seems to be to the south of east.

The upper limestone, No. 11, of the section last mentioned, makes good lime, and is much quarried and used in that way, as well as for building purposes, for which it is admirably adapted. Its compact, homogeneous structure, comparative hardness (for a limestone) and uniform pleasing gray tint, are qualities that render it a desirable material even for much of the better kinds of masonry. At present it is much quarried for caps and sills, particularly at Mr. John B. Ducols's place. In looking upon the rather rough, uneven-fractured surface of this rock, without the aid of a magnifier, no one would suspect its true nature and composition. On examining a polished or moistened surface, however, under a lens, it is seen to be composed of millions of *Fusulina* and oolitic particles, embedded in a clear calcareous base. These little bodies are so numerous that they seem to be everywhere in contact, while they are so firmly cemented together that they never separate in breaking the rock, but the fractured surface shows sections of them in every conceivable direction. Larger fossils are sometimes, but more rarely, met with in it.

At Mr. Ducols's quarry, mentioned above, the upper surface of this stratum is everywhere found, on removing the loose overlying superficial deposits, to be perfectly planed off by glacial agencies, as far in as the quarry has been opened; and it is quite probable that comparatively large areas of its upper surface have been thus polished and striated. This is a great puzzle to the quarrymen and others unacquainted with geology; but whether or not the stone-hewers trouble themselves with speculations in regard to its cause, they wisely accept the work thus ready-done to their hands; for, in cutting sills and caps from the upper part of this stratum, they always find one side already dressed more truly and evenly than they can possibly do it with a chisel. But one set of striæ was seen at this locality, and these ranged (by compass) south 10° west. No striated or polished surfaces of this kind were observed by me at any locality south of Platte River; but it is said that they have been seen at some quarries a short distance south of the mouth of that stream.

Boulders, however, of several kinds of granitic and metamorphic rocks, particularly of a kind of red quartzite, showing the action of glacial agencies here, were not unfrequently met with in the country between Platte River and Nebraska City, as well as, occasionally, south of there. These are generally not of large size, though, at one place, about three miles from Nebraska City, in a direction somewhat west of north near Bennett's mill, a huge angular mass of red quartzite was seen lying upon, or rather, projecting above, the sloping surface of the ground; for a large portion of it seemed to be embedded. It rises some 10 or 12 feet above the soil, and was estimated to contain not less than 1,000 cubic feet, and perhaps much more. This is evidently a metamorphosed sandstone, for although now an extremely hard and compact quartzite within, it shows on the slightly-weathered surface indications of planes of stratification. It presents no marks of attrition, but retains all its angles and other irregularities of surface with remarkable sharpness, showing that it must have been transported from its original position on floating ice, or, at any rate, by some agency that did not bring it in contact with other rocks. Small erratic (nearly always angular) masses of this red quartzite have been noticed by Dr. White, in Iowa, and, as he has suggested, there is scarcely any reason to doubt that they have all been transported upon or attached to masses of floating ice during the glacial period, from Southern Minnesota, the nearest point to the northward where rocks of that kind are known to occur *in situ*.

South of Platte River, the first exposure of the Upper Coal-Measure rocks of much extent seen in coming down the Missouri, is just below the town of Plattsmouth, where the following succession of strata may be seen:

Section of the beds exposed at Plattsmouth, with the names of the fossils found in each.

No.	Nature of the strata.	Thickness.
		Ft. In.
9	Slope, apparently composed mainly of Loess Drift, exposing a thickness of	80 0 10 0
8	Hard, yellowish and light-grayish limestone, in heavy beds and thinner layers, with two intercalated seams, 8 to 12 inches in thickness, of ash-colored marly clay, containing millions of *Fusulina cylindrica;* also *Retzia punctulifera, Athyris subtilita, Aviculopecten occidentalis, Entolium aviculatus, Myalina subquadrata, Macrodon tenuistriata, Allorisma subcuneata, Pinna peracuta, Phillipsia major,* &c.	12 0
7	Hard, light-grayish limestones, in thinner layers than the beds above, and containing some flinty concretions.	9 0
6	Ash-colored clay	3 0
5	Black laminated shale	1 6
4	Red clay	6
3	Yellowish incoherent sand	4 0
2	Layers light-colored and yellowish, more or less pure limestone, 10 to 12 inches in thickness, with light-colored and greenish marly clay partings; the whole passing down into light grayish and whitish marly clays, with many thin seams of limestone and numerous fossils. *Fusulina cylindrica, Rhombopora lepidodendroides, Lophophyllum, Synocladia biserialis, Polypora submarginata, Fenestella, Zeacrinus mucrospinus, Eupachycrinus verrucosus, Hemipronites crassus, Chonetes granulifera, C. Verneuiliana, Productus costatus, P. semireticulatus, P. Nebrascensis, P. longispinus, P. punctatus, P. Prattenianus, Spirifer planoconvexus, Sp. lineatus, Syntrilasma hemiplicata, Meekella striato-costata, Sp. cameratus, Spiriferina Kentuckensis, Athyris subtilita, Rhynchonella Osagensis, Retzia punctulifera, Terebratula bovidens, Aviculopecten accidentalis, Bellerophon carbonaria, Euomphalus rugorus, Nautilus ponderosus,* &c.	10 6
1	Green and reddish-brown clays; the former occupying the upper ten inches, and the remainder below being mainly brownish-red, but in places mottled with green.	10 10
	Total of Carboniferous	49 +

At the time we visited this locality, the Missouri was at nearly high-water mark, covering all below the base of the section given above. Professor Marcou, however, visited it when the river was lower, so that he saw below No. 1 about 15 feet of blue shaly clay, with, near its upper part, thin layers of ferruginous limestone; the whole being without fossils.

I found it convenient to group the beds above somewhat differently from the section given by Professor Marcou, and it will also be observed that there is a 4-foot bed of soft sandstone, and 6 inches of black laminated shale in the above section, not mentioned by him. It is not surprising, however, that these should have escaped his attention, because all the beds are so liable to slide and become so mingled and obscured from the undermining of the softer material below, that it is very difficult to make out the exact details of the section. Our opportunities, however, to make out the structure of the outcrop above high-water mark, were better than his, because several drifts and other excavations had a short time previously been made into the cliff here, by parties prospecting for coal, by which means much of the loose *débris* had been removed, so as to show more clearly the nature and arrangement of portions of the section.

To any one having even a limited knowledge of the general principles of paleontology, or of the characteristic fossils of our western Coal-Measures, it must always remain an inscrutable mystery, why Professor Marcou, after visiting this locality, and collecting most of the fossils found in these beds, should have pronounced them the remains of an eminently New Red fauna, allied to the Carboniferous, and referred the rocks to the Lower Permian. This is all the more surprising, when it is borne in mind that every species, with possibly two or three exceptions, yet known from the rocks above the mouth of Platte River referred by him to the Mountain limestone, occurs here at Plattsmouth, while even these exceptions are well-known Coal-Measure species. In addition to this, by turning to the section given on page 90, of the beds exposed three or four miles up Platte River, on the north side, it will be seen that with a few exceptions the same group of fossils found in these so-called Lower Dyassic rocks at Plattsmouth, occur there beneath the same beds called mountain limestone at Bellevue.*

But even Professor Geinitz, who was, from his personal relations with Professor Marcou, naturally inclined to favor his conclusions as much as he could, soon discovered, on examining the collections from Plattsmouth, that the so-called Lower Dyas at that locality was hopeless, because, out of the thirty-three species of fossils composing this so-called "eminently New Red fauna," he found thirty to be Carboniferous forms; and if he had been well acquainted with the fossils of our Coal-Measures, he would have undoubtedly added the other three species, as he might have done several others not included in the collections submitted to him. Indeed, so far from agreeing with Professor Marcou, in viewing these Plattsmouth beds as Lower Dyas, Professor Geinitz even goes to the opposite extreme, and places them in the upper part of the Lower Carboniferous series; though in this he is certainly in error, as I have

* I can only account for Professor Marcou's reference of this section to the Dyas, upon the supposition that the occurrence of a red bed at its base was thought by him to indicate that it belonged to a widely more recent group than the Bellevue and Omaha City outcrops; especially as he gives great weight to lithological characters. Upon this kind of evidence, however, it will be seen by turning to the section of the Omaha boring, given on page 87, that he would have to carry the Dyas down so as not only to include the beds at Omaha, referred by him to the Mountain limestone, but so as to include more than 100 feet of strata below them.

elsewhere shown,* and as any one who has studied the Western Coal-Measure fossils will at once see by examining the list of species, as they form a group especially characteristic of our Upper Coal-Measures of the Western States and Territories.

With regard to the *exact* stratigraphical relations of the section seen on the north side of Platte River, to that just below the town of Plattsmouth, of course we cannot speak very positively, as the beds cannot be traced into contact. Any one, however, who will compare collections of fossils from these two localities, must be at once convinced that the rocks belong to the same series, and must be closely related. Yet it is quite evident that the visible dip of the rocks seen along the north side of the Platte would take them somewhat beneath the Plattsmouth section, though not far below it, as the inclination of the strata here is to the northeastward, while the Plattsmouth exposure is south of east from those mentioned on the north side of Platte River.

At another locality on the Missouri, eight miles south of Plattsmouth, known as Rock Bluff, just above a village of the same name, there is to be seen perhaps the finest exposure of these Upper Coal-Measure rocks anywhere to be found in Nebraska. The lower beds here rise perpendicularly near the base from the river bank, and less precipitously above, exposing, with the exception of one uncovered interval of about 18 feet, about 120 feet of strata. The following is a section of the beds seen here:

Section of the beds exposed at Rock Bluff, on the Missouri, with a statement of the fossils found in each.

No.	Nature of strata.	Thickness. Ft.	In.
23	Loess or bluff deposit, with possibly toward the lower part some Drift..	150	0
22	Hard yellowish and grayish limestones, in rather thick layers, with some clay partings. *Fusulina cylindrica, Meekella striato-costata, Syntrilasma hemiplicata, Productus costatus, P. semireticulatus, Spirifer lineatus, Sp. cameratus, Athyris subtilita,* &c.	24	0
21	Ash-colored clay 1 foot 9 inches above, underlaid by 1 foot hard gray limestone.	2	9
20	Soft ash-colored or bluish-gray clay, 1 foot, underlaid by black laminated clay, resting upon 1 inch yellow do.	2	1
19	Very compact gray limestone, with an imperfect conchoidal fracture ...	1	6
18	Gray somewhat laminated clay, 4 feet seen, with a 4-foot space below unexposed.	8	0
17	Soft yellowish-gray sandstone, becoming almost loose sand below.......	5	0
16	Not well exposed, but apparently in part occupied by the sandstone seen above, with near the middle and at the base, some bluish clay seen.	18	0
15	Dark-bluish argillaceous limestone. *Aviculopecten occidentalis, Productus Nebrascensis, Edmondia gibbosa.*	1	9

* See Am. Jour. Sci. and Arts, vol. xliv, second ser., p. 335.

No.	Nature of strata.	Thickness.
		Ft. In.
14	Bluish and ash-colored clays, with near the middle a 3 or 4 inch band of red.	8 0
13	Hard light-gray limestone, with apparently oolitic particles. *Allorisma*.	1 6
12	Bluish, ash-colored and purple calcareous clays, with many *Campophyllum*; also *Spirifer cameratus*, *Zeacrinus hemisphæricus*, &c.	2 0
11	Hard light-gray limestone, with *Aulopora*, *Fusulina cylindrica*, *Campophyllum*, &c.	1 0
10	*a*. 7 feet bluish, ash-colored clays; *b*. 20 inches black laminated shale; *c*. 5 feet bluish clay.	13 8
9	Massive heavy-bedded, light-yellowish, rather hard limestone, with some light-colored flinty concretions. *Productus Nebrascenis*, *P. semireticulatus*, *Spirifer cameratus*, *Aviculopecten occidentalis*, *Pinna peracuta*, *Allorisma subcuneata*, *Schizodus*, *Sedgwickia granosa*, *Edmondia subruncata*, *Myalina Swallovi*, *Petalodus destructor*, *Peripristis semicircularist*, &c.	5 0
8	Very hard bluish argillaceous limestone. *Chænomya Leavenworthensis* ...	2 0
7	Black laminated shale. *Spirifer planoconvexus*	1 6
6	Light-colored clays, with chalky concretions. *Rhombopora lepidodendroides*, *Lophophyllum proliferum*, fragments crinoids, *Hemipronites crassus*, *Orthis carbonaria*, *Chonetes granulifera*, *Spirifer planoconvexus*, *S. cameratus*, *Spiriferina Kentuckensis*, *Retzia punctulifera*, *Rhynchonella Osagensis*, *Terebratula bovidens*, *Pinna peracuta*, *Solenomya*, *Nucula ventricosa*, *Euomphalus rugorus*.	5 6
5	Hard light-gray limestone, with minute rounded particles embedded....	1 3
4	Ash-colored and yellowish calcareous clay or marly material. *Fusulina cylindrica*.	0 6
3	Massive, rather compact, light yellowish limestone, with numerous *Fusulina cylindrica*; also curious cavities, apparently left by the cylindrical stems of some marine plant, often abruptly recurved like a horseshoe magnet.	5 0
2	Soft ash-colored marlite, with millions of *Fusulina*; *Fustulipora*, &c......	1 0
1	Hard, light-grayish heavy-bedded massive limestone, with great numbers of *Fusulina* embedded; also *Spirifer cameratus*, *Productus Nebrascensis*, *P. semireticulatus*, *P. punctatus*, *P. Prattenianus*, &c.	10 0
	Total of Carboniferous strata	121 0

The lower members, 1, 2, and 3, of this section, with perhaps some other beds beneath the surface of the river at the time of my visit, almost certainly represent, as suggested by Professor Marcou, the upper part of the Plattsmouth section, so that the portion of the section above these may be regarded as a continuation upwards of the section seen at Plattsmouth. Professor Marcou only mentions 60 feet of beds exposed here; but as he does not give a section of these rocks, this is probably only intended as a rough estimate of the thickness of beds exposed at and near the lower extremity of the hill. By following these outcrops about 150 yards farther up the river, however, I succeeded in ascertaining the nature, thickness, and order of succession of the various beds, with the exception of one unexposed space already mentioned, of about 120 feet of rocks here, above high-water of the Missouri.

The fact that apparently the lower beds, within 10 or 15 feet of the river at this locality, are found nearly 40 feet above the Missouri at Plattsmouth, while there is probably a fall of 5 or 6 feet in the Missouri from Plattsmouth to this locality, shows that there must be a change in the dip of the strata to the southward somewhere near Plattsmouth, as noticed by Dr. Owen. This is also further shown by the entire disappearance of this whole section of more than 120 feet of rocks beneath the level of the river between here and Nebraska City. The dip, however, is probably not due southward, but apparently somewhat east of south.

This section Professor Marcou also referred to the Lower Dyas—that is, to the upper part of the Lower Dyas—while the fossils found here, like those at Plattsmouth, he regarded as representing an eminently New Red fauna. To any person who has studied our Carboniferous fossils, however, the reasons for these conclusions must be very difficult to understand, as it will be at once observed, by a glance at the list of species found in these rocks, that they are precisely the group of forms characterizing our Coal-Measures, strangely enough at some other places referred by Professor Marcou to the Mountain limestone, as elsewhere explained. From this conclusion, however, Professor Geinitz was compelled to dissent, on examining Professor Marcou's collections from this locality, as he very properly referred these beds to the Upper Coal-Measures, to which they unquestionably belong. The strangest thing, however, is, that he should have referred these beds correctly to the Upper Coal-Measures, and the Plattsmouth section to the Lower Carboniferous, when of the thirty-four or thirty-five species of fossils found at Plattsmouth, about twenty-five also occur here at Rock Bluff, while the remaining ten or eleven Plattsmouth forms not yet found at Rock Bluff, are all, excepting two or three that are common to the Coal-Measures and Lower Carboniferous, as already stated, characteristic species of our Coal-Measures, and not yet known to occur at any horizon below the Millstone grit, in this country. He was doubtless led into this inconsistency in consequence of not having a full series of fossils from the two localities for comparison, as he seems to lay some stress upon the occurrence of numerous *Fusulina* at Plattsmouth, this fossil being generally regarded in Russia and Spain as characteristic of the upper part of the Lower Carboniferous series. He was, therefore, evidently not aware of the fact that it also occurs in vast numbers at Rock Bluff, even to the top of the section, and that, so far as yet known, it is in this country peculiar to the Coal-Measures, being most abundant in upper portion referred by Professor Marcou in part to the Dyas, and in part to the Mountain limestone, and has never yet been found in any of our rocks below the Millstone grit.

As at Plattsmouth, several short drifts and other excavations have been here made in search of coal, but, of course, without success, as there is certainly no workable bed, or even less important seam of coal, in this bluff.

Back from the Missouri about six miles, in a southwest direction from Rock Bluff, on a small stream known as Weeping Water, at a locality called Cedar Bluff, there is another very fine exposure of the Upper Coal-Measures, consisting in part of the same strata seen at Rock Bluff. Here the following beds were exposed:

Section of the rocks seen at Cedar Bluff.

No.	Nature of strata.	Thickness. Ft.	In.
23	Hard, light-gray and yellowish limestone, in rather thick layers. *Fusulina, Athyris subtilita,* crinoid columns, &c. (An excellent building-stone.)	8	0
22	Light yellowish-gray limestone, in parts hard, but generally softer than that above; and in irregular layers of various thickness.	8	0
21	Greenish-blue clay	1	6
20	Light-grayish limestone, more impure and less compact than the next limestone below. *Hemipronites crassus.*		8
19	Laminated clay, green above, darker below	1	6
18	Hard, light-gray limestone	1	8
17	Light-colored calcareous marlite, with seams of whitish hard limestone above.		8
16	Soft, yellowish sandstone, in some parts laminated; possibly also occupying some of the unexposed space below.	12	0
15	Unexposed, excepting a little greenish laminated clay about 4 feet above base.	15	0
14	Brownish-red clay	3	0
13	Light-yellowish and ash-colored clays, with some thin layers of soft impure limestone.	6	0
12	Brownish-red clay	3	0
11	Greenish clay. *Chonetes, Spirifer cameratus, Campophyllum*	1	0
10	Hard, light-bluish gray or whitish limestone, with green clay partings. *Campophyllum torquium.*	2	0
9	Greenish and yellowish clays	3	0
8	Reddish-brown clays, with one 4-inch band of bright green	3	6
7	Green crumbling clay	1	2
6	Yellowish clay with light-yellow calcareous concretions	1	0
5	Hard, bluish-gray, rather pure limestone	1	3
4	Soft, bluish argillaceous limestone, with at base 18 inches green clay	8	0
3	Light-grayish shaly clay, with a streak of black above	1	0
2	Black laminated shale	2	2
1	Dark-bluish argillaceous limestone, and indurated clay	3	3
	Total	88 +	

Having but a comparatively short time to devote to examinations here, only a few fossils were collected from the beds in place, but the following species were found among the *débris* at the base of the bluff, that had evidently fallen from the rocks cropping out above, viz: *Scaphiocrinus hemisphæricus*, *Zeacrinus mucrospinus*, *Lophophyllum proliferum*, *Campophyllum torquium*, *Hemipronites crassus*, *Chonetes*, *Productus Nebrascensis*, *P. costatus*, *Syntrilasma hemiplicata*, *Athyris subtilita*, *Rhynchonella Osagensis*, *Spirifer* (*Martinia*) *planoconvexus*, *S.* (*M.*) *lineatus*? (=*perplexus*, *McC.*,) *Spiriferina Kentuckensis*, the fragment of *Pseudomonotis* figured on Plate II, (Fig. 9), and *Euomphalus rugosus*.

This exposure was not visited by Professor Marcou, but there can be no doubt that he would have referred it to the Dyas if he had seen it, as it presents the same general lithological characters as the outcrops at Plattsmouth and Rock Bluff, with some differences in the details, and evidently belongs to the same series of Upper Coal-Measures. I am inclined to believe, however, that it represents that portion of the Rock Bluff section above bed No. 9. These beds all show a dip of a few degrees to the south, or southeast.

The upper beds here afford a good building-stone, and have been found to make good lime, while the sandstone, No. 16, is at places so soft and incoherent that it can be easily dug out and applied to most of the purposes for which sand is used. The bed of black slate at the base of the section was at one time supposed to be a sure indication that valuable coal mines might be found there, and caused the land to be held, and, if I am not mistaken, to be sold, at a high price. It is evident, however, that no coal of any value exists in this outcrop.

At Wyoming, on the Missouri, twelve miles in a direction a little west of south, below Rock Bluff, a low exposure of rocks crops out on the immediate margin of the river, showing the following beds:

Section at Wyoming, with an enumeration of the embedded fossils.

No.	Nature of strata.	Thickness.	
		Ft.	*In.*
6	Blue and red clay, entire original thickness unknown. *Productus semireticulatus*, *Syntrilasma hemiplicata*, *Athyris subtilita*, *Chonetes granulifera*, *Productus Nebrascensis*, *Spirifer* (*Martinia*) *planoconvexus*, *Hemipronites crassus*, *Spiriferina Kentuckensis*, *Rhynchonella Osagensis*, *Pleurophorus occidentalis*, *Bellerophon carbonaria*, *Euomphalus rugosus*, &c.	4	0
5	Hard gray, impure silicious limestone. *Fusulina cylindrica*, *Syntrilasma hemiplicata*, fragments crinoids, &c.	2	0
4	Light grayish ash-colored clay. *Hemipronites crassus*, *Syntrilasma hemiplicata*, *Chonetes granulifera*, *Meekella striato-costata*, *Spirifer cameratus*, *Productus longispinus*, *Spirifina Kentuckensis*, *Spirifer* (*Martinia*) *planoconvexus*, &c.	3	6
3	Massive hard yellowish, or light-gray silicious limestone, somewhat argillaceous below. *Schizodus*, *Allorisma*, *Pinna peracuta*, *Fusulina*, &c.	4	0
2	Bluish and ash-colored laminated, calcareous clays........................	5	0
1	Red gritty clay, passing down into a soft red sandstone, more or less micaceous, ripple-marked and concretionary.	9	0
	Total..	27	6

All these beds show a slight inclination to the southward or southeastward. On the upper surface of bed No. 3, I saw the same kind of curious markings, apparently of the cylindrical stems of some marine plant (often curved like a horseshoe magnet,) mentioned as occurring in bed No. 3, at Rock Bluff. The beds here at Wyoming, however, appear to hold a higher position in the series, judging from the southeastward dip of the strata here and at Cedar and Rock Bluffs, than any of the beds at either of the former localities. Professor Marcou places them on a parallel with the lower part of the section seen at Nebraska City, and I am inclined to believe they belong to that horizon, or at least very near it, in the series. He includes them, however, as a part of the Upper Permian, or Dyas, and in this Professor Geinitz concurs with him, though all the fossils yet found here occur also in the very beds referred by the latter author at Plattsmouth to the Lower Carboniferous, and all but about five of them in the same beds at Omaha and Bellevue referred by Professor Marcou to the Lower Carboniferous; while even the four species occurring here, that have not yet been found at Omaha or Bellevue, such as *Meekella striato-costata*, *Bellerophon carbonaria*, *Syntrilasma hemiplicata*, and *Pinna peracuta*, are well-known Coal-Measure species. In short, there is not the slightest reason, if we can place any reliance whatever upon fossils in identifying and determining the position of strata, for separating this outcrop from the Upper Coal-Measures.

About four miles in a nearly south direction from Wyoming, and three miles in a northwest direction from Nebraska City, at Bennett's Mill, there is to be seen along a small stream, probably 90 to 100 feet below the summit of the immediately surrounding country, an exposure showing the following beds:

Section at Bennett's Mill.

No.	Nature of strata.	Thickness. Ft.	In.
4	Hard silicious, light-gray limestone, at some places in two layers, with a clay parting between; and although compact within, showing indications of a laminated structure on weathered surfaces. *Rhombopora lepidodendroides*, *Fusulina cylindrica*, *Syntrilasma hemiplicata*, *Spirifer (Martinia) planoconvexus*, *Rhynchonella Osagensis*, *Chonetes granulifera*, *C. glabra*, *Productus longispinus*, *Spirifer cameratus*, and teeth of *Peripristis semicircularis*.	2	6
3	Light drab and ash-colored clays	3	0
2	Band of soft drab argillaceous limestone, containing *Productus Prattenianus*, *P. punctatus*, *Chonetes granulifera*, *Spirifer cameratus*, *Syntrilasma hemiplicata*, *Scaphiocrinus (?) hemisphæricus*, *Myalina subquadrata*, *Aviculopecten occidentalis*, *Pinna peracuta*, *Allorisma subcuneata*, *Pseudomonotis* sp., *Allorisma* sp. and *Chomatodus arcuatus*.	0	8
1	Light grayish crumbling indurated clay, of which about two feet were exposed.	2	0

These beds Professor Marcou and Professor Geinitz regard as occupying the same horizon as those seen at Wyoming, and the lower part of the Nebraska City section, which is doubtless very nearly or quite correct. I cannot concur with them, however, in including these rocks n the Permian, for the reasons already stated in regard to the Wyoming and other outcrops.

About two miles down the same little stream, in an east direction from Bennett's Mill, the same beds are seen at a somewhat lower level in the bed of the creek. Here the following fossils were found: *Rhombopora lepidodendroides*, *Fistulipora*, *Scaphiocrinus hemisphæricus*, *Hemipronites crassus*, *Syntrilasma hemiplicata*, *Chonetes granulifera*, *Meekella striato-costata*, *Productus semireticulatus*, *P. longispinus* (?), *Athyris subtilita*, *Spirifer cameratus*, *Spirifer (Martinia) planoconvexus* and *Platyceras*. (See Plate IV, Fig. 15, *a. b.*)

The next locality at which these rocks were examined was at the Nebraska City landing, where the following succession of beds may be seen:

Section of beds exposed at the Nebraska City landing, with an enumeration of the fossils found in each.

	Nature of strata.	Thickness.
		Ft. In.
	Loess or bluff deposit, consisting of fine light-grayish pulverulent silicious and more or less calcareous clay or marl, without distinct marks of stratification; rising back to a height of 80 to	90 0
D.	Yellowish-gray micaceous, soft sandstone, laminated or in thin ripple-marked layers, excepting 12 to 15 inches of the lower part, which is sometimes hard and compact; with fragments of plants.	10 0
C.	Drab, ash, and lead-colored, and reddish-brown clays, with, near the middle, a 9 or 10-inch hard bluish-gray argillo-calcareous layer, weathering to a rusty color. Fossils numerous, particularly near the lower part, as follows: *Rhombopora lepidodendroides*, *Lophophyllum proliferum*, *Scaphiacrinus* (?) *hemisphæricus*, *Eocidaris Hallianus*, *Synocladia biserialis*, *Fenestella Shumardi*, *Polypora submarginata*, *Glauconome trilineata*, *Lingula Scotica* (?), *Hemipronites crassus*, *Syntrilasma hemiplicata*, *Chonetes glabra*, *C. granulifera*, *Productus pertenius*, *Productus longispinus* (?), *P. Prattenianus*, *P. Nebrascensis*, *P. symmetricus*, *P. semiretriculatus*, *Rhynchonella Osagensis*, *Spirifer cameratus*, *Spirifer* (*Martinia*) *planoconvexus*, *Spiriferina Kentuckensis*, *Athyris subtilita*, *Lima retifera*, *Entolium aviculatum*, *Aviculopecten carbonarius*, *A. neglectus*, *A. Coxanus*, *A. occidentalis*, *Myalina Swallovi*, *M. subquadrata*, *Avicula longa*, *Avicula* (?) *Sulcata*, *Aviculopinna Americana*, *Pseudomonotis radialis* (? ?), *Nucula Beyrichi* (? ?), *Nucula ventricosa*, *Yoldia subscitula*, *Nuculana bellistriata*, *Macrodon tenuistriata*, *Solenomya* sp., *Solenopsis solenoides*, *Pleurophorus oblongus*, *Schizodus curtus*, *S. Wheeleri*, *Schizodus* sp., *Modiola subelliptica*, *Edmondia reflexa*, *E. Nebrascensis*, *E.* (?) *glabra*, *Prothyris elegans*, *Allorisma* (*Sedgewickia*) *subelegans*, *A.* (*S.*) *Geinitzii*, *A. reflexa*, *Dentalium Meekianum*, *Bellerophon Montfortianus*, *B. percarinatus*, *B. Marcouanus*, *B. carbonaria*, *Euomphalus rugosus*, *Orthonema subtæniata*, *Aclis Swalloviana*, *Pleurotomaria Haydeni*, *P. subdecussata*, *P. Marcouana*, *P. Grayvillensis*, *Orthoceras cribrosum*, *Nautilus occidentalis*, *Cythere Nebrascensis*, and *Cythere* sp.	39 0
B.	Several beds of hard, light-grayish, and yellowish limestones, in layers of from five to twenty inches in thickness, with soft, marly clay seams and partings. Fossils: *Fusulina cylindrica*, *Rhombopora lepidodendroides*, *Lophophyllum proliferum*, *Erisocrinus typus*, *Synocladia biserialis*, *Hemipronites crassus*, *Orthis carbonaria*, *Meekella striato-costata*, *Syntrilasma hemiplicata*, *Chonetes granulifera*, *Chonetes glabra*, *Productus longispinus* (?), *P. semireticulatus*, *P. costatus*, *P. Prattenianus*, *P. Nebrascensis*, *P. symmetricus*, *Rhynchonella Osagensis*, *Spirifer* (*Martinia*) *planoconvexus*, *S. cameratus*, *Athyris subtilita*, *Retzia punctulifera*, *Pinna peracuta*, *Myalina subquadrata*, *Allorisma subcuneata*, *Euomphalus rugosus*, *Bellerophon carbonaria*, *Phillipsia scitula*, *Cladodus mortifer* and *Deltodus* (?) *angularis*.	12 0

	Nature of strata.	Thickness.
		Ft. In.
A.	*a.* Lead-grayish and greenish clay, 4 feet *b.* Reddish-brown ferruginous, slightly gritty, indurated clay, 4 feet exposed above high-water mark.	8 0
	Total below drift	69 0

Fusulina cylindrica, mentioned among the fossils of bed B above, was not found by us in that bed immediately at the Nebraska City landing, but it occurs in this bed at Bennett's Mill and Wyoming; also in beds referred to this horizon by Professor Marcou, one and three-fourths mile and two and a half miles west of the outcrops at the landing, at elevations of from 73 to 80 feet above low-water of the Missouri. We also found it in a limestone cropping out of the bluff one mile below (southeast of) the Nebraska City landing, at a higher geological horizon than bed D, of the above section. The following other fossils, not yet found in bed B, at Nebraska City, occur in it at Bennett's Mill and Wyoming, as shown in the sections at those places: *Scaphiocrinus* (?) *hemisphæricus*, *Zeacrinus mucrospinus*, *Entolium aviculatum*, *Aviculopecten occidentalis*, *Pseudomonotis* sp., *Pleurophorus occidentalis* (?), *Peripristis semicircularus* and *Chomatodus arcuatus*.

In the foregoing section I have adopted the principal divisions recognized by Professor Marcou, and for convenience have also followed him in using letters instead of numbers, to designate the several divisions. He made other more minute subdivisions, but however well it may be to do this in first examining such exposures, with a view of keeping separate the fossils coming from each subordinate bed or seam, so soon as we know which beds are merely local modifications of color, hardness, &c., and which are more persistent and recognizable at other localities by their lithological characters and the grouping of their fossils, such minute local details had better, in comparing sections, be dispensed with, as they make the sections more complex and less easily compared.

By a glance at the list of fossils found in the bed B, it will be at once seen that with the exception of some five or six species (which are themselves well-known Coal-Measure forms), we have here all of the very same species found in the Bellevue and Omaha beds referred by Professor Marcou to the mountain limestone, and along with them a number of other species, forming altogether a group peculiarly characteristic of the western Coal-Measures, which Professor Marcou would also insist upon making mountain limestone, in Iowa, Missouri and Illinois.

This list of species, it will likewise be observed, not only includes such types as the foraminiferous genus *Fusulina*, Corals, Crinoids, Brachiopods, &c., but also Lamellibranchs, Gasteropods, and the Carboniferous and older genus *Phillipsia*, as well as Carboniferous genera of fishes. For these reasons, I must repeat that I cannot concur with Professors Marcou and Geinitz, in separating these beds from the Coal-Measures and including them in the Permian or Dyas, merely because a few such forms as *Pseudomonitis*, *Plurophorus*, and *Schizodus*, closely allied to, or possibly even in some cases identical with, European Permian species, are occasionally met with in them.

On ascending to division C, we also meet with a large proportion of the same fossils found in B, along with a considerable number not here found in the latter division. These, however, as will be seen farther on, are nearly all species that occur elsewhere in the Coal-Measures of the western States. Consequently I can see no reason whatever for drawing any important line of division between the beds included in C and those of B, or for separating either from the Coal-Measures.

The following beds were seen at Mr. Morton's and Mr. Werth's places near Nebraska City, but with the base of the outcrops elevated 73 to 80 feet above the Missouri.

*Section 1¾ and 2¾ miles due west of Nebraska City.**

No.	Nature of strata.	Thickness.
		Ft. In.
5	Slope consisting of 50 or 60 feet of Loess, at the base of which some slabs of ripple-marked micaceous sandstone were seen apparently nearly in place.	60 0
4	Bluish, greenish and drab clays, showing about 19 feet................	19 0
3	Bluish impure argillaceous limestone, some parts very hard, others decomposing and shaly. *Fusulina cylindrica*, *Scaphiocrinus* (?) *hemisphæricus*, *Rhombopora lepidodendroides*, *Hemipronites crassus*, *Chonetes granulifera*, *C. glabra*, *Spirifer* (*Martinia*) *planoconvexus*, *Spiriferina Kentuckensis*, *Athyris subtilita*, *Spirifer cameratus*, *Rhynchonella Osagensis*, *Syntrilasma hemiplicata*, *Productus costatus*, *P. semireticulatus*, *P. symmetricus*, *P. longispinus* (?), *Myalina subquadrata*, *Euomphalus rugosus*, &c.	2 0
2	Black bituminous shale, with a few inches of coal. Some of the same fossils in the shale as those in the bed above.	1 6
1	Blue laminated clay; only a few inches exposed.	

The group of fossils found in No. 3 here, it will be seen, agrees exactly, so far as they go, with that of division B of the section at the landing, with the exception of *Fusulina*, which is quite abundant here, but, as already stated, is not yet known to occur at the Nebraska City exposures, though found in division B, at Wyoming and Bennett's Mill. The rocks themselves, however, do not correspond with those composing the section at the landing, in their composition and other characters. The clays of No. 4, it is true, might correspond to a part of division C at Nebraska City, and the loose masses of ripple-marked sandstone might be supposed to have slidden from a bed agreeing with division D, but the other beds below are so different from any of those forming the lower part of the Nebraska City section, as to give rise to doubts as to their identity. At least I saw nothing there representing the black shale and coal seen at Mr. Morton's and Mr. Werth's places.

Another fact pointing to the conclusion that the beds forming the foregoing little section probably hold a higher position than those seen at the Nebraska City landing, is the occurrence of an outcrop agreeing almost exactly with the former, at a locality one mile below Nebraska City, in a southeast direction, at an elevation of 30 *feet above high water*. Here the same black shale with seams of coal is seen, with blue and ash-colored clays below, and immediately above it a hard bluish-gray argil-

* Professor Gienitz alludes to the locality at Mr. Morton's as being four miles west of Nebraska City. I am informed, however, by the engineer who ran the level between the Nebraska City landing and this outcrop at Mr. Morton's, that the distance is 1¾ miles, and the elevation 73 feet.

laceous limestone 30 inches in thickness, with *Fusulina*, *Hemipronites crassus* and fragments of other fossils, while immediately above the limestone greenish and bluish clays were seen for some 15 feet in place, with pieces of bluish argillaceous rock lying loose. Now, as this outcrop is just about one mile from the landing, and almost exactly in the direction of the dip, it cannot of course belong to any position below the beds seen at the landing, without there should be a reverse of the dip, or a fault here, of which there is no evidence whatever; while the beds themselves are entirely different from any part of that section. Hence it seems almost morally certain that this is the same limestone, black shale, coal, &c., seen at Mr. Morton's and Werth's places, brought down to within 30 feet of the Missouri at this place, by a gentle southeast dip, the distance from Mr. Morton's to this point being about two and three-quarter miles. If these are the same beds seen at Mr. Morton's, then the place they occupy here, one mile southeast of the Nebraska City outcrops, almost exactly in the direction of the dip, and yet at an actual elevation of 30 feet above the base of the outcrop at the landing, it would follow that the beds seen at Mr. Morton's and Mr. Werth's hold positions a little higher in the series than the top of the Nebraska City section, and of course that there is here a recurrence of nearly the same fauna above division C, as that found beneath it, in division B, at the landing. On the other hand, if these beds of shale, coal and limestone, seen one mile below Nebraska City, are not the same seen at Mr. Morton's, they must hold a still higher stratigraphical position; and, as they correspond to nothing in the Nebraska City section, they would at least establish the existence of the genus *Fusulina* above the horizon of division C at Nebraska City landing. It will also be seen farther on that the outcrop at Otoe City, doubtless correctly placed by Professor Marcou above the horizon of division C, contains great numbers of *Fusulina*, with various other Coal-Measure types, and, so far as yet known, none of the Permian forms.

It is to be regretted that in sinking a shaft at Mr. Morton's, to a depth of nearly 100 feet, commencing nearly on a level with the base of the outcrop alluded to, no memorandum of the thickness of the various beds and seams passed through was kept. A great number of hand specimens of the material penetrated, consisting of various-colored clays, arenaceous matter, &c., with some limestones, were carefully preserved, but, unfortunately, it was impossible to obtain any more reliable information in regard to the exact thickness and order of succession of the numerous layers penetrated than could be given from memory by the workmen; while the planking up of the shaft prevented the possibility of any information of this kind being derived from a direct examination.

It is worthy of note here, that some of the clays shown to us from the shaft, agree very well, in color and other respects, with those of division C, at the Nebraska City landing, and some of these specimens also contained a group of fossils very rarely found *together* in any other horizon than that division—such, for example, as *Chonetes glabra*, *Aviculopecten carbonarius*, *Bellerophon Marcouianus*, &c. All the fossils found in the material removed in excavating this shaft are figured together on Plate IV. No coal was struck, but at near the bottom of the shaft a black, bituminous shale, 3 feet in thickness, containing the impression of a Calamite, was penetrated. Immediately under this, a hard, white limestone, 3 feet in thickness, was also passed through.

As a means of forming some conclusions in regard to the nature of the strata beneath the surface of the Missouri at Nebraska City, the following statement of an artesian boring, made at this locality, is given. This boring was made in the valley of the river, about one mile below the landing, commencing about 13 feet above high-water mark.

*Mr. Croxton's boring at Nebraska City.**

No.	Nature of strata penetrated.	Depth.	
		Ft.	*In.*
1	Alluvial	17	0
2	Limestone	2	8
3	Gray shales	6	0
4	Sandstone	1	4
5	Limestone	2	0
6	Blue shales	2	0
7	Red shales	15	0
8	Limestone	1	0
9	Red and buff shales	7	0
10	Limestone	2	0
11	Shale	3	0
12	Limestone	1	0
13	Red sandstone	4	0
14	Gray shales	8	0
15	Red shales	5	0
16	Gray shales	20	0
17	Micaceous sandstones	8	0
18	Blue limestone	1	0
19	Buff shales	3	0
20	Gray shales	12	0
21	Red sandstone	4	0
22	Red shales	3	0
23	Gray shales with intercalations of various kinds of rock, sulphuret of iron (smutty, according to mines) and traces of petroleum.	65	0
24	Blue limestone	5	0

* The different beds passed through in this and the other artesian borings, are numbered from above downward; while those of the sections examined above the horizon of the Missouri, are numbered from below upward.

No.	Nature of strata penetrated.	Depth.	
		Ft.	*In.*
25	Coal	1	3
26	Clay and shale	13	0
27	Red shale	3	0
28	Calcareous sand	1	0
29	Blue slate rock	1	4
30	Red shales	1	0
31	Limestone with hard seam	0	10
32	Black slate	1	0
33	Brown shales	3	0
34	White limestone, very fine texture	6	0
35	White limestone	7	0
36	Gray shales	7	0
37	White sandstone, fine texture	3	0
38	Blue limestone	5	0
39	Coarse gray sandstone, (salt water)	5	0
40	Fine freestone	4	0
41	Shale	3	6
42	White limestone	6	0
43	Hard brown shale	3	0
44	Fossiliferous calcareous shale	1	0
45	Gray soft shale	1	6
46	Calcareous sand-rock	4	0
47	White lime or marl	16	0
48	Blue limestone	3	6
49	White shales or marl	3	0
50	Pyritiferous rock	2	0
51	White shales or marl	3	9
52	Crinoidal limestone	1	6
53	Brown shale or slate	2	0
54	Buff shale	2	0

No.	Nature of strata penetrated.	Depth.	
		Ft.	*In.*
55	Blue shales	10	0
56	Gray and white shales	15	0
57	Gray slate	2	0
58	Black slate	—2	0
59	Limestone	5	0
	Total	344	2

I had no opportunity to examine the borings from this well, and have to rely entirely upon the memorandum kept by those conducting the work. Although it is probable, for reasons elsewhere intimated, that the details of the section may not be exactly correct in all respects, it doubtless gives a good general idea of the strata penetrated, and is important as a demonstration that no workable beds of coal exist here beneath the level of the Missouri Valley, within nearly 350 feet of the surface. It also confirms other evidences of a southeastern dip here, because it is evident that the beds numbered 2, 3, 4, 5, 6 and 7, near the top of the boring, are the same as divisions A and B, of the section at the landing; and as they rise there nearly 20 feet above high water, and are here found entirely below that horizon, the evidence becomes quite conclusive as to the dip. It also shows that the *Fusulina* limestone cropping out of the hill 30 feet above high water, some 150 yards farther down, and in the direction of the dip, must hold a position above the whole of the section seen at the landing.

After leaving the immediate vicinity of Nebraska City, no exposures of rocks in place were observed, until we arrived at Otoe City, about five and a-half miles farther down the river, in a direction east of south. Here we saw the following beds exposed:

Section of the beds exposed at Otoe City.

No.	Nature of strata.	Thickness.	
		Ft.	*In.*
	Loess, with some Drift at the base	40	0
11	Soft, yellowish sandstone	10	0
10	Drab and ash-colored, with seams and concretions of arenaceous matter	3	0
9	Bluish laminated clays, passing gradually into the next bed below	7	0
8	Nearly black laminated clay or shale, with sometimes between it and the bed above a 6 to 8-inch bed of yellow limestone in the shale. *Productus Prattenianus*, *P. longispinus* (?), *Sp.* (*Martinia*) *planoconvexus*, *Rhynchonella Osagensis*, *Chonetes granulifera*, *Spiriferina Kentuckensis*, *Productus semireticulatus*, *Hemipronites crassus*.	—3	0
7	Very hard, dark-gray calcareous layer		4

No.	Nature of strata.	Thickness.
		Ft. In.
6	Red, green, blue, and light ash-colored clays, with near the base a 2 or 3-inch seam of black shale.	58 0
5	Soft, drab, marly material, becoming in parts hard and compact, or with hard, calcareous seams. Millions of *Fusulina cylindrica;* also, *Chonetes granulifera, Hemiphronites crassus, Spirifer cameratus, Spiriferina Kentuckensis, Syntrilasma hemiplicata, Meekella striato-costata, Productus semi-reticulatus, Myalina perattenuata.*	3 0
4	Light drab laminated clay, with a streak of black; at one place seen to swell out, so as to form a bunch of coal, 6 to 8 inches thick, with efflorescence of sulphate iron.	1 6
3	Soft, incoherent, yellowish sandstone	6 0
2	Soft, bluish, sandy shale, with large, round, and compressed oval concretions.	5 0
1	Bluish and drab clays, in parts more or less arenaceous. *Productus Prattenianus.*	15 0
	Total, exclusive of Loess and Drift	111 —

It is proper to state here, that the whole of this section was not seen in one uninterrupted exposure. All the beds from No. 1 up to, and including 8 feet of No. 6, are seen in regular succession immediately above the site of the deserted village called Otoe City. The remaining portion above was seen in another exposure about a quarter of a mile above the village, with just enough uncovered space below (allowing for a gentle inclination of the strata) to receive the beds seen immediately above the village, and, as but a moderate southeast dip was observed here, there is scarcely any reason to doubt the propriety of regarding the two exposures as directly succeeding each other in the vertical series.

At the time this section was examined, the Missouri was nearly at high-water mark. I was informed by a gentleman living in the village, however, that the bluish-drab clays forming No. 1, continue on down 12 to 13 feet farther than the base of the section as given above, and that beneath the whole there is a dark bluish-gray limestone, of which I saw some pieces containing *Hemipronites crassus.*

If the southeast dip observed near Nebraska City is continued in this direction to this place, it is quite probable that the dark-bluish gray limestone seen at low water here at the base of the Otoe City section, is the same seen cropping out over the black shale and coal, 30 feet above the Missouri, one mile below Nebraska City.* If so, it would apparently make the whole of the Otoe City section newer than the beds seen at the Nebraska City landing. This would agree nearly with the opinion expressed by Professor Marcou,† who, if I understand him correctly, thinks the lower beds here, up to No. 5 inclusive, of the foregoing section, the same as division D of the section at Nebraska City, and some 10 feet

* Professor Marcou gives the distance from Nebraska City to Otoe City as twelve miles south; in this he probably means by the curves of the steamboat channel of the Missouri River, as it is only about five and a half miles by a direct line, and in a southeast direction.

†Bull. Geol. Soc. France, XXI, new ser., p. 137.—1864.

of the bed No. 6 exposed in the bluff at Otoe City, a part of the highest bed of the Dyas. He seems not to have seen the other beds, forming the upper part of the section between a quarter and a half mile farther up the river. He also appears to have made the identification entirely from lithological characters, as neither he nor Professor Geinitz, who investigated his collections, mentions any fossils from this locality, while all of those yet known from this outcrop are, as may be seen by the names in the section, the same Coal-Measure types so often mentioned at other localities. So it will be seen we have here additional evidence to that mentioned in connection with the statement of facts observed at and near Nebraska City, that there are above the horizon of division C at Nebraska City section, repetitions of the same Coal-Measure group of fossils, seen in the beds below, including millions of *Fusulina*, with *Chonetes granulifera*, *Productus semireticulatus*, *P. Prattenianus*, *Spirifer cameratus*, *Spiriferina Kentuckensis*, *Hemipronites crassus*, *Rhynchonella Osagensis*, *Syntrilasma hemiplicata*, *Meekella striato-costata*, &c., while none of the Permian types have yet been found here!

At the village of Peru, five miles in southeast direction, below Otoe City, about a quarter to a half mile above the landing, beds of light-colored limestone, 12 to 18 inches in thickness, crop out of the side of the slope 90 to 100 feet above high water. A little lower some loose masses of limestone, that have apparently slidden from the beds above, have been dug out for building purposes. From the disintregrated limestone, or associated shaly matter, we picked up loose, many specimens of *Spirifer* (*Martinia*) *planoconvexus*, *Sp. cameratus*, *Productus Nebrascensis*, *P. longispinus*, *Athyris subtilita*, *Spiriferina Kentuckensis*, *Hemipronites crassus*, *Euomphalus rugosus*, &c. Lower in the hill indications of red clays were seen, though not well exposed, the whole hill being probably composed of the same, or at any rate, a part of the same, beds exposed at Otoe City.

Less than a quarter of a mile, however, below the village of Peru, there is an abrupt exposure of yellowish and light-gray, soft, somewhat micaceous sandstone, with large, round and compressed or oval concretions of arenaceous matter, of considerable hardness. There are also very curious irregularly and obliquely arranged seams and isolated masses of dark-bluish shaly matter and clay. These appear as if the sandstone had been very irregularly eroded in places during its deposition, and the shaly matter deposited in the depressions, and then more sand upon it again. Fragments of coal were also seen embedded in the sandstone, along with stems of Calamites, and broken-up leaves of ferns. The sandstone can scarcely be said to be stratified, but appears massive with the exception of some oblique marks of deposition, and the intercalated seams of shaly matter. The latter are not continuous for any distance, but often end very abruptly, or in other cases become much attenuated, and again swell out to a foot or so in thickness. They do not appear to conform to the bedding of the sandstone, but cut obliquely across it at various angles, and yet their laminated structure, and fragments of plants, show they were deposited in water. This exposure of sandstone rises abruptly from the edge of the river at high water, to an elevation of about 60 to 65 feet, and almost certainly holds a lower stratigraphical position than the outcrops of limestone seen 90 to 100 feet above the Missouri, a short distance above the village. Its positior is doubtless nearly the same as that of the lower part of the Otoe City section, though it is more arenaceous here, and perhaps thicker. Just over it some red clay was seen, apparently a little slidden from its proper position.

At the time of our visit, the Missouri being high, no beds below this sandstone were exposed above water. Dr. Owen, however, mentions seeing beneath evidently the same bed, either here or near here, a black bituminous shale, and brown encrinital limestone.

As an additional evidence that this sandstone holds a lower position in the series than the limestones mentioned high in the hill above the village, we may mention that at a locality about two and a half miles west of south from here, and a mile or so back from the river, at a higher elevation than the sandstone, there are outcrops of light-colored limestone agreeing exactly with those seen cropping out in the hill above the village. In the rubbish thrown out at a quarry in these beds, we found *Spirifer* (*Martinia*) *planoconvexus*, *Sp. cameratus*, *Chonetes granulifera*, *Pleurophorus* sp., fragments of *Myalina*, *Pleurotomaria* sp., *Bellerophon carbonaria*, *B. Montfortianus*, &c.

At Brownville, about five miles below Peru, in a direction east of south the following outcrops were seen:

Sections of the various beds exposed at Brownville.

No.	Nature of strata.	Thickness.
		Ft. In.
14	Loess rising back with the slope from 30 or 40 to	100 0
13	Dark-bluish, very fine unctuous clay, becoming nearly black below, and weathering to drab color. *Aviculopecten Whitei*, *Spirifer* (*Martinia*) *planconvexus*, *Productus pertenuis*, small *Polyphemopsis*, *Pleurotomaria*, *Macrocheilus*, *Myalina perattenuata*, and *Nuculana* sp. (Seems to be merely the remains of a thicker bed.)	2 0
12	Yellowish-gray, granular or sub-oolitic limestone; massive, but showing a disposition to divide into two layers.	3 0
11	Unexposed	10 0
10	Whitish, soft argillaceous limestone, 6 to 8 inches thick	0 8
9	Red, purple, and greenish clays	10 0
8	Whitish and yellowish impure limestone, rather massive. *Allorisma subcuneata*, stems of crinoids, *Bellerophon percarinatus*, *Productus Nebrascensis*, &c.	3 0
7	Purple clay	1 0
6	Soft, whitish limestone	6 0
5	Bluish clay	5 6
4	Black shale and seams impure coal, with impressions fern leaves	1 0
3	Blue clay, with fragments of coal and iron pyrites	20 0
2	Black, hard rock, with crystals calc spar	2? 0
1	Soft, yellowish micaceous sandstone, with irregular seams and alternating laminæ of black and greenish, more or less carbonaceous and sandy material, with fragments of coal. Many broken leaves of ferns, pieces of *Calamites*, &c. *Neuropteris hirsuta*, *Neuropteris Loschii* (identified by Professor Lesquereux), and Coprolites of some Selachian fish, as determined by Professor Agassiz.	57

Of the lower member of this section only about 25 feet were exposed above high water of the Missouri, when examined. I was informed, however, by an intelligent gentleman who sunk a shaft into it through several of the overlying beds, that they penetrated it to a depth of 57 feet without passing through it. Several persons also stated that when the Missouri is low, branches and trunks of trees are seen in this rock, which, judging from their descriptions, are probably *Lepidoderon* and *Sigillaria.*

This is doubtless the same sandstone seen just below Peru, about five miles in a direction west of north from here. If so, there must be a moderate dip in this direction, or the bed is not so thick, as it only rises at this place about 25 feet above high water, while its summit there was 65 feet above the river, and there must be 3 or 4 feet fall in the river in that distance.

Of the beds filling the interval between Nos. 1 and 4, of the foregoing section, I only saw the upper part of No. 3, but they were described to us and their thickness given by the gentleman who sunk the shaft already mentioned, through them.

A little below the landing at Brownville, and not more than 100 yards in a southeast direction from where the sandstone No. 1 of last section is seen, showing a thickness of 25 feet above high water, 5 or 6 feet of greenish fine arenaceous clay was seen at the base of the hill, not more than 6 or 7 feet above the level of the river. Above this greenish clay, about the same thickness of reddish-brown clay occurs. Just over the latter fragments of whitish argillaceous limestone were lying, as if they had slidden down from a decomposing bed of the same. Among these great numbers of *Fusulina* and *Chonites granulifera* were found loose; also fragments *Spirifer cameratus*, *Productus semireticulatus*, &c. Above this some beds of red clay, and in parts green, are imperfectly exposed, and still higher, or about 50 feet above high water of the Missouri, some thin beds of whitish, crumbling, impure limestone, and light-drab clay are imperfectly exposed. Among the loose fragments of the limestone were found *Chonetes granulifera*, *Productus semireticulatus*, *P. longispinus*, *P. Nebrascensis*, and some stems of crinoids. Immediately above the crumbling limestone some indications of a thin bed of black shale were observed, while the remainder of the hill (some 60 to 70 feet above) seemed to be composed entirely of Loess.

It is probable all these beds of clay and limestone had slidden some from their proper horizons, as it would require a greater dip than we have reason to believe exists here, to bring the whole of the 25 feet of sandstone seen above high-water mark, just above the Brownville Landing, beneath the base of the hill here, only about 100 yards farther down the river.

Between one and two miles below Brownville, along the same line of bluffs, the following exposures were seen:

Section one and a half miles below Brownville.

No.	Nature of strata.	Thickness.
		Ft. In.
8	Loess rising back with the slope from 30 or 40 to	10 0
7	Bluish, fine clay, a few inches only remaining; apparently same as No. 13 at Brownville.	0 0
6	Yellowish, granular sub-oolitic limestone	0 10

No.	Nature of strata.	Thickness.
		Ft. In.
5	Ash-colored clay, with whitish calcareous seams, more or less indurated, with some blue and red clays below.	8 0
4	Yellowish-gray, impure, sub-oolitic limestone	2 0
3	Red and bluish clays	10 0
2	Whitish, impure limestone. *Productus Prattenianus*	0 10
1	Red, bluish, and ash-colored clays, with loose fragments of whitish limestone strewed along the upper part of slope from the bed above; also loose *Productus longispinus* (?), *P. Nebrascensis*, *Sp. cameratus*. Near base great numbers of *Chonetes granulifera;* with *Myalina subquadrata*, *Bellerophon percarinatus*, &c. The whole of this bed not very clearly exposed.	43

These exposures indicate a continuation of the upper part of the Brownville section, the lower sandstone being probably mainly below the level of the Missouri here. It is worthy of note that the beds 4 and 6 here agree exactly in structure, color, and composition, with the two layers composing No. 12 of the Brownville section, but here they are separated by about 8 feet of clays, represented at Brownville, in same position, by a mere seam, which is at other places wanting. This illustrates the changes liable to occur in the beds of these rocks in comparatively short distances.

At Aspinwall, seven miles below Brownville, in a direction slightly east of south, we had an opportunity to examine various outcrops, from which the following section was constructed:

Sections of beds exposed at Aspinwall.

No.	Nature of strata.	Thickness.
		Ft. In.
15	Loess varying in thickness from 30 to	40 0
14	Grayish and drab clays, with thin seams of light-colored calcareous matter.	6 0
13	Massive, rather hard, yellowish limestone. Numerous *Myalina perattenuata;* also *Aviculopecten occidentalis*, *Hemipronites crassus*, *Allorisma subcuneata*, *Productus Nebrascensis*, *Edmondia Aspinwallensis*, *Euomphalus rugosus*, *Nautilus occidentalis*, &c.	3 0
12	Bluish and drab clays, with one or more 6 to 8-inch seams soft, grayish, impure limestone.	20 0
11	Yellowish limestone	1 6
10	*a.* Bluish clay, 5 feet *b.* Red clay, 5 feet *c.* Bluish laminated clay, 3 feet	13 0
9	Bluish and whitish impure limestone. *Productus longispinus* (?), *Chonetes granulifera*, *Athyris subtilita*, *Hemipronites crassus*, *Spirifer* (*Martinia*) *planoconvexus*, *Meekella striato-costata*, *Productus pertenuis*, *Productus semireticulatus*, *Spiriferina Kentuckensis* and a tooth of *Xysrtodus* (?) *angularis*.	4 0

No.	Nature of strata.	Thickness. Ft. In.
8	Coal	0 6
7	Clay	2 0
6	Bluish indurated clay	4. 0
5	Hard, bluish, impure limestone	1 6
4	Black shale	4 0
3	Coal	1 10
2	Bluish clay	15 0
1	Sandstone, above low water	3 0
	Total below Loess	78 10

At the time of my examination, No. 1, and part of No. 2, of this section were concealed by high water. No. 3 was seen rather imperfectly a little below its true horizon, owing to the slipping of a part of the outcrop from undermining by the river. Nos. 4, 5, 6, 11, and 12 were more or less perfectly exposed, and 13 was very clearly seen. All the information in regard to the thickness, order of succession, &c., of the other beds not seen by me, was obtained from an intelligent miner who has made various excavations into the hill, and stripped off the loose material at places, so as to be able to see all the beds in their proper positions. Most of these openings were covered again by slides, and the falling in of the clay, loose rocks, &c., at the time of my visit. The fossils from No. 9 were found in fragments of the rock thrown out of a shallow shaft.

The miner working here says all these beds have a perceptible dip up the river, north or northwest, as he has determined by sinking a shaft at a point a little north of the main exposure, and he thinks the whole section passes beneath that seen at Brownville. The dip, however, may be local, as the beds here above No. 8 appeared to me very like those composing the imperfect exposures seen a short distance below Brownville. He says the bed of coal No. 3 and associated rocks, agree exactly with those seen at Clorinda, Page County, Iowa, where he has been engaged in mining coal.

It will be observed that this bed of coal is the thickest one yet seen in any natural exposure, or struck in any of the shafts or borings in these rocks. As near as can be determined from specimens taken from the rather weathered outcrop, it seems to be of good quality, and preparations are being made to mine it. It is not easy to determine yet, whether it can be profitably mined. It certainly could not, I should think, in a district where thicker beds are accessible, as it will probably be necessary to remove most of the overlying rather soft black shale in working it, in order to have a solid roof that can be timbered up, if not to give working space. Another disadvantage is that it cannot be conveniently mined at the outcrop along the bluff, owing to the tendency of all the beds of the lower part of the section to slide down, in consequence of the river washing away the soft underlying beds. Consequently, they will have to sink a shaft 70 to 80 feet in depth, back from the river some little distance. In a country like this, however, with such a vast area of fertile soil, capable of sustaining a dense population, where there is at present a deficiency of wood, it is probable that this coal may be profitably mined, unless better and more extensive beds may be found within workable distances by sinking deep shafts in this region, which seems doubtful.

The rocks seen here at Aspinwall were also, like those at Brownville, Peru, and other localities along this part of the Missouri, referred by Professor Marcou to the Permian (= Dyas), but upon what evidence he does not say, nor is it apparent to any one who regards fossils as any guide in identifying rocks, as those found here consist of the same forms constituting the group so often mentioned as characterizing the Coal-Measures of the Western States.

At Arago, about fourteen miles southeast from Aspinwall, I saw at the base of the hill, near 12 feet above the Missouri, some 7 inches of yellow limestone, containing *Productus longispinus*, and a small *Myalina;* and over this 18 inches of hard, compact bluish limestone, containing *Productus punctatus*, *Chonetes*, *Pinna*, *Entolium aviculatum*, &c. Above the latter some bluish clays were seen, and some large masses of impure bluish limestone that had slidden from a higher bed, with numerous *Fusulina*, were lying along the slope.

Two miles above Rulo, and some six or seven miles in a direction east of south from Arago, the following outcrop was examined:

Section two miles above Rulo on the Missouri.

No.	Nature of strata.	Thickness.	
		Ft.	*In.*
7	Loess with perhaps some Drift seventy to	80	0
6	Massive yellow limestone	5	0
5	Gray and yellowish impure limestone and drab clays	4	6
4	Bluish and drab arenaceous clay with fossil ferns. *Neuropteris hirsuta* and *N. Loschii.*	7	0
3	Coal	0	6
2	Indurated clay, called soapstone by the miners. (Not seen)	0	4
1	Bluish laminated sandstone, very soft, with thin streaks of black, and looking very much like No. 1 of the Brownville section.	8	0

I did not see the whole of No. 1 of this section, it being partly covered by the high water of the Missouri, though I was informed by a miner at work in a drift immediately over it, that he had seen as much as 8 feet of it exposed. The whole of this exposure seemed to me to have bodily slipped a little below its true horizon, probably from the washing away of the soft sandstone beneath by the river. This appearance is also confirmed by the statements of the miners, who informed me that the coal ends and all the beds change abruptly at the end of the drift, 40 to 50 feet in. The thickness, composition, and order of succession of the beds, however, can be very clearly seen. Of course any attempt to mine so thin a seam of coal by drifting in and removing a portion of the clays above, will hardly be remunerative, and there is little or no reason for believing that it will be found materially to increase in thickness farther in the hill.

At another locality, one and a quarter miles south of Rulo, and at an elevation of about 25 feet above high water of the Missouri, a shaft and boring were sunk to a depth of about 76 feet. The shaft was filled with water at the time of our visit, but the miner who sunk it, Mr. St. Louis, gave the following statement of rocks perforated:

Shaft and boring one and one-fourth miles south of Rulo.

No.	Nature of strata penetrated.	Depth.	
		Ft.	*In.*
1	Yellow indurated clay, called soapstone by the miners	18	0
2	Yellow limestone	3	0
3	Blue clay. In this clay, lying near the shaft, we found the following fossils: *Rhombopora lepidodendroides*, an incrusting species of *Fistulipora; Polypora submarginata, Hemipronites crassus, Productus Nebrascensis, P. Prattenianus, Chonetes granulifera, Syntrilasma hemiplicata, Spirifer (Martinia) planoconvexus, Sp. cameratus, Nucula* (?) sp., *Pleurophorus* sp., *Bellerophon Kansasensis, Bellerophon* sp., *Naticopsis* sp., *Pleurotomaria perhumerosa*, and several undetermined species of *Murchisonia*.	12	0
4	Hard, gray limestone	11	0
5	Blue clay	17	0
6	Limestone	3	0
7	Blue clay bored into below the limestone	12	0
	Total	76	0

In another shaft, sunk only about 30 yards south of the above, and commencing at a horizon about 20 feet lower, we were informed a bed of coal 16 inches thick was struck at a depth of 17 feet. It seems difficult to account for this bed not having been struck in the other shaft, sunk so near, to a depth of 76 feet. The miner thought that it was to be accounted for by the northern dip of the strata, but this is scarcely possible, as no evidence whatever of such high inclination of the strata was observed in some beds of yellow *Fusulina* limestone seen cropping out of the slope a little above the horizon of the deeper shaft, though there seems to be good evidence of a moderate northern dip here. If there was no mistake in all the facts given, there is probably a fault here, though it may be the case that the disagreement between the beds observed in the two shafts was produced by a sliding down of some of the beds penetrated in the 17-foot shaft at some distant period, when the Nemaha (in the valley of which this latter shaft was sunk) ran close along the base of the hill at this place, as often takes place in these rocks.

At the mouth of the Great Nemaha, a mile or two farther down the Missouri, Dr. Hayden saw an exposure (the same mentioned by Dr. Owen) of soft sandstone rising 20 or 30 feet above the river, with above it a thin (5 or 6-inch) seam of coal connected with arenaceous shales, containing the same ferns found over the bed of coal two miles above Rulo, and at Brownville.

The elevation of this coal and sandstone here above the Missouri shows there is quite a perceptible rising of the strata in this direction, the same coal being only about 8 feet above the river two miles above Rulo, though it had apparently slidden somewhat below its true horizon at the latter place. I am inclined to believe this sandstone under the coal the same bed seen at Peru and Brownville, and at the base of the section at Aspinwall, though it may be another holding a lower position. If it is the same, there can be little doubt but the exposures here near Rulo hold a position in the series above the horizon of the Nebraska City section. On these points, however, more detailed examinations than we had an opportunity to make are desirable.

With the completion of the observations here in the vicinity of Rulo, ended our examinations of the rocks seen along the Missouri in Nebraska, between Omaha City and the southern boundary of the State. As some deep borings, however, had been made at St. Joseph, Missouri, and Atchison, Kansas, it seemed desirable that all the facts revealed by these borings should be noted and taken into consideration in connection with those observed along the Missouri in Nebraska above; consequently these two places were visited on my return eastward.

The distance between Rulo, in the southeast corner of Nebraska, and St. Joseph, in Missouri, is, by an air line, about thirty-three miles, the direction from the former to the latter being nearly due southeast. Along this interval I made no examinations, though it is evident from the observations of Dr. Owen, Professor Swallow, and particularly those of Mr. Broadhead, that Coal-Measure rocks, characterized by the same fossils already so often mentioned, form, with the overlying Loess, the hills bounding the Missouri Valley, as above Rulo. It is probable, however, that the rise of the strata below Rulo brings up a considerable thickness of rocks between that place and St. Joseph that are not seen at or near Rulo. If I understand Mr. Broadhead's section correctly, he recognizes about 700 feet of strata altogether, as rising up between the sandstone at the mouth of Great Nemaha and St. Joseph, Missouri. From his description of these rocks, I have little doubt that they include the whole or a part of the Nebraska City section, and all of the beds bored through at that place, composed in part of the Rock Bluff and Plattsmouth sections. There may, however, be here a considerable thickness of intercalated strata not represented as far north as Nebraska City, as there are evidences of a thickening of the series toward the south. The fact that all of these rocks rise up from beneath the sandstone at Nemaha, which may be the same seen at Peru, believed to hold a higher stratigraphical position at least than the division C at Nebraska City, favors the view expressed, while the few fossils mentioned by Mr. Broadhead in these rocks at least show them to belong mainly to the same Coal-Measures series. These are, *Fusulina cylindrica*, *Chætetes*, *Hemipronites crassus*, *Syntrilasma hemiplicata*, *Productus Nebrascensis*, *P. Prattenianus*, *Athyris subtilita*, *Spiriferina Kentuckensis*, *Myalina subquadrata* and *Aviculopecten carboniferus*. Mr. Broadhead used different, but synonymous, names for several of these fossils.

While at St. Joseph I visited a place about one mile, a little west of north from the city, where a shaft had been sunk (commencing, perhaps, some 70 or 80 feet above the Missouri) to a depth of about 100 feet, in search of coal. Judging from the material thrown out, for more than half the distance it had penetrated through Loess, of which there are heavy deposits here. Beneath this they struck fine, rather bright blue indurated clay, more or less laminated, and showing some minute specks of mica on the partings. This evidently belongs to the coal series, and not to the Drift or Loess. From the quantity thrown out, it was probably penetrated to a depth of about 40 feet, and of course below the level of the Missouri, without passing through it, or striking any coal or beds of limestone, or other hard rock.

At another locality two miles south of St. Joseph the deep boring, to which allusion has already been made, was sunk. For the following statement of the various beds penetrated, and for the privilege of examining specimens of the borings from most of the beds, I am under obligations to T. B. Weakley, esq., of St. Joseph, one of the gentlemen at whose expense the boring was made.

Section of a boring two miles south of St. Joseph, 60 feet above high water of the Missouri.

No.	Nature of strata penetrated.	Depth.	
		Ft.	*In.*
1	Loose surface material. [Loess and soil]	21	0
2	Blue clay, (with specks of mica)	74	0
3	Sandstone	16	0
4	Soapstone. [Indurated clay]	10	0
5	Limestone	4	0
6	Soapstone. [Indurated clay]	2	0
7	Slate	2	0
8	Coal	1	2
9	Bituminous shale	6	0
10	Limestone	7	6
11	Slate	5	0
12	Limestone	6	0
13	Black slate	2	0
14	Limestone	4	6
15	Black shale or slate	5	6
16	Limestone	1	0
17	Slate	14	0
18	Limestone	4	6
19	Hard soapstone. [Indurated clay]	3	0
20	Sandstone	2	6

No.	Nature of strata penetrated.	Depth.	
		Ft.	*In.*
21	Soapstone. [Indurated clay]	2	6
22	Limestone	4	9
23	Dark-bluish slate	24	0
24	Bluish-gray limestone	12	0
25	Slate	6	0
26	Bituminous shale	1	6
27	Blue slate	8	0
28	Limestone	4	0
29	Slate and shale	20	0
30	Light-gray or drab limestone	6	0
31	Green slate	20	0
32	Bluish limestone and clays	40	0
33	Coal	1	0
34	Blue soapstone. [Indurated clay]	3	8
35	Bluish-gray slate or indurated clay	4	6
36	Light-bluish gray and darker limestone	16	6
37	Light-blue soapstone. [Indurated clay]	3	0
38	Coal	2	0
39	Light-drab limestone	9	0
40	Marble. [Compact limestone]	4	0
41	Limestone	3	0
42	Slate	2	6
43	White limestone	2	6
44	Blue marley clay	14	0
45	Bluish-drab soapstone. [Indurated clay]	7	0
46	Slate	38	0
47	Limestone	10	6
48	Soapstone. [Indurated clay]	15	0
49	Light-gray limestone	20	0
	Total	496	7

The bed of blue clay (No. 2 of the boring) is almost certainly the same struck in the shaft, one mile above St. Joseph, as the borings from it agree well with the clay thrown out at the shaft. It is probably the most extensive bed of that character in the whole series, and may possibly represent the 40-foot bed, No. 19, called "blue slate," in the Omaha boring.

Between St. Joseph and Atchison, Kansas, I had no opportunity to make any examinations; but Mr. Broadhead's explorations of these rocks along the Missouri side of the river show a continuation of the same great series of Coal-Measures without any workable beds of coal. The distance between these two places is, by a right line, about seventeen miles, and the direction from St. Joseph to Atchison very nearly due southwest.

At Atchison, and along the bluffs between there and "Riverside," the residence of Mr. George Scarborough, just above the village of Sumner, and three miles below Atchison, I had an opportunity to make out very satisfactorily (with the exception of one unexposed interval of 17 feet) a continuous section of Coal-Measure beds, altogether more than 175 feet in height above the Missouri.

During my examinations here, I was materially aided by Mr. Scarborough, whose taste for geological and botanical investigations had previously led him to explore the surrounding country, so as to be able to direct my attention to the best exposures, and otherwise to assist me in making out the following section of the rocks at this locality.

Section of the rocks exposed at Riverside, Kansas, and along the river bluff between there and the Atchison Landing.

No.	Nature of strata.	Thickness. Ft.	In.
22	Slope, probably Loess, to the summit of highest point, known as Prospect Hill.	27	0
21	Rough, yellowish limestone, splitting into thin pieces on weathered surfaces; great numbers of *Fusulina*.	2	0
20	Blue clay below, yellow above	2	0
19	Two layers yellowish-gray limestone with 18 inches clay between	5	6
18	Ash-colored clays with *Chonetes granulifera*	2	6
17	Black, thinly laminated bituminous shale	2	6
16	Hard, impure, yellowish-gray limestone	2	0
15	Blue clay	6	0
14	Soft, yellowish limestone	3	0
13	Grayish limestone weathering to yellow, in two layers, upper 1 foot, lower 20 inches. *Fusulina*.	2	8
12	Bluish and ash-colored, more or less laminated and sandy clays, with near the top a 2½-inch seam of impure coal, or carbonaceous matter.	11	6

No.	Nature of strata.	Thickness. Ft.	In.
11	Yellowish and gray or bluish-gray sandstone; some parts very soft, but a few layers very hard and breaking almost like quartzite; the soft part sometimes laminated, and in others in 10 to 12 inch layers.	23	0
10	Unexposed	17	0
9	Hard, bluish-gray, impure limestone. *Productus Nebrascensis, P. pertenius, Chonetes granulifera, Athyris subtilita.*	2	6
8	Unexposed space	21	0
7	Hard, bluish-gray, impure limestone, some parts argillaceous, others a little arenaceous. *Productus Nebrascensis, P. Prattenianus, Athyris subtilita.*	4	0
6	Bluish laminated clay, with near the top two or three hard calcareous layers.	5	0
5	Bluish and light-grayish limestones, weathering to yellowish, in irregular layers, with thin gray and yellowish partings. Many *Fusulina, Productus Nebrascensis, P. Prattenianus, Spirifer cameratus, Sp. lineatus* (?), *Athyris subtilita, Pinna peracuta, Avicula longa, Allorisma subcuneata, Syntrilasma hemiplicata, Schizodus,* and teeth of *Cladodus.*	16	0
4	Yellowish and drab laminated clays	2	6
3	Black, regularly laminated shale	2	6
2	Hard, bluish-gray, argillaceous limestone, weathering to yellowish in places. *Fusulina cylindrica, Hemipronites crassus, Chonetes glabra, Productus Prattenianus, P. longispinus* (?), *Edmondia subtruncata, Aviculopecten occidentalis, Bellerophon crassus,* &c.	1	10
1	Bluish and drab-colored, more or less laminated, clay, with near the top many *Chonetes granulifera;* also *Myalina* sp.	42	0
	Total Coal-Measures	175	6

Eight or ten feet of No. 1 of this section were under water at the time the examinations were made, and the nature of that part was learned from Mr. Scarborough and other persons at Atchison. At the Atchison Landing, the top of the heavy stratum of limestone, No. 5, which forms a well-marked horizon, is a little more than 65 feet above high water of the Missouri, while at Mr. Scarborough's place, (Riverside), three miles below Atchison, in a direction east of south, its upper layers are 100 feet above high water. Allowing 2 feet for the slope of the river surface in that distance, it would leave a difference of 37 feet in the actual elevation of the various corresponding beds at Atchison and Riverside, showing a dip here to the northwestward of a little more than 12 feet to the mile.

At first I was unable to reconcile this with the boring two miles below St. Joseph, at which point it seemed probable this conspicuous bed of limestone should have been struck in the boring near the level of the Missouri. On reflecting, however, that the direction from Atchison to the locality where the boring below St. Joseph was made, although up the river, is not north, but a little east of northeast, while the dip of the strata here is northwest, or possibly a little west of northwest, it will be seen that the heavy bed of blue clay struck in the St. Joseph boring, at an elevation of 39 feet above the Missouri, almost certainly holds a stratigraphical position a little below the horizon of the 16-foot bed of limestone mentioned above. This conclusion is strengthened by the occurrence of a heavy bed of bluish and drab laminated clay beneath the 16-foot bed of limestone at Atchison, as well as by the fact that another boring at Atchison, commenced 22½ feet above high water of the Missouri, and about 12 feet below the above-mentioned limestone, after passing through 15 feet of loose earth and fragments of rock (debris of the limestone in the hill above), struck shale, or laminated clay, and passed through a thickness of 11 feet of it. The "conglomerate rock," No. 3, 5 feet in thickness, mentioned in the memorandum of the Atchison boring, beneath the laminated clay, is doubtless the same sandstone, No. 3, of the St. Joseph boring, with some of the very hard concretions so often seen in the sandstones of this series. The want of exact agreement of thickness and composition of this and other beds penetrated, is easily accounted for by the changes so common in the lithological characters of the beds composing this series, and the different methods of grouping the subordinate beds, at the two borings.

For the following statement of the boring at Atchison, I am under obligations to George W. Glick, esq., of Atchison, the president of the company that sunk the boring:

Boring at Atchison, Kansas, commencing 22½ feet above high-water mark of the Missouri; made by the Atchison Coal Company, 1865–'66.

No.	Nature of strata penetrated.	Depth.	
		Ft.	*In.*
1	Loose earth and fragments of rock	15	0
2	Shale, (laminated clay)	11	0
3	Conglomerate rock, [so called by the workmen]	5	0
4	Bituminous shale with a little coal	19	0
5	Sandy mixed shale	82	0
6	Hard limestone	8	0
7	Gray shale	1	6
8	Sandstone, very hard	2	1
9	Sand and shale	7	0
10	Shale, clayey and sandy	27	0
11	Dark-brown clay, with small seam of coal	8	0
12	Sandy shale	1	0
13	Gray shale with sand-rock at bottom	7	0
14	Bituminous shale	6	0
15	Hard lime-rock	21	0
16	Soapstone. [Indurated clay]	—1	6

No.	Nature of strata.	Depth.	
		Ft.	In.
17	Very hard rock	3	0
18	Shale with a little coal	3	0
19	Sandy ironstone	1	6
20	Gray shale	9	0
21	Lime-rock		6
22	Shale	5	0
23	Lime-rock	1	0
24	Gray shale	4	0
25	Lime-rock	2	0
26	Shale with coal	1	0
27	Lime-rock	3	0
28	Hard lime-rock	10	0
29	Soapstone. [Indurated clay]	2	0
30	Lime-rock	4	0
31	Sandy shale	2	0
32	Hard lime-rock	0	4
33	Sandy shale	2	0
34	Sand-rock	5	0
35	Hard shale	3	0
36	Sandstone	0	6
37	Shale	10	0
38	Hard rock	1	6
39	Shale	3	0
40	Lime-rock	2	0
41	Shale, (hard)	4	0
42	Sand-rock	0	6
43	Lime-rock	4	6

If I am not mistaken, this boring, as already suggested, commenced at very near the same geological horizon as that at St. Joseph, and, as far as it went, merely explored, at least in part, the same ground as that penetrated at St. Joseph, allowing for the thinning out of some unimportant beds, and the thickening or change in the composition of others. If so, the money expended on the Atchison boring added little or nothing to the information developed by that at St. Joseph. If it is, therefore, desired to make further explorations at either of these localities, I would advise that the St. Joseph boring be carried 500 or 600 feet deeper, and if a workable bed of coal should be struck there, it may reasonably be expected to exist at not far from the same depth beneath Atchison.

Having thus given a somewhat detailed sketch of the various natural sections of these rocks seen along the Missouri, in Nebraska, below Omaha City, together with statements of the thickness, order of succession, and included fossils of the numerous subordinate beds, along with the facts revealed by the several shafts, drifts, borings, and other excavations examined, it will be interesting to consider the vertical range of the different types of organic remains so often mentioned, and their relations to those of our Carboniferous rocks elsewhere, with the view of throwing some light upon mooted questions respecting the age of some of these Nebraska strata.

In order to facilitate such a review of the range of the fossils already mentioned in detail in their proper places in the several sections given, the following tabular list of all the species has been prepared. With regard to the abbreviations, Kans., Mo., Io., Ill., W. Va., and Ky., inserted in the columns after the names, it is perhaps scarcely necessary to explain that they are used instead of the names of the several States, Kansas, Missouri, Iowa, Illinois, West Virginia, and Kentucky.*

The other columns designate localities in Nebraska at which the various species have been found. For the species found at Nebraska City two separate columns are used, to show which forms occur in the division B and which in C, the line of separation between these two divisions being more strongly marked upon paleontological grounds *at this particular locality* than others, though in some parts of the West the fossils of both divisions are, with few exceptions, known to range through the whole series.

As it would be inconvenient, as well as unnecessary, for the proposed objects of comparison, to give a separate column for each of the numerous localities at which the fossils enumerated were found, only a few of the more prominent localities are thus noted, and these are the particular localities at which the outcrops occur that have been by Professors Marcou and Geinitz referred in part to the Lower Carboniferous, and in part to the Coal-Measures and Permian (= Dyas). In the column on the right, some additional localities are also occasionally noted, and of these it is proper to remark that the outcrops at Otoe City, Aspinwall, Brownville, and Peru are believed to hold a higher position in the series than division C, at Nebraska City. That at Otoe City was so regarded by Professor Marcou, in which opinion I am, as elsewhere stated, inclined to agree with him, and there are equally good grounds for believing the beds seen at Peru, Aspinwall, and Brownville are in part or entirely upon the same horizon, or very near it, as these Otoe City outcrops.

* The comparatively small number of Illinois, Missouri, Iowa, Kansas, and Nebraska species, noted in the Kentucky column, is due to the fact that little attention was devoted, during the progress of the Kentucky survey, to collecting fossils. So far as the Coal-Measure fossils of Kentucky are known, however, the remains of both animals and plants agree very closely with those of Illinois, as might be expected from the exact correspondence shown by Lesquereux to exist between the various beds composing this series in these two States.

Tabular list, illustrating the geological and geographical range of the fossils of Eastern Nebraska.

No.	Names of species.	COAL-MEASURES IN OTHER STATES.						ESPECIAL LOCALITIES IN NEBRASKA.						Miscellaneous localities in Nebraska.
		Kans.	Mo.	Io.	Ill.	Ky.	W. Va.	Omaha.	Bellevue.	Platts-mouth.	Rock Bluff.	Nebraska City, Wyoming, Bennett's Mill.		
												B	C	
	PLANTS.													
1	*Neuropteris hirsuta*, Lesqx.[1]				Ill¶	Ky.								Brownville and two miles above Rulo.
2	*Neuropteris Loschii*, Brgt.				Ill¶									Do.
3	*Calamites*													Brownville.
	ANIMAL REMAINS.													
4	*Fusulina cylindrica*, Fischer	Kans.	Mo.	Io.	Ill¶			*	*	*	*			Nebraska City, above bed C; Otoe City; Brownville, Rulo, &c.
5	*Rhombopora lepidodendroides*, Meek	Kans.	Mo.	Io.	Ill¶				*	*	*	*	*	North side Platte, Rulo.
6	*Fistulipora*, undt. sp	Kans.	Mo.	Io.	Ill¶				*	*	*	*		Rulo.
7	*Syringopora multattenuata*, McCh				Ill¶									3 miles up Platte River, north side.
8	*Lophophyllum proliferum*, McCh., (sp.)		Mo.	Io.	Ill¶					*	*	*	*	Do.
9	*Campophyllum torquium*, Owen, (sp.)		Mo.	Io.	Ill¶						*			Cedar Bluff.
10	*Scaphiocrinus? hemisphæricus*, Shm.	Kans.	Mo.	Io.	Ill¶			*				*	*	3 miles up Platte, north side; Cedar Bluff.
11	*Zeacrinus mucrospinus*, McCh	Kans.	Mo.	Io.	Ill¶					*				Cedar Bluff.
12	*Erisocrinus typus*, M. & W			Io.	Ill				*			?		
13	*Archæocidaris triserratus*, M				Ill			*						Three miles up Platte River.
14	*Eocidaris Hallianus*, Geinitz												*	
15	*Fenestella*, sp				Ill¶				*	*				
16	*Fenestella Shumardi*, Prout												*	
17	*Polypora submarginata*, Meek				Ill¶					*		*	*	
18	*Polypora*, sp	Kans.		Io.									*	
19	*Synocladia biserialis*, Swallow[2]	Kans.	?	Io.	Ill¶					*		*	*	Three miles up Platte River.
20	*Glauconome trilineata*, Meek												*	
21	*Glauconome Americana*, Swallow, (sp.)	Kans.		Io.	Ill							*		
22	*Lingula Scotica*, Davidson?				Ill								*	
23	*Orbiculoidea*, sp				Ill¶									Morton's shaft, 1¾ miles west of Nebraska City.
24	*Chonetes Verneuiliana*, N. and P	Kans.	Mo.	Io.	Ill¶					*				Three miles up Platte River.
25	*Chonetes granulifera*, Owen		Mo.	Io.	Ill¶			*	*	*	*	*	*	Cedar Bluff, Otoe City, Rulo, &c.
26	*Chonetes glabra*, Geinitz	Kans.	?	Io.								*	*	Morton's shaft, Nebraska City.
27	*Productus pertenuis*, Meek	Kans.			Ill¶							*	*	Morton's.
28	*Productus Nebrascensis*, Owen	Kans.	Mo.	Io.	Ill¶	Ky.	W. Va.	*	*	*	*	*	*	Platte River, Brownville, Rulo, &c.
29	*Productus symmetricus*, McCh	Kans.	Mo.	Io.	Ill¶				*			*	*	Platte River and other localities.
30	*Productus punctatus*, Martin, (sp.)	Kans.	Mo.	Io.	Ill¶				*	*	*	*		Morton's shaft.
31	*Productus costatus*, Sowerby?	Kans.	Mo.	Io.	Ill¶	Ky.		*	*	*	*	*		

32	*Productus semireticulatus*, Martin, (sp.)	Kans.	Mo.	Io.	Ill¶.		W. Va.	*	*	*	*	*	*	Otoe City, Brownville, Aspinwall, &c.
33	*Productus longispinus*, Sowerby	Kans.	Mo.	Io.	Ill¶.	Ky.		*	*	*	*	*	*	Do.
34	*Productus Prattenianus*, Norw	Kans.	Mo.	Io.	Ill¶.	Ky.	W. Va.	*	*	*	*	*	*	Otoe City, Rulo, &c.
35	*Hemipronites crassus*, M. and W	Kans.	Mo.	Io.	Ill¶.	Ky.	W. Va.		*	*	*	*	*	Platte River, Otoe City, Aspinwall, Rulo.
36	*Meekella striato-costata*, Cox, (sp.)	Kans.	Mo.	Io.	Ill	Ky.				*	*	*		Otoe City, Aspinwall.
37	*Syntrilasma hemiplicata*, Hall, (sp)	Kans.	Mo.	Io.	Ill					*	*	*	*	Otoe City, Rulo, &c.
38	*Orthis carbonaria*, Swallow	Kans.	Mo.	Io	Ill¶						*	*		Platte River.
39	*Rhynchonella Osagensis*, Swallow	Kans.	Mo.	Io.	Ill¶				*	*	*	*	*	Platte River, Otoe City.
40	*Spirifer cameratus*, Morton	Kans.	Mo.	Io.	Ill¶.	Ky.	W. Va	*	*	*	*	*	*	Otoe City, Brownville, Rulo.
41	*Sp. (Martinia) plano-convexus*, Shum	Kans.	Mo.	Io.	Ill¶.	?	W. Va.		*	*	*	*	*	Platte River, Otoe City, Brownville.
42	*Sp. (Martinia) lineatus*, Martin, (sp.)	Kans.	Mo.	Io.	Ill¶.				*	*	*			
43	*Spiriferina Kentuckensis*, Shum	Kans.	Mo.	Io.	Ill¶	Ky.		*	*	*	*	*	*	Otoe City, Aspinwall.
44	*Retzia punctulifera*, Shum	Kans.	Mo.	Io.	Ill¶.	Ky.				*	*	*		
45	*Athyris subtilita*, Hall, (sp.)	Kans.	Mo.	Io.	Ill¶.	Ky.	W. Va.	*	*	*	*	*	*	Aspinwall.
46	*Terebratula bovidens*, Morton	Kans.	Mo.	Io.	Ill¶.					*				Platte River.
47	*Lima retifera*, Shumard	Kans.			Ill¶.								*	
48	*Entolium aviculatum*, Swallow, (sp.)		Mo.	Io.	Ill¶.					*		*	*	Do.
49	*Nucula Beyrichi*, V. Schaur ??				Ill¶.							*	*	
50	*Nucula ventricosa*, Hall	Kans.	?	Io.	Ill¶.		W. Va.				*		*	
51	*Yoldia subscitula*, M. and W?												*	
52	*Nuculana bellistriata*, Stevens, (sp.)	Kans.		Io.	Ill¶		W. Va.						*	Brownville.
53	*Macrodon tenuistrata*, M. and W			Io.	Ill¶.								*	
54	*Schizodus curtus*, M. and W	Kans.			Ill¶.							*	*	
55	*Schizodus Wheeleri*, Swallow, (sp.)		Mo.	Io.	Ill¶.						*		*	Platte River.
56	*Schizodus*, sp		?		Ill¶.								*	
57	*Avicula longa*, Geinitz, (sp.)	Kans.		Io.	Ill¶.								*	
58	*Avicula ? sulcata*, Geinitz												*	
59	*Pseudomonotis*, sp. [3]	?			Ill					*		*		Cedar Bluff.
60	*Pseudomonotis radialis*, Phill., (sp.) ??				Ill								*	
61	*Myalina Swallovi*, McChesny	Kans.	Mo.	Io.	Ill¶.						*		*	
62	*Myalina subquadrata*, Shum	Kans.	Mo.	Io.	Ill		W. Va		*	*		*	*	Platte River.
63	*Myalina perattenuata*, M. and W	Kans.			Ill									
64	*Aviculopecten occidentalis*, Shum	Kans.	Mo.	Io.	Ill¶.					*	*	*	*	Otoe City, Brownville, Aspinwall.

[1] These plants were identified by Professor Lesquereux, who informs me that *N. Loschii*, although found in the Permian of Europe, ranges through the whole thickness of the Coal-Measures in Illinois; while *N. hirsuta*, which has been thought by some to be identical with a European Permian form, was first described by Prof. L. from the Coal-Measures of Pennsylvania, and he says it ranges throughout the whole area and thickness of the Illinois Coal-Measures. The fact that these two ferns occur at Morris, Ill., associated with a curious plant (?), thought to belong to a genus (*Palæoxyris*) only known in Europe in the Trias, might lead those unacquainted with all the facts to suspect that these Morris beds could scarcely be so old as the Carboniferous. It must not be forgotten, however, that these beds belong to the lower part of the Illinois Coal-Measures, and that we find there, directly associated with these plants, the well-known European Coal-Measure genus of *Crustacea*, *Anthracopalemon*, together with three other genera of *Crustacea* (*Palæocaris*, *Euproops*, and *Eurypterus?*) which, although not certainly identical generically, with European Carboniferous forms, are certainly far more nearly allied to the same than to anything yet known from the Permian. Such facts only show that several types of both animals and plants appeared at considerable earlier periods here than in Europe. This is even more strikingly the case in the Cretaceous flora of this country than in that of the Carboniferous, which fact led one of the highest European authorities on fossil botany, at first, to refer a collection of plant remains from our Cretaceous to the horizon of the Miocene.

[2] This species ranges down into the Chester and St. Louis limestones, of the Lower Carboniferous, in Illinois.

[3] A species of this genus occurs at Leavenworth City, Kansas, far down in the Coal-Measures, in a bed referred by Professor Swallow to the upper part of the middle Coal-Measures. Dr. White also finds it at about the middle of the Upper Coal-Measures in Iowa, and I have seen an imperfect specimen of the genus from the Lower Coal-Measures of West Virginia.

¶ This mark in the fourth column after the letters "Ill." indicates that the species is known not only in the *Upper*, but also in the *Lower*, beds of the Coal-Measures of Illinois. All of those from West Virginia were found in the Lower Coal-Measures.

Tabular list, illustrating the geological and geographical range of the fossils of Eastern Nebraska—Continued.

No.	Names of species.	COAL-MEASURES IN OTHER STATES.						ESPECIAL LOCALITIES IN NEBRASKA.						Miscellaneous localities in Nebraska.
		Kans.	Mo.	Io.	Ill.	Ky.	W. Va.	Omaha.	Bellevue.	Platts-mouth.	Rock Bluff.	Nebraska City, Wyoming, Bennett's Mill.		
	ANIMAL REMAINS—Continued.											B	C	
65	*Aviculopecten neglectus*, Geinitz, (sp.)				Ill								*	
66	*Aviculopecten carboniferous*, Stv., (sp.)		Mo.	Io.	Ill¶		W. Va.						*	Morton's shaft, Nebraska City.
67	*Aviculopecten Whitei*, Meek			Io.	Ill¶							*	*	Morton's shaft; Brownville.
68	*Aviculopecten Coxanus*, M. and W				Ill¶								*	
69	*Aviculopinna Americana*, Meek			Io.									*	
70	*Pinna peracuta*, Shumard	Kans.	Mo.	Io.	Ill					*	*	*		
71	*Modiola ? subelliptica*, Meek	Kans.											*	
72	*Pleurophorus oblongus*, Meek				Ill¶								*	
73	*Pleurophorus occidentalis*, M. and H				Ill							* ?		Otoe City.
74	*Edmondia ? reflexa*, Meek				Ill¶								*	Morton's shaft.
75	*Edmondia ? glabra*, Meek				Ill								*	
76	*Edmondia ? Nebrascensis*, Geinitz												*	
77	*Edmondia subtruncata*, Meek	Kans.									*			
78	*Edmondia Aspinwallensis*, Meek						W. Va.							Aspinwall.
79	*Solenomya*, sp				Ill¶								*	
80	*Solenomya*, sp									?	*			
81	*Prothyris elegans*, Meek				Ill¶								*	
82	*Chænomya Leavenworthensis*, M. and H.	Kans.			Ill¶						*			
83	*Allorisma* (*Sedgwickia*) *reflexa*, Meek												*	
84	*Allorisma* (*Sedgwickia*) *Geinitzii*, Meek				Ill¶								*	
85	*Allorisma* (*Sedgwickia*) *subelegans*, M				Ill¶								*	
86	*Allorisma* (*Sedgwickia*) *granosa*, Shum	Kans.			Ill						*			
87	*Allorisma subcuneata*, M. and H	Kans.	Mo.	Io.	Ill¶				*	*	*	*		Brownville and Aspinwall.
88	*Solenopsis solenoides*, Geinitz, (sp.)				Ill¶						*		*	
89	*Dentalium Meekianum*, Geinitz				Ill¶								*	
90	*Bellerophon carbonarius*, Cox	Kans.	Mo.	Io.	Ill¶	Ky.	W. Va.			*		*	*	
91	*Bellerophon Montfortianus*, N. and P	Kans.	Mo.	Io.	Ill¶		W. Va.		?				*	Morton's shaft.
92	*Bellerophon Marcouanus*, Geinitz				Ill¶								*	
93	*Bellerophon percarinatus*, Con	Kans.	Mo.	Io.	Ill¶		W. Va.						*	Brownville.
94	*Bellerophon Kansasensis*, Swallow	Kans.			Ill¶									Rulo.
95	*Euomphalus rugosus*, Hall	Kans.	Mo.	I.	Ill¶		W. Va.		*		*	*	*	Cedar Bluff, Aspinwall.
96	*Platyceras Nebrascensis*, Meek				Ill¶							*		Morton's shaft, Nebraska City; two miles below Bennett's Mill.
97	*Macrocheilus intercalaris*, M. and W				Ill							*		2¾ miles southwest of Nebraska City.
98	*Macrocheilus primogenius*, Con. ?			Io.	Ill¶		W. Va.		*					
99	*Orthonema subtæniata*, Geinitz, (sp.)			Io.									*	
100	*Aclis* (?!) *Swalloviana*, Geinitz, (sp.)				Ill ?								*	
101	*Pleurotomaria Haydeniana*, Geinitz												*	

102	*Pleurotomaria perhumerosa*, M.				Ill									Morton's shaft, Nebraska City; Rulo.
103	*Pleurotomaria sphærulata*, Con.	Kans.	Mo.	Io.	Ill¶.	Ky.			*					
104	*Pleurotomaria inornata*, Meek.				Ill¶.									Morton's shaft, Nebraska City.
105	*Pleurotomaria Grayvillensis*, N. and P.	Kans.	?	Io.	Ill¶.	Ky.	W. Va.						*	
106	*Pleurotomaria Marconanus*, Geinitz												*	
107	*Pleurotomaria subdecussata*, Geinitz												*	
108	*Murchisonia Nebrascensis*, Geinitz												*	
109	*Orthoceras cribrosum*, Geinitz			?	Ill¶.		W. Va.						*	
110	*Nautilus occidentalis*, Swallow	Kans.	Mo.		Ill¶.		W. Va.						*	Aspinwall.
111	*Nautilus ponderosus*, White & St. John			Io.	?					*				
112	*Cythere Nebrascensis*, Geinitz			Io.									*	
113	*Phillipsia*, sp.	?		Io.						*			*	
114	*Phillipsia scitula*, M. and W.				Ill¶							*		
115	*Phillipsia major*, Shumard	Kans.			Ill¶.					*				
116	*Deltodus? angularis*, N. and W.			Io.	Ill									Do.
117	*Cladodus mortifer*, N. and W.	Kans.			Ill						*	*		
118	*Petalodus destructor*, N. and W.				Ill						*			
119	*Diplodus compressus*, N. and W.				Ill						*			Rulo, Nebraska.
120	*Peripristis semicircularis*, N. & W., (sp.)				Ill			*	*					Morton's shaft.
121	*Chomatodus arcuatus*, St. John[1]											*		
122	*Xystrodus? occidentalis*, St. John			Io.									*	

[1] This species was found at Bennett's Mill, near Nebraska City, by Professor O. St. John.

The inseparable linking together of all of these Nebraska rocks under consideration, by their organic remains, as well as their identity with the Upper Coal-Measures of Iowa, Illinois, and other neighboring States, is so clearly exhibited in the foregoing list, as to render any extended comments upon these points almost unnecessary. Some concluding remarks, however, may assist in making these facts more manifest to those who may not have the time, or inclination, fully to analyze the list.

In the first place, it may be remarked, that as Professor Geinitz, the paleontological authority to whom Professor Marcou submitted his collection, now concedes that the Rock Bluff and Plattsmouth sections (which had been by the last-mentioned author referred to the Permian) really belong to the Carboniferous, the only questions that need be discussed here are, whether Professor Geinitz was right in referring the Plattsmouth beds to the Lower Carboniferous or Mountain limestone series, instead of to the Upper Coal-Measures, as he correctly referred the Rock Bluff section; and whether or not he and Professor Marcou are right in placing the Wyoming and Nebraska City outcrops on a parallel with the Permian or Dyas of Europe?

In regard to the first question, it is only necessary to repeat, that out of thirty-six species known from Plattsmouth, twenty-five (including millions of *Fusulina*), as may be seen by the list, also occur at Rock Bluff; while the remaining eleven Plattsmouth species not yet found at Rock Bluff, with one or two exceptions, may all (like the others) be seen to occur in the Upper and many of them in the Lower Coal-Measures of Illinois, Iowa, and other neighboring States. In addition to this, of all the species yet known from both of these localities, only some four or five (and these are forms, in this country, common to both the Coal-Measures and Lower Carboniferous) have ever been here found at any horizon below the Millstone grit. It is true, a larger proportion than this of these fossils agree in their affinities with European Lower Carboniferous forms; but the fact that the rocks in which they occur, in Kansas, shade without any defined physical or paleontological break, into the Permian above, *ought to be* convincing evidence that these Plattsmouth beds cannot be properly included in the Lower Carboniferous series.* This is also clearly shown by ther identity with the Upper Coal-Measure beds of Iowa and Illinois, which in the latter State, can be traced without interruption southward to where they and other still Lower Coal-Measure strata, containing, along with most of the same animal types, the remains of a well-marked Coal-Measure flora, rest directly upon extensive deposits, 300 or 400 feet in thickness, of Millstone grit; while beneath the latter, we have a great series of massive limestones and other strata, corresponding completely, not only in their physical characters and position, but also in the affinities of their entire group of fossils, with the Mountain limestone of Europe.

By a glance at the foregoing list, it will also be seen that the same reasoning applies, with equal force, against Professor Marcou's reference of the outcrops above the mouth of Platte River, at Bellevue and Omaha, to the Mountain limestone; since of the twenty-six species of fossils now known from these localities, twenty-four are the common forms of our Coal-Measures, while none of them, excepting a few that are well

* Since this was written, Dr. White, the able State geologist of Iowa, has, in his report, shown in the clearest manner possible, by sections across that State from the Mississippi River to the Missouri at Plattsmouth, that the rocks along the latter stream, at and near Plattsmouth, really belong to the Coal-Measures, and that there are no Lower Carboniferous rocks there within one thousand feet or more of the surface.

known to be common to the Coal-Measures and Lower Carboniferous rocks of the West, have ever been found below the horizon of the Millstone grit. Indeed, all but three or four of them (which are themselves well-known Coal-Measure species) also occur in the Plattsmouth beds, referred by Professor Marcou to the Lower Dyas. He has, it is true, made an effort to explain away this latter fact, on the ground that most of the fossils known to him from Omaha and Bellevue are *Brachiopoda*, which he thinks cannot be relied upon in the identification of strata, owing to their great vertical range. We now know, however, from these very beds at Bellevue, as may be seen by the foregoing list, the lamellibranchiate genera, and Coal-Measure species, *Myalina subquadrata*, and *Allorisma subcuneata*, both of which occur in the very beds referred by him elsewhere to the Upper Dyas. Again, we have the crustacean genus and Coal-Measure species, *Phillipsia major* of Shumard, both from these so-called Mountain limestone beds of Bellevue, and from the so-called Lower Dyas of Plattsmouth. Still further, we have from the Bellevue beds the well-known *vertebrate* Coal-Measure species, *Peripristis semicircularis*, which, strangely enough, supposing Professor Marcou correct, also occurs at Nebraska City and Bennett's Mill, in the so-called Upper Dyas, and at Rock Bluff, in the upper part of the so-called Lower Dyas (see Plates III and IV). So even if we admit that the testimony of the *Brachiopoda* may be, in some cases, less reliable than that of a few other types, we cannot ignore it when directly corroborated by that of the lamellibranchiates, trilobites, and fishes. I can, therefore, see no reason whatever for separating these beds from the Coal-Measures.

We come now to consider the propriety of Professors Marcou and Geinitz's reference of division B of the Nebraska City section, and at Wyoming and Bennett's Mill, to the Upper Dyas, or more properly to the Dyas at all. As shown in the list we know altogether from these beds about forty species of fossils, of which twenty, or one-half, are common to the same and the outcrops above Plattsmouth, referred by Professor Marcou to the Mountain limestone. Thirty of them also occur in the Plattsmouth and Rock Bluff strata, the first of which are placed by Professor Geinitz in the Lower Carboniferous, and the latter in the Upper Coal-Measures; while thirty-seven or thirty-eight, nearly the whole, consisting of *Fusulina*, polyzoa, corals, crinoids, brachiopods, lamellibranchiates, trilobites, and fishes, are found in the Upper, and, in part, the Lower Coal-Measures of Iowa, Illinois, and other neighboring States; and none of them, excepting a very few forms that are common to our Coal-Measures and Lower Carboniferous rocks, have ever been found in the latter in this country. Hence it seems to me clearly manifest that there is no reliable paleontological evidence for separating these beds from the Upper Coal-Measures. The only fact that gives the slightest show of reason for such a conclusion is the occurrence in them of the genera *Pseudomonotis* and *Synocladia*, with species resembling Permian forms of the genera *Schizodus*, *Pleurophorus*, *Nuculana*, &c., but it is exceedingly doubtful whether *any* of even these fossils are specifically identical with any known Permian form. The *Synocladia* is certainly not so, while it, and all the others, are well known to range far down into unquestionable Coal-Measures of the West, even below the horizon of the rocks referred by Professor Marcou to the Mountain limestone above Plattsmouth. For instance, a bed of limestone at near the level of the Missouri, at Leavenworth, Kansas, in which a *Pseudomonotis*, nearly allied to *P. speluncaria*, is known to occur, has been by Professor Swallow, who carries the line between the Upper Coal-Measures and Permian in that State several hundred feet lower than any others in this country, placed some 700 or 800 feet below the base of the Permian, in the Middle Coal-Measures. Dr. White, the State geologist of Iowa, also finds this genus near the middle of the Upper Coal-Measures in that State. Professor Marcou would doubtless go around this difficulty by including all the strata in the Permian, as far down as a single one of these Permian types is known to range, regardless of all the other associated fossils. But in this he would be met by a still greater difficulty in having to include in the Permian the very same beds he at other places refers to the Mountain limestone, and at some places, as at Leavenworth City, even still lower strata than those so referred by him.

But if these few fossils were known beyond doubt to range no lower in the series than division B of the Nebraska City section, and it could be positively demonstrated that they are all specifically identical with well-known European Permian species, should they alone set aside all the overwhelming weight of evidence of all the other thirty-odd associated Carboniferous species, including such Carboniferous and older genera as *Fusulina, Lophophyllum, Campophyllum, Solenopsis,* true *Orthis,** *Phillipsia, Cladodus, Chomatodus, Peripristis,* &c.? I should think not.

There being, therefore, no reason whatever for separating division B, with its numerous undoubted Coal-Measure species, from the Carboniferous, the next question to be considered in this connection is, whether there are any sufficient reasons for drawing an important division here between the beds B and C. For the solution of this question we must again appeal to the imbedded organic remains; and by examining the foregoing list, it will be seen that of about 70 species of fossils already known from division C (and some 8 or 9 others found in higher beds and not yet known to occur in C), about 32, as already stated, come in this region directly up from division B and lower beds, while we also now know that some 35 or more of the others, together with nearly all of those found in division B, occur at various horizons in the true Coal-Measures of Illinois and other Western States—most of them, indeed, there ranging into the lower part of that series. In fact, from what we now know of the range of species in our western Coal-Measures, I do not hesitate to express the opinion that nearly or quite all of the species found in division C, at Nebraska City, will yet be discovered in beds holding positions below the middle of the Coal-Measures of Illinois and other Western States.

From all of the facts, therefore, now determined, it must, I think, be clearly evident that all of these strata under consideration along the Missouri, that have been by some referred in part to the Mountain limestone, in part to the Permian or Dyas, and in part to the Coal-Measures, really belong entirely to the true Coal-Measures; unless the division C, at Nebraska City, and some apparently higher beds below there on the Missouri, may possibly belong to the horizon of an intermediate series between the Permian and Carboniferous, for which, in Kansas, Dr. Hayden and the writer proposed the name Permo-carboniferous. This latter distinction, however, it should be remembered, is, as we have always explained, even in Kansas, merely an arbitrary one, not founded upon any well-defined physical or paleontological break between these upper beds and the Upper Coal-Measures.

It is true that in first announcing the existence of Permian rocks in

* I am of course aware that the genus *Hemipronites* (= *Streptorhynchus*) occurs in the Permian, and is sometimes, by those who are not very particular about generic distinctions, called *Orthis;* but I am not aware of any true *typical Orthis,* of the type of *O. Michilini,* and *O. Carbonaria,* having been yet found in the Permian of the Old World, though this genus may possibly occur at that horizon there.

Kansas (Trans. Albany Inst., vol. IV., 1858), we also, upon the evidence of a few fossils from near Otoe and Nebraska Cities, resembling Permian forms, referred these beds to the Permian; but on afterwards finding that these fossils are there directly asssociated with a great preponderance of unquestionable Carboniferous species; and that there is also in Kansas a considerable thickness of rocks between the Permian and Upper Coal-Measures containing, along with comparatively few Permian types, numerous unmistakable Carboniferous forms, we abandoned the idea of including these Otoe and Nebraska City beds in the Permian. And all subsequent investigations have but served to convince us of the accuracy of the latter conclusion.

In his work on the fossils of the rocks under consideration,* Professor Geinitz has, without indicated doubts, identified, mainly from divisions B and C at Nebraska City, the following European Permian species:

Guilielmites permianus, Gein.†
Synocladia virgulacea, Phill. sp.
Polypora biarmica, Keyserling.
Stenopora columnaris, Schlot. sp.
Productus horridus, Sowerby.
Productus horrescens, de Vern.
Strophalosia horrescens, de Vern.
Camarophoria globulina, Phill. sp.
Avicula speluncaria, Schlot. sp.
Avicula pinnæformis, Geinitz.
Aucella Hausmanni, Goldf. sp.
Pleurophorus simplus, Keyserling.
Pleurophorus Pallasi, de Vern.
Nucula Beyrichi, v. Schaur.
Nucula Kazanensis, de Vern.
Arca striata, Brown.
Schizodus obscurus, Sowerby.
Schizodus Rossicus, de Vern.
Schizodus truncatus, King.
Solenomya biarmica, de Vern.
Serpula planorbites, Munster.

In regard to nearly all of these, and several others of Professor Geinitz's identifications of these fossils, however, I have been reluctantly compelled to differ with him, after a very careful study of extensive collections. Indeed, so far from these fossils being in all cases specifically identical with the foreign forms to which he has referred them, I have already elsewhere shown that some of them do not even belong to the same genera, or families;‡ while very few, if any of them appear, to me really identical, specifically, with European Permian species. For instance, the shell he has identified with *Camarophoria globulina*, is a true *Rhynchonella*, and at least one of those he referred to *Strophalosia horrescens* is a true *Productus*; while he has identified two distinct species with the so-called *Clidophorus Pallasi*, de Vern., neither of which belongs to the same genus as the type of the Russian species, which is figured and described as an edentulous shell, by de Verneuil, under the name *Modiola Pallasi*. But in all of these cases, where I differ with Professor Geinitz respecting the relations of these Nebraska fossils, I have endeavored to give the student the means of forming his own conclusions, by figuring on the same plates for comparison the foreign species to which they have been, as I believe, erroneously referred.

Of all the various fossils yet known from the Nebraska rocks referred by Professors Marcou and Geinitz to the Permian, there are really very few that seem to me to be so closely similar to European Permian forms, that no entirely satisfactory specific distinctions have yet been detected in their external characters. These are the forms referred by Professor Geinitz to *Nucula Beyrichi*, *Leda Kazanensis*, *Schizodus Rossicus*,§ *Avicula* (*Pseudomonotis*) *speluncaria*, and *Pleurophorus Pallasi*. The first four of

* Carbonformation und Dyas in Nebraska, 1866.

† In the bed at Nebraska City, from which the specimens referred by Prof. Geinitz to this species were obtained, we only found, after repeated and diligent examinations, mere fragments of plants too much broken up to be clearly identified even generically.

‡ See Am. Jour. Sci. and Arts, vol. XLIV, new series, p. 170, and p. 331, 1867.

§ *S. Rossicus* is said by Eichwald (Lethæa Ross. 1, p. 999) to occur both in the Carboniferous and Permian rocks of Russia.

these are, I acknowledge, *very* similar to the Permian species to which he has referred them, while the fifth seems to be as nearly like *Pleurophorus costatus.* To these we may perhaps also add the *Schizodus* represented by fig. 2, on plate X, which is very closely allied to *S. obscurus*, though more compressed. The fact, however, that we know comparatively little of the hinges and interior of these shells, while they are mainly such types as usually present few reliable external characters for specific distinction, and belong for the most part to genera in which species are frequently very similar, must weaken our confidence in these specific identifications, under such circumstances. But when we take into consideration the fact that precisely the same forms are found in the Western States far down in beds acknowledged by all to belong to the Coal-Measures, and which have even at some places been referred by Prof. Marcou to the Mountain limestone, and especially when we bear in mind, the numerous unquestionable Carboniferous species with which they are here directly associated in the beds under consideration, the impropriety of basing important conclusions upon them must be apparent.

I am not, however, wishing to conceal the fact that several of the *genera* found in these supposed Permian rocks, in Nebraska, and their equivalents in Iowa, Illinois, &c., are believed to be generally regarded as not in Europe dating back beyond the Permian epoch. But in considering these, we must not overlook the other genera equally characteristic of the Carboniferous, found in the very same rocks. The following lists include both types of genera:

A. CARBONIFEROUS GENERA.		B. GENERA BELIEVED NOT TO DATE BACK OF THE PERMIAN IN EUROPE.
1. *Fusulina*, 1 or 2 sp.	10. *Petalodus*, 1 sp.	
2. *Lophophyllum*, 1 sp.	11. *Cladodus*, 1 or 2 sp.	
3. *Campophyllum*, 1 sp.	12. *Peripristis*, 1 sp.	1. *Synocladia*, 1 sp.
4. *Eupachycrinus*, 1 sp.	13. *Chomatodus*, 1 sp.	2. *Aulosteges*,* 1 sp.
5. *Erisocrinus*, 1 sp.	14. *Xystrodus*, 1 sp.	3. *Pseudomonotis*, 2 sp.
6. *Orthis*, 1 sp.	15. *Deltodus*, 1 sp.	4. *Aviculopinna*, 1 sp.
7. *Solenopsis*, 1 sp.	16. *Antliodus*, 1 sp.	5. *Bakevellia*, 1 sp.
8. *Prothyris*, 1 sp.		6. *Placunopsis*,† 1 sp.
9. *Phillipsia*, 2 sp.		7. *Lima*, 1 sp.

It will thus be seen that while we have in the whole series of Upper Coal-Measures and Permo-carboniferous rocks of Kansas, Nebraska, Iowa and Illinois, seven genera, consisting of one of *Polyzoa* and five or six of true *Mollusca*, that have not, so far as known to the writer at this time, been found in Europe at any horizon below the Permian, and two of these, it is believed, not even that low in the series *there*, we find in the same horizon, and for the most part directly associated with these, sixteen other genera, consisting of one of *Foraminifera*, two of Corals, two of *Crinoidea*, three of *Mollusca*,‡ one of Trilobites, and seven of Fishes, not known to occur anywhere in more recent rocks than the Carboniferous. Seven of these, however, viz., *Erisocrinus*, *Eupachycrinus*, *Prothyris*, *Deltodus*, *Peripristis*, *Antliodus*, and *Xystrodus*, being new genera, it might be supposed ought not to be counted against the Permian. But it is worthy of note that even the typical species of

* Dr. White has discovered at about the horizon of the middle of the Rock Bluff section, in the Upper Coal-Measures of Iowa, attached to the coral *Campophyllum*, a minute Brachiopod, apparently possessing the characters of *Aulosteges*.

† A species presenting at least all the external characters of the genus *Placunopsis*, has been found in the Upper Coal-Measures of Illinois (see Proceed. Chicago Acad. Sci. 1, March, 1866, p. 13), and the same has since been discovered by the Geol. Survey of Ohio, in the Lower Coal-Measures of the latter State.

‡ The genus *Bellerophon*, of which we have four or five species in these rocks, has also, I believe, not been found in the Permian of Europe, though one species is known from the Trias (St. Cassian beds).

the genus *Erisocrinus*, occurs at Bellevue, in the very Coal-Measure beds erroneously referred, by Professor Marcou, to the Mountain limestone, as well as in the equivalent Upper Coal-Measure beds of Illinois. It is also certainly represented at far lower horizons by several lower Carboniferous species, two of which occur as low as the Burlington limestone; while the only other known species of *Eupachycrinus* is from the horizon of the Keokuk group. Again, the only other known species of the genus *Prothyris* was discovered by Professor Winchell in the very oldest member of the whole Carboniferous series in Michigan; and of the nine described species of *Antliodus*, and seven of *Deltodus*, all of the former, and all but one of the latter, are Lower Carboniferous forms, which is also the horizon of the only two known European species thought by Newberry and Worthen probably to belong to *Antliodus*. Again, the only known European species, believed by St. John to belong to the new genera *Peripristis* and *Xystrodus*, are Mountain limestone species, one of which is the type of the genus *Xystrodus*. The genus *Chomatodus* is likewise, I believe, essentially, if not exclusively, a Mountain limestone group in Europe. At any rate, of the ten described species of this country, nine are from the Lower Carboniferous, and only one (an Upper Coal-Measure species of Illinois) was found at any higher horizon. The genus *Petalodus* seems also to be nearly or quite confined to the Mountain limestone of Europe, though two or three species are known from the Upper Coal-Measures of Illinois, Kansas, Nebraska, and Pennsylvania. *Fusulina* is also, I believe, usually considered in Europe mainly, if not exclusively, a Lower Carboniferous genus, or at any rate one not ranging into the Permian.

On the other hand, of the six or seven genera mentioned in the foregoing list, that are supposed to be in Europe confined to the Permian and later rocks (none of which types, it will be observed, rank even among the higher lamellibranchiate *Mollusca*), the genus *Aviculopinna*, as may be seen by the figures and description, seems to be closely allied to Professor McCoy's Carboniferous and older genus *Pteronites*. Again, the genus *Bakevellia*, although most generally regarded in Western Europe as a Permian type, is said, by Eichwald, also to occur in the Carboniferous of Russia; and it seems, also, to be represented in Nova Scotia, according to high authority, at or below the base of the enormously developed Coal-Measures of that region. And almost certainly the very species of *Synocladia* alluded to, also occurs in the Chester and St. Louis limestones of our Lower Carboniferous series.

So it is evident that whether we look at the *species* or the *genera* found in the rocks under consideration, we see that there is an overwhelming weight of evidence against the conclusion that they represent the Permian of the Old World. Indeed, looking at the *genera* of animal remains alone, without regard to the plants or the position of these beds above the horizon of the Millstone grit and Mountain limestone, and by taking into consideration the greater weight due the evidence of such highly organized types as Trilobites and Fishes, it would not be difficult to bring forward a better argument in favor of the conclusion that this whole series belongs to the horizon of the European Mountain limestone, than that it represents the Permian. The mingling, however, of the few Permian types mentioned in the same beds with the numerous Carboniferous forms, already enumerated, through a considerable thickness of these and higher strata, together with the physical structure of our entire Carboniferous system, demonstrate, it seems to me, as clearly as facts can, that these rocks belong to the Coal-Measures, and that here we have no abrupt break between the Carboniferous and Permian.

REMARKS ON THE PROBABILITY OF FINDING VALUABLE BEDS OF COAL WITHIN PROFITABLE WORKING DISTANCE OF THE SURFACE IN EASTERN NEBRASKA.

It is well known that along the valley of Des Moines River, and some of its tributaries in Iowa, 75 to 100 miles east of the Missouri, extensive and valuable beds of coal exist. This coal is of good quality, in beds varying from a foot or so to 7 feet in thickness, and is regarded by Dr. White, the State Geologist, as belonging to the Lower Coal-Measures. In going to Nebraska in the spring of 1867, I visited, with that gentleman, several localities where these Lower Coal-Measure rocks are seen, and as the facts observed there have a bearing on the coal question in Eastern Nebraska, a few remarks on these Des Moines Valley beds are necessary to a clear understanding of what will follow in regard to coal in Nebraska.

Dr. White had previously thoroughly explored this region and the country through from here to the Missouri, and knew all the localities where the best exposures are to be seen, and I am under great obligations to him for conducting me through from Iowa City to Nebraska City; and it is but just that I should state here, that all the facts observed by me, in the geology of Iowa, were previously known by him, and have been published in an interesting paper from his pen, in the American Journal of Sciences and Arts, vol. XLIV, 1867, and subsequently in his able report published by legislative authority.

The Lower and Middle Coal-Measures of the Des Moines Valley had also been, to a considerable extent, explored by Professor A. H. Worthen, when connected with a former survey of Iowa, and described at length in the report of that survey, published in 1858. They consist mainly of sandstones, shales, coal, and some thin, impure limestones, the maximum thickness of the whole, in this region, being supposed by Dr. White to be about 800 feet. They only come to the surface in the region of the Des Moines River, and northeast of there, excepting in some of the deeper valleys somewhat farther west, and are believed by Dr. White to include all the workable beds of coal in the State. Although in some places quite fossiliferous, they are generally not so much so here as the Upper Coal-Measures occupying all the country from near the Des Moines, westward to the Missouri.* As there is no Millstone grit here, the lower beds in this region rest directly down upon the Lower Carboniferous rocks.

The Upper Coal-Measures west of the Des Moines, as already shown by Dr. White, are not separable from the Middle series by any physical or abrupt paleontological break, a number of the species of fossil being common to both, and the whole having here a uniform and slight general dip to the west of south.

The last locality at which we saw any of the beds of the Middle series in going southwestward from the Des Moines Valley was near Winterset, Madison County, in the rather deep valley of Middle River, a tributary of the Des Moines. Here we saw only one of the very upper beds of the Middle series, in the bottom of the valley. This valley, like all the others in this region, is merely one of erosion, cut down to a depth of perhaps 250 feet below the common elevation of the country by the

* In some of the upper beds of the Middle series, however, we found *Sperifer cameratus*, Morton, *S. lineatus*, *S. planoconvexus*, Shumard, *Productus Nebrascensis*, Owen, *P. Longispinus*, Sowerby ?, *P. Punctatus*, Martin (*sp.*), *Hemipronites crassus*, M. & W., *Chonetes mesoloba*, N. & P., together with *Gervilla longa*, Geinitz (a true *Avicula*), and *Aviculopecten carboniferus* Stevens (= *Pecten Hawni*, Geinitz).

stream now flowing in it. The hills on each side, to an elevation of about 220 feet, or perhaps more here, are composed of the succeeding higher series of the Coal-Measures, consisting of some comparatively heavy beds of light-yellowish limestones, with bluish, drab, and reddish clays, and sandy, micaceous shale, some black laminated shale, and a few inches of impure coal, the whole being overlaid by more or less heavy deposits of Drift.

In this Upper series of Coal-Measures, as already intimated, only separated as a matter of convenience, and not by any natural break from the Middle, we found the following fossils, nearly or quite all of which had been previously collected here by Dr. White: *Fusilina cylindrica*, *Rhombopora* and *fistulipora*, sp., *Eocidaris Halli*, *Hemipronites crassus*, *Meekella striatocostata*, *Productus costatus?*, *P. Nebrascensis*, *P. Longispinus*, *Spirifer cameratus*, *S. planoconvexus*, *S. lineatus*, *Spiriferina Kentuckensis*, *Chonetes Verneuiliana*, *Athyris subtilita*, *Terebratula bovidens*, *Retzia punctulifera*, *Macrodon tenuistriatus*, *Myalina perattenuata*, *M. subquadrata*, *Aviculopecten occidentalis*, *Bellerophon crassus*, and an undetermined *Phillipsia*. Dr. White also found in one of the upper limestones a tooth of a *Cochliodus*.

This, it will be seen, is, as far as it goes, with the exception of one or two species, almost precisely the same fauna characterizing the Upper Coal-Measures along the Missouri, already so often mentioned, where the whole series, however, is much more developed, the entire thickness of the upper series here, near Winterset, being estimated by Dr. White as probably not exceeding 200 to 250 feet.

On ascending from this valley, we saw no more exposures of the Middle series, the southwestward inclination of the strata taking these beds beneath the bottoms of the deepest valleys. At various places, however, south of there, on Grand River, in Union County, and westward to the Missouri, somewhat above Nebraska City, the Upper series was seen more or less exposed, and readily recognized by the various fossils already mentioned. As these fossils, however, have nearly all a great vertical range in this Upper series, and beds at different horizons, especially in Kansas and Nebraska, where these rocks are greatly developed, are known to resemble each other very closely in lithological characters, it is not easy to be always sure of the *exact* horizon in the series, of particular beds seen at different localities. Dr. White, however, is quite confident that he can clearly identify the subordinate beds of most of the Winterset exposure, already mentioned, in the bluffs of the Missouri, on the Iowa side, twelve miles northeast of Nebraska City. Whether these beds are exactly equivalent or not with those at Winterset, there is not the slightest reason for doubting that they belong to the same Coal-Measure series, as do all the intervening exposures between these two localities.

On our arrival at Nebraska City, parties there interested in sinking a shaft, at Mr. Morton's farm, one and three-fourth miles west of the landing at that place, in search of coal, knowing that we had been examining the country across from the coal region of the Des Moines Valley to the Missouri, naturally wished to know how deep we thought they would probably have to penetrate to find coal. After taking into consideration the gentle dip of the strata from the outcrops of the Lower productive Coal series near the Des Moines, and the thickness of the Upper nearly or quite barren series at Winterset, Iowa, and making some allowance for the thickening of the latter in a westerly or southwesterly direction, of which we had seen some evidence, Dr. White and the writer told the Hon. J. Sterling Morton that we thought it possible the productive series might be struck within about 500 feet of the surface of the Missouri Valley at Nebraska City. We apprised him, however, of the hazard and uncertainty generally attending such mining enterprises, and advised that before proceeding farther with the shaft (at that time carried to a depth of 80 or 90 feet) that they should first sink an artesian boring 600 to 800 or 1,000 feet if coal should not be sooner struck. We were not at that time, however, nor until on the eve of starting from Nebraska City for Omaha City, aware of the fact that a boring had already been sunk in the Missouri Valley, a short distance below Nebraska City, to a depth of about 344 feet without striking any workable bed of coal.

On taking this important fact, and others observed in the exposures seen between Nebraska. City and Omaha City, into consideration, together with those developed by the borings at Omaha, we were led to the conclusion that it is exceedingly improbable that sufficiently thick beds of coal exist at Nebraska City within profitable working distance of the surface. As we came back from Omaha, a few weeks later, Dr. Hayden and I called to advise Mr. Morton to suspend work on the shaft, and, if he wished to make further exploration, first to sink Mr. Croxton's boring 500 or 600 feet deeper, if coal should not be sooner struck. Mr. Morton, however, was not at home, and we left the suggestions with his family. After my return to Washington I also wrote to him to the same effect.

Of course it is not within the scope of human wisdom to decide *positively* and *beyond doubt*, by any means short of actual boring, whether or not workable beds of coal could be struck there at depths that could be profitably wrought. The weight of evidence, however, from all the facts now known, is undoubtedly against success in such an undertaking. Some of the reasons for this conclusion are stated below.

In the first place, although some facts were observed indicating a slight northerly or northeasterly inclination of the strata on the north side of Platte River, there is little reason to doubt that the boring at Omaha City, about thirteen miles farther north, penetrates nearly, if not quite, the whole 400 feet, into lower strata than those forming the Plattsmouth section, where at low water about 60 feet of rocks can be examined inch by inch, and certainly contain no workable coal. From near this point there is evidently a gentle south or southeastward dip, so that at Rock Bluff, about six miles by a right line in a direction a little east of south from Plattsmouth, the upper part of the section at the latter place is brought so low as to leave little room to doubt that in these two sections we have an opportunity to examine in detail about 150 or 160 feet of strata (with the exception of a few feet at Rock Bluff, almost certainly occupied by sandstone) without finding any coal.

On going twelve miles farther down the river, in a nearly south direction, to Wyoming, we find the whole of the Rock Bluff section has, between these points, passed beneath the level of the Missouri, so that here at Wyoming we observe a low exposure, not agreeing with any part of the Rock Bluff section, but apparently occupying nearly the horizon of the lower part, or divisions A and B, of the Nebraska City section. It is possible, however, that there may be 40 or 50 feet of other unexposed strata between the top of the Rock Bluff section and the Wyoming outcrop. From these exposures and the borings at Nebraska City and Omaha City, we have pretty good data for the belief that there are no workable beds of coal beneath the Missouri level at Nebraska City for a depth of at *least* 600 or 800 feet. Consequently the Upper barren measures almost certainly thicken much more rapidly in coming southwestward from Winterset, Iowa, than had been supposed. Dr. White's numerous local sections, however, observed at various points west and southwest of Winterset, through to the Missouri, seem to show very clearly that this thickening of the Upper series is not due to the deposition of newer strata upon the equivalents of those seen at Winterset, but to a thickening of the same beds seen there, or with probably, in part, to the intercalation of other similar strata between or beneath the lower members of the same.

At Nebraska City, and below there, at Otoe City, Brownville, and Aspinwall, there are, perhaps, altogether, near 150 to 200 feet of additional strata, holding probably all a position above the geological horizon of the top of the boring at Nebraska City. The various natural exposures, shallow shafts, drifts, and other excavations in these beds, show that they almost certainly contain no thicker bed of coal than the 22-inch bed seen at Aspinwall; so that these beds may, altogether or in part, also be added to the entire *explored* thickness of strata almost certainly without any seam of coal thicker than that cropping out at Aspinwall.

Somewhere below Aspinwall, a gentle dip of the strata to the north or northwest takes place, and causes lower beds to rise to the surface as we descend the Missouri from the southeast corner of the State. But, from what we saw near Rulo, it seems rather improbable that any beds come to the surface between that place and St. Joseph, in Missouri, belonging to much lower geological horizons than those penetrated by the borings at Nebraska City and Omaha, though it is possible some beds may exist there not represented under the latter places. At any rate, the observations of Mr. Broadhead, and those of Dr. Owen and Professor Swallow, along the Missouri, render it exceedingly improbable that any workable beds of coal exist in any of the strata holding a geological position between those seen at Rulo and the horizon of the top of the boring at St. Joseph, in Missouri. And as the boring at the latter place was carried to a depth of 480 feet, without striking a workable bed of coal, and no such beds crop out between there and Atchison, Kansas, while none such were struck in a boring at the latter place, extending to a depth of about 300 feet, the evidence becomes very strong that we have here a great series of nearly barren Upper measures, containing no beds of coal of more than about 2 feet in thickness.*

It is worthy of note in this connection, that Mr. Broadhead had arrived at the same conclusion in regard to the Coal-Measures of northwestern Missouri, from the examination of numerous natural exposures before any of these borings had been made. By examining a paper on the Coal-Measures of that region, published by him in the Proceedings of the St. Louis Academy of Sciences in 1865, it will be seen that he gives a continuous section of beds (with some unexposed spaces) of nearly 2,000 feet, before reaching a bed of coal as much as 2½ feet in thickness; all of those noted above being from a few inches to 2 feet in thickness. It is quite probable that all the numerous thin beds and seams of shale, clay, limestone, sandstone, &c., composing this long section, are not always here placed by him in their *exact* order of superposition, since it is generally impossible to see more than comparatively a few of them exposed in any one natural section, while it is very difficult, in a series of rocks like this, so liable to change their lithological characters in short distances, to be always right in identifying the same beds at different localities,

* Dr. White's observations in Iowa have also led him to the conclusion that the Upper and most of the Middle Coal-Measures of that State, contain no bed of coal of more than two feet in thickness.

especially as most of their fossils have a great vertical range. There is little room for doubting, however, that this section gives a good general approximation to the true structure of the series in that region. It evidently includes most, if not all, of the series referred by Professor Marcou and Professor Geinitz to the Dyas, Coal-Measures, and Mountain limestone; also the Upper and some of the Middle Coal-Measures, as well as probably some of the Lower Permian, of Professor Swallow's Kansas section.

There are a few leading facts that ought to be impressed upon the minds of those who may wish to seek coal in Eastern Nebraska. The first of these is, that the exposed rocks here belong to the upper, member of the great coal-bearing series, and that, although of considerable thickness, all observations and experience, not only here in Nebraska, but also in Kansas, Missouri, and Iowa, as already stated, render it almost a moral certainty that this Upper series, in this region, contains none but thin and unreliable beds of coal. Consequently if good workable beds exist here, it may be regarded as a nearly settled question that they lie *deep* beneath the level of the lowest valleys, in the Lower or Middle Coal-Measures; and if ever made available it must be by deep shafts. It would not be prudent, however, to attempt to reach these lower beds by directly sinking shafts; the proper course is first to demonstrate their existence here, and their depth, by artesian borings.

In making such borings here, they should, of course, commence in the deepest valleys. It ought also to be remembered, that it is simply throwing away money to make two or more such borings near each other, for in a country where the strata are so slightly inclined, and so free from faults as this, a single boring settles the question for or against the existence of coal, to the depth that it penetrates, for a considerable area of the surrounding country; since regular workable beds of coal are, with rare exceptions, not confined to one spot, but extend with the other strata for more or less considerable distances. Indeed, it is very probable that one boring at Omaha City, or six or eight miles up Platte River, where the rise of the strata would give a somewhat lower start, and another somewhere in the valley of the Great Nemaha, if carried to a depth of 1,200 to 2,000 feet, in case coal should not be sooner found, would, with the boring already made at Nebraska City, and the numerous natural exposures examined, settle the question in regard to the existence of workable beds of coal within accessible depths, in Eastern Nebraska.

Again, it ought to be generally known that money or labor expended in sinking shafts or excavating drifts into strata, every bed and seam of which may be examined in detail, and seen to be barren of coal in natural exposures in the immediate vicinity, will be certainly without reward. Much hard labor has also been squandered in Nebraska by excavating drifts into beds of black shale, with the delusive hope that if followed a hundred yards or so into a hill, they may be found to change into valuable beds of coal. Or, in thus following a 2 or 3-inch seam of coal, expecting it to be found much thicker farther in. Although such a result is not always certainly beyond the range of *possibilities*, it may be regarded as so nearly the case in these rocks as practically to amount to about the same thing. Indeed it may be almost taken for granted, from the numerous sections already examined inch by inch, and the facts developed by the borings, shafts, and other excavations already made, that no more valuable bed of coal than the 22-inch bed at Aspinwall, will ever be found here by such superficial openings.

In regard to finding workable beds of coal within accessible depths in Eastern Nebraska by deep borings, I would remark, in conclusion, that although not prepared to discourage entirely all hope of success, it is proper to state that all the known facts are unfavorable; not only on account of the great thickness of the Upper nearly barren series, but because it is by no means *certain* that we would strike the productive measures here, even after going entirely through the Upper nearly barren series, since the lower beds *may* thin out in this direction. The great importance, however, of a good supply of coal to a country so rich in other natural resources, and at present so scantily supplied with wood, certainly renders it desirable that some of the borings already commenced should be sunk to greater depths before abandoning all hope of success. That at Omaha City, for instance, which probably starts at a lower geological horizon than any other yet commenced in the State, should be at least continued on to a depth of 1,000 to 2,000 or more feet, in case coal should not be sooner found; for it is scarcely possible to estimate, at this time, at what depths it might be profitable, at some future time if not at present, by the aid of improved implements and skill, to mine coal here, directly on the great thoroughfare of trade and travel between the Atlantic and Pacific shores.*

---

* This report was prepared between three and four years back, and as I have not since visited Nebraska, I am at this time (April, 1871) unacquainted with any of the facts bearing on the question in relation to the probability of finding workable beds of coal there, that more recent borings or other explorations may have brought to light.

As some persons seem to be surprised that it should be thought possible to mine coal at the great depth of 2,000 feet, it may not be out of place to mention here, that coal is now being profitably raised in England from the enormous depth of 2,418 feet, while some of the best English mining engineers believe that within 20 years it will be there raised from shafts 3,000 feet in depth.

DESCRIPTIONS OF FOSSILS.

PROTOZOA.

FORAMINIFERA.

Genus FUSULINA, Fischer.

FUSULINA CYLINDRICA, Fischer.

Pl. I, Fig. 2; Pl. II, Fig. 1; Pl. V, Fig. 3, *a*, *b*; and Pl. VII, Fig. 8, *a*, *b*.

Fusulina cylindrica, Fischer, 1837, Oryct. Moscou, p. 126, Pl. 18, Figs. 1-5; Murch. de Vern. and Keys., 1845, Geol. Russ., 11, p. 16, Pl. 1, Fig. 1, *a*, *b*, *c*, *d*, *e*; Owen, 1852, Geol. Survey Wisconsin, Iowa, and Minn., p. 131, &c.; Meek & Hayden, 1859, Proceed. Acad. Nat. Sci., Philadelphia, p. 24; Dana, 1863, Man. Geol., p. 164, Fig. 193; Meek & Hayden, 1864, Paleont. Up. Mo., p. 14. Pl. 1, Fig 6, *a*, *b*, *c*, &c.; Meek, 1864, Paleont. California I, Part I, p. 4, Pl. 2, Fig. 1; Geinitz, 1866, Carb. und Dyas in Neb., p. 71, Tab. V, Fig. 5.

Shell small, fusiform or sub-cylindrical, more or less ventricose in the middle, somewhat obtusely pointed at the extremities, which generally have the appearance of being a little twisted. Surface smooth, excepting the septal furrows, which are moderately distinct, more or less regular, and a little curved as they approach the extremities. Aperture apparently linear, and not visible as the specimens are generally found. Volutions 6 to 8, closely coiled, the spaces (near the middle) being rarely more than twice the thickness of the shell walls. Septa from 20 to about 33 in the last or outer turn of adult specimens, counting around the middle; comparatively straight near the outer walls, but strongly undulated laterally within; foramina passing through the walls, moderately distinct in well-preserved specimens—as seen under a high magnifying power, in transverse sections near the middle of the shell, somewhat radiating, and numbering in the outer turn of a medium-sized specimen, from 12 to 20 between each two of the septa.

Varying considerably in size and form; length or greater diameter of medium-sized individuals, 0.27 inch; thickness, or shorter diameter, 0.08 inch.

In following the general practice of referring this to the Russian species, I am not only governed by comparisons with figures and descriptions of the latter, but I have had an opportunity to make direct comparison with specimens of *F. cylindrica* kindly sent to me by Colonel Romanowski, of the Russian mining-engineers's department, from the Ural Mountains. It is true these Russian specimens are not in a condition to show very clearly in polished sections the minute details of their internal structure under the microscope, but so far as I have been able to determine from the comparison, they seem to agree well with the American form.*

I have never yet seen any specimens presenting the form of that figured by Professor Geinitz under Fischer's name, *F. compressa*. If this compressed bicarinate appearance is not due to accidental pressure, I should think it a distinct species; though the specimens of these little foraminiferous shells have often been subjected to various accidents,

* I was interested to find among the Russian specimens sent by Colonel Romanowski, the globose species of *Fusulina* described by me in the California report, from the Carboniferous rocks of that State, under the name *F. robusta*, with which the Russian authorities had correctly identified it.

and besides they vary so much naturally, that the separation of the species is generally attended with great difficulty.

Locality and position.—One of the specimens figured is from a locality two and a half miles west of Nebraska City; it also occurs at Bennett's Mill, Wyoming, &c., all in beds referred by Professor Marcou and Professor Geinitz to division B of the lower part of their Upper Dyas: likewise at a higher position one mile southeast of (or below) Nebraska City, and at lower positions in the Coal-Measure on the Missouri, at Rock Bluff, Plattsmouth, Bellevue, Omaha City, &c. In short, it ranges all through the Coal-Measures, and is widely distributed in the Western States and Territories.

RADIATA.

POLYPI.

Genus RHOMBOPORA, Meek.

Small ramose corals, with non-septate, short tubular cells, radiating obliquely outward and upward on all sides from an imaginary axis; cell-mouths rhombic or rhombic-oval, and very regularly arranged in longitudinal and oblique spiral rows, the former of which are sometimes separated by more or less flexuous longitudinal ridges; interspaces usually rather thick, and not pierced by transverse pores, but occupied by very minute non-septate longitudinal cells that are closed and represented at the surface by minute granules or spinules.

Type, *Rhombopora lepidodendroides*, Meek.

The little corals composing this group are closely allied to some of those included by Professor Hall in his genus *Trematopora*, if not really congeneric with a part of the same. They differ, however, from the typical species of that group, such as *T. tuberculosa*, in not having the interspaces between the cells septate, but occupied by very minute tubes or pores (closed at the surface), as well as in the rhombic form, and very regular arrangement of the cell-mouths. *Trematopora aspera*, Hall, has much the external aspect of our genus, excepting in its elongate oval instead of rhombic cell-mouths.

Millepora rhombifera, Phillips, *Vincularia ornata*, Eichwald, and *Favosites serialis*, Portlock, may, with much confidence, be included in this genus. The species of this group with which I am acquainted are from the Carboniferous rocks, through the whole of which the genus ranges, and I suspect that it is also represented in the Devonian and Silurian. None of the species I have yet seen seem to be incrusting or have hollow tubular stems, but there may be such among the undescribed species. Although some species of this genus have been referred to Goldfuss' genus *Vincularia*, they are widely removed from the typical Cretaceous species of that genus.

RHOMBOPORA LEPIDODENDROIDES, Meek,

Pl. VII, Fig. 2, *a*, *b*, *c*, *d*, *e*, *f*.

? *Stenopora columnaris* (*pars*), Geinitz, 1866, Carb. und Dyas in Neb., p. 66; (not Schloth. 1813).

Ramose, slender, cylindrical or slightly compressed, and bifurcating at regular, distant intervals; divisions nearly straight between the points of bifurcation, where they diverge at angles of about 70° to 80°; composed of small, short, nearly round, tapering tubes, that ascend from an imaginary axis, obliquely outward, with more or less curve to the surface, near which they are separated by interspaces, which in cross-sections show the minute cellular structure; calices arranged very regularly in quincunx, so as to form vertical and oblique rows; distinctly rhombic at the surface, where their margins are roughened by small, prominent, node-like grains, placed one at each corner, with smaller granules along the edges between.

Entire size unknown; thickness of one of the largest branches, 0.07 inch; length between bifurcations, 0.40 inch; number of calices in 0.20 inch, measuring longitudinally, 12; number of calices in 0.05 inch, measuring in the direction of the oblique rows, 4.

The rhombic form of the cell-mouths, and their very regular arrangement, are often well marked in this species. These characters, however, and the granules along their margins, are only to be seen in well-preserved specimens, for when slightly worn the granules are removed, and the angular character of the calices becomes nearly obsolete. When the surface is ground a little obliquely, the calices present an oval outline, as seen in Fig. 2 *e*, of Plate VII, while the thin interspaces present a cellular structure, as if occupied by minute vertical cells; one of these cells is larger than the others between the ends of the calices and seems to correspond to the larger grainlike node between the ends of each two of the calices. In longitudinal sections I have been entirely unable to detect transverse plates in the calices or interspaces. In old specimens the cell-walls seem to have become proportionally thicker and the cells consequently smaller.

As this is the only little coral we found in division C, at Nebraska City, among extensive collections, I am led to think it must be one of those referred by Professor Geinitz to *Stenopora columnaris* of Schlotheim. However that may be, I have no doubt, after a careful comparison with ramose specimens sent under the latter name from Posneck, Thuringia, that the Nebraska fossil is clearly distinct. In the first place, it differs in the regular rhombic form of the apertures of its well-preserved calices, and in the uniform arrangement of a larger granule at each of their corners, and one or more rows of very minute ones along each margin between. The specimens of *S. columnaris* mentioned above, have granulated margins to the calices; but these grains are irregularly arranged, while the calices themselves never present the rhombic outline seen in our fossil, but are always oval at the surface.

In division B, at Nebraska City, the coral under consideration is often almost completely covered by the following described incrusting species. As Professor Geinitz includes under the one species *Stenopora columnaris* both incrusting and ramose forms, it is probable he included this as one of the varieties of that species. But Edwards and Haime consider the species *columnaris* as peculiarly an incrusting coral. At any rate, the species here described differs so materially from that incrusting it, that the two can be distinguished almost at a glance by external characters alone, while internally, also, they seem to me to present structural differences of generic importance. These differences are so readily seen in the figures as to render any detailed comparisons unnecessary.

Locality and position.—Division C, of Nebraska City section; also in division B of the same at that place, where, as above stated, it is generally incrusted with another coral. It likewise occurs in the latter horizon at Bennett's Mill, Wyoming, &c., and at lower horizons in the Upper Coal-Measures at Rock Bluff, and Plattsmouth, as well as at various positions in the Upper Coal-Measures of Nebraska, Kansas, Iowa, Missouri, and Illinois.

Genus FISTULIPORA, McCoy.

FISTULIPORA NODULIFERA, Meek.

Pl. V, Fig. 5, *a*, *b*, *c*, *d*.

Stenopora columnaris Geinitz, 1866, Carbonf. und Dyas in Nebraska, p. 83; (not *S. columnaris*, Schoth., sp.

Corallum incrusting crinoid columns, small branching corals, and other objects, so as often to assume cylindrical, irregular, or false ramose forms; surface more or less nodulous, but without granules or spinules; cell-mouths circular, small, or generally less than 0.01 inch in diameter, and, when not worn, with prominent smooth margins, generally directed a little obliquely outward from the centers of the prominences on the latter; interspaces usually slightly less than or about equal to the diameter of the cell-mouths, but sometimes (particularly on the prominences) wider; composed of from one to three ranges of minute vesicles, usually a little wider than high, and separated by very thin diaphragms; diaphragms of principal cells not seen.

This coral is quite common through most of the Coal-Measures of the West. It often incrusts crinoid columns, and, wherever associated with the more delicate little coral for which I have proposed the name *Rhombopora lepidodendroides*, it incrusts the slender stems and branches of that species so completely as entirely to conceal the latter, and thus to appear as if it were itself ramose. Transverse or longitudinal sections, however, reveal the little stems, with an entirely different structure within. I have seen no examples of it ramose, however, where not thus modified by the very distinct coral upon which it so often grows.

I can scarcely doubt that this is one of the species referred to *Stenopora columnaris* by Professor Geinitz, as it is common in the same beds, and at the same localities from which he cites that species, some of the forms of which it nearly resembles in general appearance, as well as in its mode of growth. I cannot believe, however, that it is nearly allied to that species, because it shows no traces whatever, on perfectly preserved specimens, of the granules or spinules of the interspaces, characterizing the forms included by Professor Geinitz in Schlotheim's species, where well preserved. On the contrary, it seems to me to have the structure of *Fistulipora* of McCoy; while the European forms referred by Professor Geinitz to the species *columnaris* are thought to belong, in part at least, to the genus *Labechia*, by Edwards and Haime.

Among the described species of *Fistulipora*, this coral seems to agree most nearly, in the size of its cell-tubes and the structure of its intermediate vesicular tissue, with *F. minor* of McCoy. Still its cell-mouths are smaller and more closely arranged, there being from six to seven in the space of one line, while in Professor McCoy's species, four occupy the same space. It also never appears to grow to the thickness of that species, and consequently has much shorter tubes. The nodes or prominences of its surface are comparatively large, like those seen on *Monticulipora*, d'Orbigney, and on these the cell-mouths are most widely separated, and usually directed slightly outward from the middle of each node, in which case the upper margin or lip of each cell is slightly more prominent than the other.

I do not know that this nodose character has been observed in the genus *Fistulipora*, and hence the coral under consideration would seem to differ from the typical species of that genus about as *Monticulipora* differs from *Chætetes*. These nodes, however, are not always developed, and seem to have no regular arrangement. Sometimes they seem to be merely represented by slightly more scattering spaces of cell-mouths, without any elevation.

Locality and position.—Nebraska City, Bennet's Mill, and Wyoming, from bed B of the Nebraska City section. It also occurs at various other localities and positions in the Coal-Measures of Nebraska, Iowa, Illinois, and other Western States.

Genus SYRINGOPORA, Goldfuss.

SYRINGOPORA MULTATTENUATA, McChesney.

Pl. 1, Fig. 5 *a*, *b*, *c*, *d*.

Syringopora multattenuata, McChesney, 1860, Descriptions New Paleozoic Fossils, p. 75; also 1865, Pl. 2, Fig. 4, *a*, *b*, illustrations of same.

Corallum forming large masses; corallites cylindrical, long, slender, flexuous, more or less radiating, varying in their distance apart from once or twice their own breadth to close contact; connecting tubes numerous, slender, transverse, irregularly distant, but generally rather close; epitheca thick and rather strongly wrinkled. Septa unknown; tabulæ very obliquely and irregularly arranged.

Size of corallum unknown, but fragments indicate a diameter of 5 to 7 inches. The diameters of the corallites are quite uniform, and about 0.07 to 0.08 inch. They are rarely more than twice their own diameter apart, and often so closely compacted together as to become more or less angular, in which cases the connecting tubes are of course obsolete, but connecting openings take their places. In these compact examples, when the corallum is broken parallel to the corallites, these little transverse openings give it almost the appearance of a *Favosites*. This species seems to be related to *S. geniculata*, Phillips, but differs in having the corallites often closely compacted, and perhaps generally slightly smaller.

Locality and position.—Four miles up Platte River, north side, in the Upper Coal-Measures; it is also common at the same horizon in Illinois, eight miles south of Springfield.

Genus LOPHOPHYLLUM, Edwards & Haime.

LOPHOPHYLLUM PROLIFERUM, McChesney, (sp).

Pl. V, Fig. 4 *a*, *b*.

Cyathaxonia prolifera, McChesney, 1860, Descr. New Pal. Foss., p. 75; also 1865, Fig. 1. Pl. 2, illustrations same.

Cyathaxonia, sp. Geinitz, 1866, Carb. und Dyas in Neb., pp. 65 and 66, Tab. V, Figs. 3 and 4. Comp. *C. tortuosa*, Michelin, 1846, Icon. Zooph., p. 258, Pl. 59, Fig. 8.

Corallum elongate-conical, more or less curved, or sometimes nearly straight, tapering to a pointed base; epitheca very thin, with more or less distinct encircling wrinkles and striæ of growth, crossed by longitudinal striæ; rarely sending off a few spines near the base. Calice nearly or quite circular, moderately deep; columella prominent, compact in texture, compressed above, with its longer axis coincident with the general curve of the corallum; septa from about 30 to 50, every alternate one generally considerably shorter than the others, which latter extend to the columella, near which they are sometimes a little tortuous.

The largest of the specimens from which the foregoing description was drawn up is about 0.82 inch in length, and 0.42 inch in breadth at the larger end; the species, however, sometimes attains quite double this length.

I have not seen specimens of this species showing the septal fossula very clearly. Those from Nebraska are generally broken and distorted, but on comparison with examples from the Coal-Measures of Illinois, they are found to agree well in all their known characters. By examining longitudinal sections, I have ascertained that there are in this coral rather distant tabulæ or plates, extending outward with a strong downward curve from the columella, and hence that it does not even belong to the family *Cyathaxonidæ*, but seems to agree well with *Lophophyllum*.

Professor Geinitz compares this species to *Cyathaxonia tortuosa* of Michelin, from the Carboniferous rocks of Tournay, to which it seems to be very similar in external characters; but if that species is a *Cyathaxonia*, of course the form under consideration must be widely different.

Locality and position.—The specimens described are from division B of the Nebraska City section. Professor Geinitz also figures it from division C of the same section, and it occurs at Rock Bluff, Nebraska; and at Springfield, Lasalle, and numerous other localities in the Coal-Measures of Illinois, as well as in the adjoining States, and Texas.

Genus CAMPOPHYLLUM, Edwards & Haime.

CAMPOPHYLLUM TORQUIUM, Owen, sp.

Pl. 1, Fig. 1, *a*, *b*, *c*, *d*.

Cyathophyllum vermiculare?, Owen, 1852, Report Geol. Survey Wisconsin, Iowa, and Minnesota, p. 133, Pl. IV, Fig. 2; (not Goldfuss, 1826).
——— *flexuosum?*, Owen, 1852, ib., Fig. 3, *a*, *b*; (not C. (*Madrepora*) *flexuosum*, Linn., 1767, nor Goldfuss, 1826).
——— *torquium*, Owen, 1852, ib., explan. Pl. IV, Fig. 2.

Corallum simple, attaining a rather large size, elongate-conical, and often variously geniculated or bent when two or three inches in length, but becoming nearly straight, subcylindrical, and considerably elongated in the larger half of adult individuals. Epitheca thin, with small encircling wrinkles and strong undulations of growth, showing no traces of septal costæ when unabraded, but, where even slightly worn, exposing the regularly disposed septa, and thin intervening dissepiments distinctly. Calice circular or slightly oval, comparatively shallow, with thin margins, from which its sides slope rather steeply inward for some distance, and then descend very abruptly into a deeper, narrow, central depression; provided at the outer side of the general curve of the corallum with a moderately distinct septal fossula, formed by the shortening of one of the primary septa, and the bending down of the tabulæ at that point. Principal septa from 30 to 48, extending from about one-half to two-thirds of the distance from the exterior toward the center, stout and usually nearly straight inside of the vesicular outer zone, but becoming distinctly more attenuated (as seen in transverse sections) and somewhat curved or a little flexuous in crossing the vesicular area, where they alternate with an equal number of very short, thin ones; tabulæ very wide or occupying about two-thirds of the entire breadth, as seen in longitudinal sections, and passing nearly horizontally across, with a more or less upward arching; dissepiments thin and forming numerous obliquely ascending, small vesicles, in transverse sections seen to pass across between the costæ with an outward curve.

Entire length unknown, but individuals incomplete at both extremities, 5 inches in length and 1.60 inches in breadth, have been met with. These were probably not less than 7 or 8 inches in length when entire. Individuals of this size show at the thickest part 9 costæ in a space of 0.50 inch.

Young tortuous individuals of this coral were referred by Dr. Owen with doubt to *Cyathophyllum tortuosum*, Goldfuss, and large straighter fragments of adult specimens, from the same beds, were doubtfully referred by him to *C. vermiculare* of Goldfuss. After a careful examination, however, of a large number of specimens, I am led to the conclusion that they have the same internal structure, and really belong to one species. Although the large straighter specimens resemble quite nearly in their general appearance Goldfuss' figure of *C. vermiculare*, they are found to differ widely in their internal structure, *C. vermiculare* having from 64 to 100 septa, the larger of which extend in to the center. In the same way the smaller flexuous specimens resemble *C. flexuosum* of Goldfuss, with which they agree, more nearly in internal structure. They differ, however, in having the tabulæ much more closely crowded, and arching upward, or, in other words, curved downward on each side, as seen in longitudinal sections.

In some respects our coral seems nearly related to *Caninia sub-ibicina* of McCoy (as illustrated in his Brit. Pal. Foss., Pl. 3 I, Figs. 35, 35 *a*), but it has not near so many septa, that species having as many as about 130 in the two series.

Locality and position.—Very abundant in beds 11 and 12 of the Rock-bluff section; also common at about the same horizon at Cedar Bluff. Dr. White likewise finds it near this horizon in the Coal-Measures of Iowa; and it also occurs in the Coal-Measures of Illinois.

ECHINODERMATA.

Genus ERISOCRINUS, Meek & Worthen.

ERISOCRINUS TYPUS, M. & W.

Pl. 1, Fig. 3, *a*, *b*.

Erisocrinus typus, Meek & Worthen, 1865, Am. Jour. Sci., Vol. XXIX, p. 174; also 1866, Geol. Rep. Ill., Vol. II. p. 317, Fig. 33.
Philocrinus pelvis, Meek & Worthen, 1865, Am. Jour. Sci., Vol. XXIX, p. 350.
Erisocrinus Nebrascensis, Meek & Worthen, March, 1865, ib., p. 174.

Fig. 1.

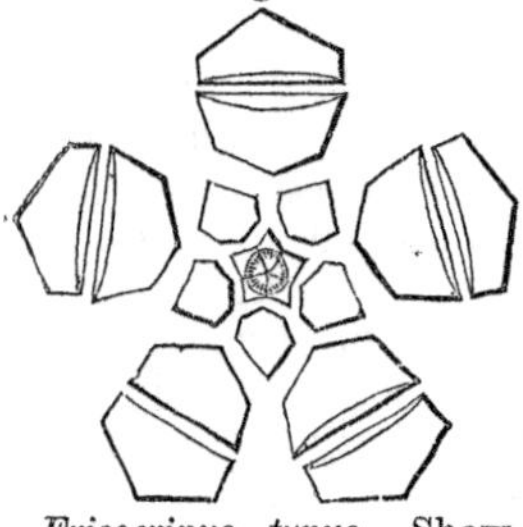

Erisocrinus typus.—Showing the structure of the body from the base to the second radials inclusive. (From Illinois State Geological Report.)

Body below the summit of the first radials, basin-shaped, rounded below, and obscurely subpentagonal in outline as seen from above or below; composed of thick, smooth, slightly convex plates. Basal pieces small, occupying a shallow concavity of the under side, about half hidden by the column, all pentagonal in external form. Subradial pieces considerably larger than the basal, and all equally hexagonal in form. First radial pieces four or five times as large as the subradials, wider than long, equal, and all pentagonal; supporting upon their broadly and evenly truncated superior sides the second primary radials, which are of nearly the same size and form as the first, but have their sloping sides above instead of below, while they each support above, two first brachials, or a series of secondary radials yet unknown. Surface smooth.

Breadth of body below the summit of the first primary radials, 0.72 inch; height of same, 0.35 inch.

From the slightly larger proportional size of the subradial pieces, I was at one time led to believe this specifically distinct from the Illinois specimens upon which the genus and species *E. typus* were founded. Further comparisons of additional specimens from the original locality near Springfield, however, showed clearly that no separation could be made on this character.

When Professor Marcou announced, some time back, that he had found a crinoid at Nebraska City, nearly related to *Encrinus*, I very naturally supposed he alluded to the form here under consideration, knowing it to be not unfrequently met with in these rocks in Nebraska, as well as in Illinois; while its body, below the summit of the first radial pieces, has almost exactly the structure of the corresponding parts of *Encrinus*, excepting that it has a well-developed series of basal pieces within the series usually regarded as basals in *Encrinus*.* It was not until the publication of Professor Geinitz's memoir, in which the form alluded to by Professor Marcou was figured, that I was aware the species mentioned by him is one more widely removed from *Encrinus*, by having a large, well-developed anal piece resting down upon one of the first subradials, a character modifying the whole structure of the parts of the body above.

The form mentioned by Professor Marcou, and figured by Professor Geinitz (*Poteriocrinus hemisphæricus*, Shumard), *Zeacrinus mucrospinus*, McChesney, and that here under consideration, as well as perhaps several others only known by fragments, are all more or less nearly related. Indeed, the separate first primary radials, and, perhaps, some of the other parts of these crinoids, when found disconnected, can scarcely, if at all, be distinguished specifically. The truncated subradial, however, on the anal side of *Poteriocrinus hemisphæricus* (not a true *Poteriocrinus*) can always be distinguished from any of the parts of the other forms mentioned; while the curiously developed spine-like second radials of *Zeacrinus mucrospinus* can always be readily recognized.

All of these crinoids are peculiar to the Coal-Measures of the West, where they are widely distributed.

Locality and position.—The specimen figured on Pl. 1, was found by Dr. Hayden, eight or nine years since, at Bellevue, Nebraska. The species was first described from the Upper Coal-Measures at Springfield, Illinois. We also have fragments of apparently this crinoid from division B, at Nebraska City.

Genus SCAPHIOCRINUS, Hall.

SCAPHIOCRINUS? HEMISPHÆRICUS, Shumard, sp.

Pl. V, Fig. 1, *a*, *b*; and Pl. VII, Fig. 1, *a*, *b*, *c*.

Poteriocrinus hemisphæricus, Shumard, 1858, Trans. St. Louis Acad. Sci. 1, p. 221.
Cyathocrinus inflexus, Geinitz, 1866, Carb. und Dyas in Neb., p. 62, Tab. IV, Fig. 20, *a*, *b*, *c*, and Fig. 28, *a*, *b*, *c*.

Body below the summit of the first radials sub-hemispherical, with the

* I mentioned in 1865 (Am. Jour. Sci., XXIX, 174), seeing a series of minute rudimentary pieces in a true *Encrinus*, belonging to Smithson's private collection at the Smithsonian Institution, within the range usually regarded as basals. As these, however, are not so developed as to be seen externally, and assume the character of true basals, and our crinoid seems to differ in the structure of its superior parts, it cannot be properly referred to that genus, though evidently closely allied to it.

under side deeply concave. Base very small, pushed or inverted entirely within the cavity of the body, and nearly or quite hidden externally; column facet round and deeply sunken. Sub-radial pieces of moderate size, very nearly equal, having a general pentagonal form, excepting the one on the anal side, which is a little truncated above by the anal piece, so as to give it a general hexagonal outline; each, however, has an additional very obscure mesial angle at its connection with the base, and all are strongly incurved below, to form the concavity of the under side. First radial pieces nearly twice as large as the sub-radials, twice as wide as high, and equally pentagonal, the upper sides being longer than either of the others, and all evenly truncated. Second primary radials (at least the two on the anal side) comparatively narrow, but still wider at the base than high, rounded on the outer side, a little constricted in the middle, and pentagonal in form, the two upper sloping sides supporting the first divisions of the arms, which are composed, at least for the first three ranges, each of a single series of wedge-formed pieces. First anal piece comparatively small, a little concave, resting upon one of the sub-radials, and connecting with the first radial on each side, above which it projects; supporting upon its slightly incurved upper edge a second piece, the form of which is unknown. Surface smooth, or only with traces of minute granules.

Fig. 2.

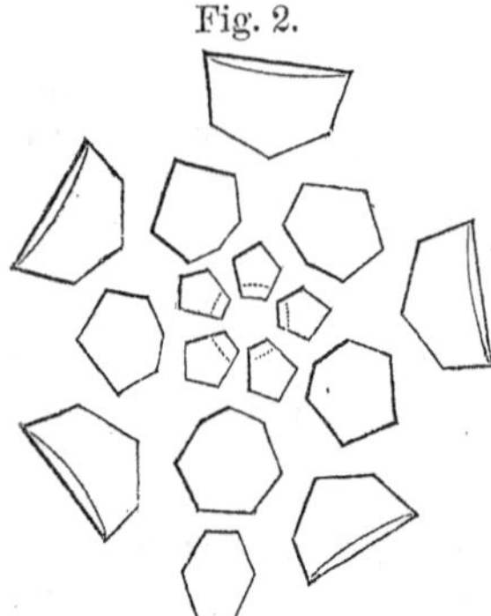

Scaphiocrinus? hemisphæricus.—Diagram showing the structure of the body from the base to the first radial pieces, inclusive.

The specimen from which the above description was made out is 0.66 inch wide, and 0.27 inch high to the summit of the first radial pieces.

In this, as in the several known species of *Erisocrinus*, and some other allied forms of the Coal-Measures, the first radial pieces are very thick on the upper edge, which has a neatly crenated ridge, with a furrow on each side of it, along the outer margin. These first primary radial pieces have exactly the form of the corresponding parts in *Erisocrinus*, and, as we have reason to believe, of one or more undefined genera of these rocks, and when found detached cannot be distinguished. The sub-radial piece, however, on the anal side of this crinoid, can at once be distinguished from any of those of *Erisocrinus* by the truncation of its upper side (see Fig. 28, *a*, *b*, *c*, Tab. IV, of Geinitz's Carb. und Dyas). On the inside of the cup of this and the allied forms alluded to, the sunken basal pieces are seen to rise so as to form a kind of pyramidal or conical protuberance.

Until the vault of these crinoids is known it will be difficult to determine whether they are more nearly allied to *Poteriocrinus* or to *Cyathocrinus.* From all analogy, however, I am led to think they will be found to possess the large prolonged trunk or proboscis of the *Poteriocrinus* group, instead of the merely vaulted summit and lateral tube of *Cyathocrinus.*

When we compare the structure of this crinoid, as far as yet known, with that of *Scaphiocrinus simplex*, Hall, the type of that group, it will be seen to agree in all respects, aside from mere specific characters, such as its concave under side, slight difference of form, &c., though it differs more widely from several of the other species that have been referred to that group, in having but a single anal piece, composing a part of the walls of the cup. Hence I have been led to place it provisionally in the

Scaphiocrinus group, generally regarded as a sub-genus under *Poteriocrinus.* It is worthy of note, however, that it is still a matter of doubt whether *Scaphiocrinus* will not be found a synonym of *Graphiocrinus*, de Kon., when it is known whether or not that type has minute basal pieces within those supposed to be such.

Locality and position.—Professor Geinitz figures a fine example of this species from division C of the Nebraska City section; and we have it from Bennett's mill, included by him and Professor Marcou in division B of the Upper Dyas. We also have it from Omaha, &c., from a lower position in the Coal-Measures referred by them to the Lower Carboniferous. It also occurs in the Coal-Measures of Iowa, Kansas, Illinois, &c.

Genus ZEACRINUS, Troost.

ZEACRINUS ? MUCROSPINUS, McChesney

Pl. V, Fig. 2 *a*, *b*, *c*.

Zeacrinus mucrospinus, McChesney, 1860, New Paleozoic Foss., p. 10; also, 1865, illustrations of same, Pl. 4, Fig. 7, *a*, *b*.
Actinocrinus, sp., Geinitz, 1866, Carb. und Dyas, Tab. IV, Fig. 29, *a*, *b*; (not *Actinocrinites*, Miller, 1821).

As pointed out by me some time back (Am. Jour. Sci., Vol. XLIV, new series, p. 176), the spines figured in Professor Geinitz's work, cited above, are those of a type belonging to the same family as *Poteriocrinus*. This curious crinoid has a cup almost exactly of the same form as the last, being sub-hemispheral, rounded below, and provided with a very small deeply sunken base, almost entirely hidden by the column. Its anal series of plates, however, is different from that of the last, these pieces being here more numerous and arranged as in *Zeacrinus*. Its most remarkable peculiarity, however, consists in having its second primary radial pieces produced outward into long spines. These spines Professor G. mistook for those of some species of the *Actinocrinites* group, like *A. Gouldi* of Hall. The spines of the latter, however, belong to an entirely different part of the crinoid, being modified vault-pieces, and hence have the head, or larger end, of a very different form. Consequently those under consideration can be distinguished at a glance from the spines of any of the species allied to *Actinocrinites*. In addition to this, the two types occupy distinct geological horizons, no *Actinocrinoid* being known in this country as high in the series as the Coal-Measures.

I have reproduced on Plate V, Professor Geinitz's figures of these spines, because we found no good specimens of them from any of the localities near Nebraska City. For comparison I have also given a copy of Professor McChesney's figure of one of these pieces of his typical specimen. That figured by Professor Geinitz, however, has the larger end somewhat obscured by adhering foreign matter, or from wearing so as partly to obliterate the markings on the upper side (Fig. 2 *a*) supporting the two first divisions of the arms.

Locality and position.—Professor Geinitz figures the spines of this species from Bennett's Mill, three miles northwest of Nebraska City, from beds referred by him and Professor Marcou to the lower part (division B) of their Upper Dyas. We have them from Plattsmouth, Cedar Bluff, Omaha, and numerous other localities and positions in the Coal-Measures of Nebraska and Kansas. They likewise occur in the same horizon in Iowa, Illinois, Missouri, Arkansas, &c. But they never, so far as known, occur at any horizon below the Coal-Measures; though Professor Worthen and the writer have described an allied representative species from the Chester Limestone of the Illinois Lower Carboniferous series. No analogous crinoid has ever been described in Europe from the Permian.

Genus EUPACHYCRINUS, M. & W.

EUPACHYCRINUS VERRUCOSUS, White & St. John.

Hydreionocrinus? verrucosus, White & St. John, 1869, Trans. Chicago Academy of Sciences. Vol. 1, p. 117, Fig. 1.
Eupachycrinus verrucosus, White and St. John, M. S.

"Body below the top of the first radial plates basin-shaped, with its base slightly depressed, more than twice as wide as high, and composed of somewhat massive, convex plates, which are joined by close sutures, which latter are in the middle moderately deep, caused by the beveling of the edges of the body-plates. Base moderately large, pentagonal, having a little more than one-third its diameter covered by the upper joint of the column. Width and height of subradial pieces about equal, strongly convex from base to top; three of them pentagonal, and two of them hexagonal, there being properly no angle at the base of any of them. First radial plates all pentagonal, not quite so large as the subradials, nearly twice as wide as high, all of them truncated above, so that the top of the calyx is seen to be nearly level when the arms are removed; the facets to which they were attached so broad as to occupy more than half the semi-diameter of the calyx. First anal piece nearly half as large as a subradial, quadrangular, resting between the superior sloping sides of two of the subradials, an inferior sloping side of one of the first radials, and the second anal piece, all of it being below the summit of the calyx. Second anal piece pentagonal, not quite so large as the first, but projecting considerably above the top of the calyx, and bordered by the first anal, one subradial, two first radials, and surmounted by two other small anal pieces. Surface marked by distinct verrucose elevations; and the whole surface, including the sutures, presenting a fine granulated appearance under the magnifier."

Fig. 3.

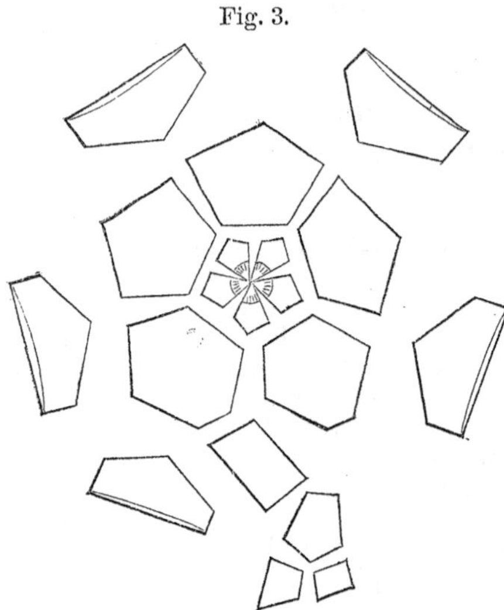

*Eupachycrinus verrucosus.**—Diagram showing structure of the body, from base to first radials inclusive.

Fig. 4.

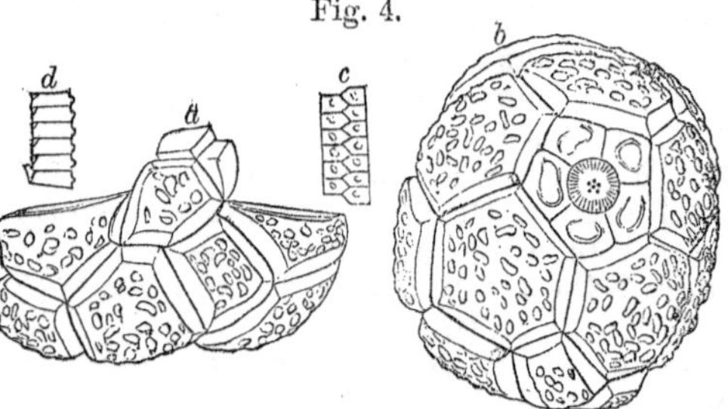

Eupachycrinus verrucosus.—*a*, an outline posterior view of the body; *b*, view of under side of same a little distorted; *c*, a view of the dorsal side of a piece of one of the arms; *d*, a lateral view of same, showing the first joint of one of the pinnulæ at the left lower corner.

* I am under obligations to Professor C. A. White, the State geologist of Iowa, for the use of the type specimen from which these cuts were made.

Height of body to the top of the first radial pieces, 0.55 inch; breadth of same, about 1.13 inches.

This crinoid agrees nearly, in the number and arrangement of the parts of its body, with the genus *Poteriocrinus*, from the typical species of which, however, its body differs materially in its depressed form and concave base; but it differs more especially in the structure of its arms, which, instead of being each composed of a single series of pieces, consist each of a double series of interlocking pieces. It, therefore, belongs to the same group as *Graphiocrinus 14-brachialis* of Lyon (Owen's Kentucky Geol. Rep., Vol. 3, Pl. 1, Fig. 2, 2*a*, and 2*b*), for which Mr. Worthen and the writer proposed the name *Eupachycrinus*, in Vol. 3, Illinois Geol. Rep., p. 177. At one time we had been led to think Mr. Lyon's type might possibly belong to *Hydreionocrinus*, de Koninck. Facts, however, more recently ascertained, in regard to the structure of some analogous American forms, seem to indicate that the part in *Hydreionocrinus* supposed to be composed of the united arms is probably only the ventral portion of the body extended upward.* If this is so, it most probably had free arms; though even in that case we have no evidence that they possessed the structure of those of the type under consideration; and if its *arms* are united to form a kind of extension of the body, as supposed by Professor de Koninck, then the species here described would certainly be widely removed, because the indentations for the attachment of pinnulæ, along their inner sides, show that its arms were free. It also has relations to *Zeacrinus* and *Scaphiocrinus*, from the typical forms of which, however, it differs materially in having its arms composed of a double series of pieces. Specifically this crinoid is closely allied to *Hydreionocrinus tuberculatus* (*Erisocrinus tuberculatus*, Meek & Worthen, from the Coal-Measures of Illinois), which proves to have the structure of arms characterizing this group, instead of that seen in those of *Erisocrinus*. The species under consideration, however, is more delicate, and differs in the peculiar verrucose style of its ornamentation, instead of having a distinctly nodose surface. It also differs in having its sutures more channeled by the beveling of its plates, than the species *tuberculatus*.

Dr. White and Mr. St. John have placed this species in MS. in the genus *Eupachycrinus*.

Locality and position.—The only known specimen of this species was found by Dr. White in the bed No. 2 of the Plattsmouth section, at Plattsmouth, Nebraska.

Genus ARCHÆOCIDARIS, McCoy.

ARCHÆOCIDARIS ? TRISERRATA, Meek.

Pl. 1, Fig. 6 *a*, *b*, *c*.

Of this species I only know the primary spines, but these are so peculiar that they will alone probably be sufficient to distinguish it. They are moderately long, rather slender, and usually a little arched near the base, where they present a nearly or quite circular section. Farther up they soon become compressed with a rhombic section, the lateral margins being sharp and regularly serrated, the little teeth-like projections being inclined outward or toward the apex of the spine, and apparently sometimes alternately arranged. Of these there were probably 25 to 30 on each side of the spine, and 9 to 12 of them may be counted in the

* See remarks of the writer and Mr. Worthen on this subject in Proceed. Acad. Nat Sci., Philad., Ap. 1870, p. 29.

space of half an inch. On the middle or concave side of the curvature of the spine there is a third serrated carina, apparently extending about two-thirds of its length from near the base toward the apex, the serrations becoming more distantly separated toward the outer extremity. On the other side three or four rows of much smaller granules, or minute, elongated nodes, extend from the shank of the spine, and gradually become obsolete near the middle. The ring at the head is moderately well defined, and, when well preserved, is seen to be faintly milled. The perforation of the abruptly contracted articulating extremity is about one-third as wide as the diameter of the shank. Surface, at least near the shank, where not worn, marked with exceedingly minute longitudinal striæ, entirely invisible without the aid of a good magnifier.

Length, apparently about 2.30 inches. No entire specimens, however, were found.

This species is evidently allied to *A. biangulatus*, Shumard (Trans. St. Louis Acad., 1, p. 223), to which I was at first inclined to refer it. As Dr. S., however, only mentions the two lateral rows of serrations of the primary spines, and says nothing about the others on the upper and lower side, and particularly as he gives the entire number of the serrations on each side as 12 to 14, his species should be distinct.

Locality and position.—Upper Coal-Measures, Omaha, Nebraska; and same position (bed No. 8) in an exposure seen on Platte River, three miles above its mouth. I have also seen a specimen of it from Vermillion County, Illinois, where it was found by Mr. Brodhead, fifty feet below coal-bed No. 6 of the Illinois section. The specimen from Illinois, like that here described, consists of a single spine. Until other parts of the fossil can be known, it is not possible to determine whether or not it is a true *Archæocidaris*.

Genus EOCIDARIS, Desor.

EOCIDARIS HALLIANUS, Geinitz.

Pl. VII, Fig. 9, *a*, *b*, *c*, *d*.

Eocidaris Hallianus, Geinitz, 1866, Carb. und Dyas in Neb., p. 61, Tab. V, Fig. 1 *a*, *b*, 2 *a*, *b*.

Of this delicate little species I have seen only fragments of the spines from Nebraska City and other localities in Nebraska. Entire spines, however, occur on Grand River, Union County, Iowa, in a bed containing many of the same fossils associated with it at Nebraska City, and holding a position about the middle of the Upper Coal-Measures of that region.

Locality and position.—Division C of Nebraska City section.

MOLLUSCA.

POLYZOA.

Genus FENESTELLA, Lonsdale.

FENESTELLA, sp.

Pl. 1, Fig. 4 *a*, *b*.

Fenestella plebeja, Geinitz, 1866, Carbon. und Dyas in Neb., p. 58, Tab. V, Fig. 8 *a*, *b*, (not McCoy, 1844).

I have no specimens of this species for study and illustration, and therefore merely reproduce Professor Geinitz's figures to call attention to the fossil as one of the forms to be looked for in these rocks. I must differ very decidedly from Professor Geinitz, however, in regard to the identity of this species with *F. plebeja* of McCoy, supposing both forms to have been nearly correctly figured, since they are represented so widely different that very few would think it necessary even to compare them. The differences are too strongly marked to be the result of any errors in the drawings, or due to peculiarities of different varieties of the same species, as may be seen by comparing the published figures of the two forms (see figure of *F. plebeja* copied on Plate VII, Fig. 11, from Professor McCoy's work on the Carboniferous fossils of Ireland, and that here given from Professor Geinitz's Carb. und Dyas in Neb). It is also worthy of note here that Professor McCoy's description agrees exactly with his figure, the fenestrules being represented in both as "rectangular, and from two to three times as long as wide," while in Professor Geinitz's enlarged figure of the Nebraska fossils they are represented without any traces of angles, and of a broad-oval or subcircular form, but slightly longer than wide.

We can never hope for any approximation to precision in paleonlotogy or geology, while such hasty identifications as this are insisted upon by geologists.

Locality and position.—Bellevue and Plattsmouth, in Upper Coal-Measures.

FENESTELLA SHUMARDI, Prout ?.

Plate VII, Fig. 3 *a*, *b*, *c*.

Fenestella Shumardi, Prout, 1858, Trans. St. Louis Acad. Sci., Vol. 1, p. 232.

Polyzoum growing apparently in flabelliform expansions, and composing an extremely fine, delicate net-work; branches very slender, of uni form size, rather flattened, and comparatively coarsely striated on the nonporiferous side, bifurcating at rather regular intervals of from 0.20 to 0.25 inch, the divisions diverging but slightly; fenestrules oblong, or about once and a half to nearly twice as long as wide, distinctly quadrangular, especially as seen on the non-poriferous side, and about equaling the breadth of the branches; dissepiments extremely slender or scarcely more than one-fourth as thick as the branches, not widened at the end on the non-poriferous side, but often somewhat expanded by a pore at one or both ends on the other side. Poriferous side with a mesial carina *apparently* sometimes bearing minute projecting points, and on each side of this angle about two and sometimes three comparatively large pores, generally arranged so that there is one at each end of each dissepiment, and another between these opposite each side of each fenestrule.

Entire size of polyzoum unknown; number of fenestrules in 0.25 inch, measuring longitudinally, 13 to 14; ditto, measuring transversely, 15 to 16; breadth of branches, 0.02 inch.

As near as can be determined from Dr. Prout's description alone, this very delicate little species seems to agree quite well with his *F. Shumardi*, from the Carboniferous limestone (probably of the age of the Coal-Measures) at the Organ Mountain, New Mexico. Still, as it is exceedingly difficult to determine the relations of such fossils with precision from descriptions alone, it is quite possible that the form under consideration may be found distinct on comparison of specimens. If so, I would propose to designate it by the name *F. perelegans*.

Compared with foreign described species, this is perhaps most nearly related to *F. plebeja* of McCoy (Carb. Foss. Ireland, Pl. XXIX, Fig. 3); but on comparison it will be seen to have its fenestrules only once and a half to nearly twice as long as wide, instead of from two to three times as long as wide, while it has only two or three pores instead of four or five opposite each side of each fenestrule. Its pores are also proportionally larger, while the entire polyzoum forms a more delicate net-work, as may be seen by comparison with a figure of Professor McCoy's species given on Plate VII.

Locality and position.—Division C of the Upper Coal-Measure section at Nebraska City, Nebraska. It has also been discovered by Dr. Newberry in the Lower Coal-Measures of Ohio.

Genus POLYPORA, McCoy.

POLYPORA SUBMARGINATA, Meek.

Pl. VII, Fig. 7 *a*, *b*.

Polypora marginata, Geinitz, 1866, Carb. und Dyas in Neb., p. 69, Pl. V, Figs. 11 *a*, *b*, and 12 *a*, *b*; (not McCoy 1844, Synopsis Carb. Fossils Ireland, p. 206, Pl. XXIX, Fig. 5).
Compare *Fenestella polyporata*, Portlock, 1843, Geol. Londonderry, p. 323, Pl. XXII A, Fig. 1; (not Phillips).

Polyzoum growing in flabelliform (or infundibuliform?) expansions; longitudinal branches bifurcating at more or less regular intervals, and having their lateral margins sharply carinate; dissepiments about half as wide as the branches; fenestrules oblong oval, usually about twice as long as wide, or sometimes proportionally a little wider or longer, their breadth being about the same as the branches. Non-poriferous side with the branches convex, and a little prominent along the middle, flattened toward the carinate lateral margins, and in good specimens finely and beautifully striated longitudinally; dissepiments rounded and sometimes showing faint traces of striæ. Poriferous side with branches provided with a mesial row of small nodes or distinct granules, giving them a subcarinate appearance, and five rows of alternating pores, those of the middle row being placed one between each two of the nodes, while the other four rows are placed two on each side of the middle range; pores sometimes with a slightly raised, very narrow lip, most prominent around the outer side of the lateral ones, which, however, scarcely imparts any irregularity to the outline of the sharp lateral margins; dissepiments obtusely subcarinate along the middle, and sometimes showing traces of striæ, as on the other side.

Entire size of Polyzoum unknown; number of fenestrules, measuring longitudinally, in 0.50 inch, 6; same, measuring transversely, 9 to 11 in same space.

Professor Geinitz refers this species to *P. marginata* of McCoy; but if we are to be guided by Professor McCoy's figures and description, it seems to me we must regard it as being clearly distinct, as may be seen by Figs. 13 *a*, *b*, *c*, copied from Professor McCoy's Synopsis of the Carboniferous Fossils of Ireland. On comparing these with the figures of our fossil, it will be seen that he represents the fenestrules as being proportionally longer, and the dissepiments much more slender. The most important differences, however, may be seen in the enlarged figures of the poriferous sides of the two, as represented by Figs. 7 *b* and 13 *c*. In our species, for instance, there is, when well preserved, a row of distinct little nodes along the middle of each branch, neither represented in the figures nor mentioned in the description of McCoy's species. In addition to this, he mentions in his description, and shows in his figure, waved lines running along between the pores, not seen in the Nebraska species; and he only shows 3, or rarely 4, pores in each row opposite each fenestrule, while in our species there are from 4 to 5, or rarely 6, but generally 5. Surely such differences as this cannot be ignored if we mean to distinguish species at all.

It is true that in the figure given by Professor Geinitz, he shows no traces of the row of little nodes along the middle of the poriferous side; this, however, is due to the fact that his specimen had been subjected to some wearing or abrasion. At any rate, I have before me worn specimens showing the pores very clearly, and having the little nodes entirely removed on some parts, and more or less preserved on other portions of the same; while in others they are as distinct as seen in our enlarged figure. It may be argued that McCoy's figure may have been taken from a worn specimen also; but the fact that it showed the peculiar waved striæ so distinctly between the pores, seems entirely incompatible with the conclusion that it could have suffered such abrasion.

Our species would seem to be more nearly allied to a form referred by General Portlock to *Fenestella polyporata* of Phillips, and cited by Professor Geinitz as a synonym of *P. marginata*, McCoy, particularly as General Portlock mentions seeing on some specimens a row of minute granules along the middle of the branches. The form figured by General Portlock, however, is certainly different from Phillips's species, which is clearly figured with but two rows of pores to each branch, while Portlock's shows four, the one being a *Fenestella*, and the other a true *Polypora*. As General Portlock merely quotes Phillips's meager, unsatisfactory description, and gives a figure very unlike Professor McCoy's, I would neither feel warranted in identifying his species with *P. marginata*, nor in connecting our fossil with either of them.

Locality and position.—The specimens from which our Fig. 7 of Pl. VII were drawn, came from division C of the Nebraska City section, at that place.

POLYPORA, (sp. undetermined).

Pl. VII, Fig. 6.

Polypora biarmica, Geinitz, 1866, Carb. und Dyas in Neb., p. 68, Pl. V, Fig. 13, *a*, *b* (—Keyserling ? 1846).

As we found no specimens of this species in division C of the Nebraska City section, I have merely given copies of Professor Geinitz's figures of a specimen from that locality and position. In order that others may have some means of forming their own conclusions in regard to its relations to *P. biarmica* of Keyserling, I have also copied Count Keyserling's figures of his original typical specimen, published in his Petschora Land, Pl. 3, Fig. 10, and 10 a. I can only say these figures look to me exceedingly unlike, and I should not be willing to identify the Nebraska form with the Russian, if the figures are even approximately correct representations of the fossils. In addition to this, the form figured by Professor Geinitz occurs in the unquestionable Coal-Measures of Kansas and Iowa.

Locality and position.—Division C of the Nebraska City section.

Genus SYNOCLADIA, King.

SYNOCLADIA BISERIALIS, Swallow.

Pl. VII, Fig. 5, *a*, *b*, *c*, *d*, *e*.

Synocladia virgulacea ?, Swallow, 1858, Trans. St. Louis Acad. Sci., I, p. 179; (not Phillips, 1829).
——— *biserialis*, Swallow, *Ib.*
? *Septopora Cestriensis*, Prout, 1858, Trans. St. Louis Acad. Sci., I, p. 448, Pl. XVIII, p. 2, 2 *a*, *b*, *c*.
Synocladia virgulacea, Geinitz, 1866, Carb. und Dyas in Neb., p. 70, Pl. V, Fig. 14; (not Phillips, sp.).

Polyzoum infundibuliform? or composed of broad overlapping, or folded, rapidly-spreading flabelliform expansions, attached by a small root-like base. Primary branches, larger than the others, which are alternately and irregularly given off at various angles on each side, and themselves variously divided and subdivided, or rather increased by lateral and intercalated branchlets. Dissepiments, smaller than the branches, usually more or less arched or geniculated, and occasionally giving origin to intermediate branches. Fenestrules, often transversely oblong or irregularly subquadrangular, and usually wider than the branches. Non-poriferous side, with branches and dissepiments rounded, and finely and regularly striated, as well as provided with a few very scattering, irregularly disposed, round, and apparently superficial dimorphous pores, nearly as large as the true pores or cells of the other side; never provided with projecting root-like processes. Poriferous side, with branches and dissepiments, more or less distinctly carinated along the middle, the carina being occupied by a row of little pointed nodes, sometimes rising into short little spine-like projections, with generally on or near the base of each, one or two minute dimorphous pores; true pores or cells moderately large and rounded, with a slightly raised margin; arranged in two rows, one on each side of the carina of each branch and dissepiment, to the lateral margins of which they usually impart a more or less undulated outline; sometimes on the larger primary branches, and some of the larger dissepiments, an occasional odd pore or cell, not properly belonging to either of the two rows, is placed along near the lateral margins, but these never form a third continuous row.

The entire size of the polyzoum is not known, but some of the specimens indicate a diameter of 4 or 5 inches.

This species was regarded by Professor Swallow as a variety of *S. virgulacea*, for which he proposed the name *biserialis*. Professor Geinitz also referred it to *S. virgulacea* of Phillips, without even regarding it as being sufficiently distinct to require separation as a variety. To me, however, it seems to present differences of too much importance to admit of being included even as a variety of *S. virgulacea*. In order the more clearly to illustrate these differences, I have given on Plate VII, Fig. 12 *a*, a copy of a part of Professor King's enlargement of the poriferous side of *S. virgulacea* for comparison with Fig. 5, *d*, an enlargement of a portion of the same side of the Nebraska fossil. On comparing these, it will be seen that *S. virgulacea* has no mesial carina and is provided with three or more rows of pores or cells (Professor King says, in his description, from 3 to 5), arranged in longitudinal furrows, the middle one of which is bounded on each side by a slight carina, with small, minutely poriferous nodes; the entire arrangement being very different from the characters *constantly* presented by the Nebraska species. In addition to this, so far as yet known from the examination of numerous specimens, the latter never shows any traces of the root-like process of the non-poriferous side, so often seen on *S. virgulacea* (See Pl. VII, Fig. 12, *b*); while it constantly shows on this side a few scattering, irregularly-arranged dimorphus pores, neither mentioned in any of the descriptions, nor illustrated in any of the figures of the European species I have seen. For these reasons I can but regard it as a clearly distinct species. Like the European *S. virgulacea*, the species under consideration varies in the arrangement of its branches, fenestrules, and dissepiments, to some extent, though it is constant in its other characters.

Locality and position.—The specimens figured on Pl. VII are from division C, of the Nebraska City section; those on Pl. IV are from a shaft at that place, sunk some distance below this horizon, probably below division B, in which it also occurs. We also have it from a lower position at various localities in the Upper Coal-Measures of Kansas and Nebraska, Iowa, &c.; while in Kansas it ranges up into the Permo-carboniferous, and possibly into the Permian. In Illinois it occurs in both the Upper and the Lower Coal-Measures; and we have there also found an undistinguishable form in the St. Louis and Chester limestones, of the Lower Carboniferous series. (See Proceed. Acad. Nat. Sci., Philad., March, 1870, p. 15.)

Genus GLAUCONOME, Goldfuss.

GLAUCONOME TRILINEATA, Meek.

Pl. VII, Fig. 4, *a*, *b*, *c*, *d*.

Compare *Glauconome grandis*, McCoy, 1844, Carb. Foss. Ireland, p. 199 Pl. XXVIII, Fig. 3.

Polyzoums with main stem long, slender, and, as far as known, of equal breadth the entire length, not quite twice as wide as the lateral branches. Lateral branches, long, rather rigid, sometimes alternating, and, in other examples, nearly opposite, distant from each other slightly more than the breadth of the main stem, with which they range at an angle of about sixty degrees; like the main stem, showing no taper as far as known; sometimes themselves giving off (at least on the lower side) a few distant lateral subdivisions, at some distance out from the main stem. Non-poriferous side, with main stem and branches minutely and very regularly striated longitudinally. Poriferous side, with a mesial ridge along the main stem, consisting of three slightly-raised lines, the middle one of which is larger than the others; on each side of this ridge is a row of pores, three of which are placed between each two of the lateral branches, and about their own breadth from the margin, to the outline of which they do not impart any irregularity; lateral branches with, on each side of a linear ridge, two rows of alternating pores, very slightly smaller than those of the main stem, and imparting sometimes a slightly undulated outline to the margins.

Entire length unknown; a specimen imperfect at both extremities 1.24 inches in length, sending off on each side six lateral branches in a space of 0.27 inch.

This may not be distinct from Professor McCoy's species, which is certainly a very closely allied form. As the main stem of our fossil, however, shows on the poriferous side, three longitudinal mesial-raised lines or ridges, as seen under a good magnifier, and nothing of the kind is mentioned in the description or shown in the figures of *G. grandis*, I do not feel warranted in identifying the Nebraska species with it. There are, also, some differences in the angle at which the lateral branches are given ff, as well as in the number of pores, that indicate a specific difference

The presence of a secondary branchlet seen on one of our specimens may be another indication of further differences, though as this is only given off farther out from the main stem than the lateral branches are preserved in the specimen figured by Professor McCoy, they may really exist in his species.

Locality and position.—Division C, at Nebraska City.

BRACHIOPODA.

Genus LINGULA, Bruguiere.

LINGULA SCOTICA, *var.* NEBRASCENSIS.

Pl. VIII, Fig. 3, *a*, *b*.

? *Lingula Scotica*, Davidson, 1860, Monogr. Carb. Brach., Scotland, p. 62, Pl. 5, Fig. 36, 37, and 37 *a*.

Shell attaining nearly a medium size, compressed, subovate or ovoid-subtrigonal, a little longer than wide, the greatest breadth being near the anterior margin; front rather broadly rounded, or at maturity somewhat more nearly straight along the middle, rounding rather abruptly into the antero-lateral margins; sides convex in outline and converging (near the beaks) at an angle of about 90°; beaks rather obtusely pointed and a little convex. Surface polished and ornamented by prominent, linear concentric ribs or striæ, separated by flat spaces three or four times as wide as the ribs, in which traces of much finer, obscure irregular concentric striæ can be seen by the aid of a magnifier. Internal cast nearly smooth, excepting three or four radiating linear impressions near the beak, apparently formed by little ridges on the inner side of the shell, but which are sometimes made visible on the outside, apparently in consequence of accidental pressure upon the thin shell.

Length, 0.55 inch; breadth, 0.49 inch.

This shell is so very nearly like *L. Scotica*, Davidson, from the Carboniferous rocks of Scotland, that I cannot, with the present means of comparison, regard it as more than distinguishable as a variety of that species. Its chief differences consist in its greater breadth in proportion to its length, and the more convex outline of its lateral margins, together with the larger size of its concentric lines. The former characters cause the beak to be rather distinctly less attenuated. As I have seen but a single specimen, I have no means of determining whether it is the shorter or longer valve from which the figure and description have been prepared.

Locality and position.—Lower part division C, Nebraska City.

Genus ORBICULOIDEA, d'Orbigny.

ORBICULOIDEA, sp

Pl. IV, Fig. 3.

I only know this species from an impression of the outside of the under valve in the clay matrix, with adhering portions of the shell, showing the inside. The outline of the valve is broad ovate, the posterior end being the narrower. It is rather convex in the region of the apex, and flat, or even slightly concave, between that and the anterior margin. The apex is situated between the middle and the posterior side, but nearer the former. The surface is ornamented by very regular, distinct concentric striæ, some of which bifurcate as they pass around the sides from the posterior toward the rather broadly-rounded front.

The remaining portion of the shell, between the apex and the posterior margin, shows very clearly that there is not a slit passing through the valve, as in true *Discina*, though there is evidently a deep furrow on the outside of the valve passing from the apex a little more than half-way to the posterior margin, at which point there was a small round or oval foramen. On the inside a prominent smoothly-rounded ridge (*d*, of Fig. 3) corresponds to the deep furrow on the outer side. The shell is quite thin and has the usual dark semi-corneous appearance of *Discina*. It is possibly an undescribed species, but not having the means of fully characterizing it, I propose no name for it at present.

Anterio-posterior diameter, 0.64 inch; breadth, 0.55 inch.

Locality and position.—The specimen figured and described was found in sinking a shaft on Hon. J. S. Morton's place, one and three-quarters miles west of Nebraska City landing.

Genus PRODUCTUS, Sowerby.

PRODUCTUS COSTATUS, Sowerby ??sp.

Pl. VI, Fig. 6, *a*, *b*.

? *Producta costata*, J. D. C. Sowerby, 1827, Min. Conch., Pl. 560, Vol. VI, p. 115; Phillips, 1836, Geol. Yorks., 11, p. 213, Pl. VII, Fig. 2.
——— *costellata*, McCoy, 1844, Synopsis Carb. Foss., Ireland, p. 108, Pl. XX, Fig. 15.
——— *costatus*, de Verneuil, 1845, Geol. Russ. et Ural Mts., Vol. II, p. 268, Pl. XV, Fig. 13, *a*, *b*; de Koninck, 1847, Monogr. Prod. Pl. VIII, Fig. 3, and Pl. X, Fig. 3; Davidson, 1860, Monogr. Scott. Carb. Brach., p. 44, Pl. II, Figs. 22–24; also Monogr. British Carb. Brach., p. 152, Pl. XXX, Figs. 2–9; Shumard, 1855, Missouri Geol. Report, p. 216; Hall, 1858, Iowa Geol. Report, Vol. I, Part II, p. 712, Pl. XXVIII, Figs. 3 and 4.
——— *Portlockianus*, Norwood & Pratten, 1854, Journ. Acad. Nat. Sci. Philad., III, p. 15, Pl. 1, Fig. 9, *a*, *b*, *c*.
——— sp., Prof. Henry D. Rogers, 1858, Geological Report, Pennsylvania, Vol. II, p. 833, Fig. 687.

Shell of medium size, wider than long, very convex; hinge margins about equaling the greatest breadth of the valves. Ventral valve exceedingly gibbous, and very strongly incurved, with a deep-rounded sinus extending from near the beak to the front, to which it imparts a sinuate outline; umbo prominent and distinctly incurved, so as to pass somewhat within the hinge margin; ears well-defined, arched, and rather distinct from the abrupt swell of the umbo, from which they are sometimes separated by a small ridge or fold. Dorsal valve flattened in the visceral region, and more or less abruptly curved or geniculated toward the front and anterior lateral margins, the former of which usually shows a small mesial ridge. Surface of both valves ornamented with distinct, rather unequal, depressed, and rounding radiating costæ, which sometimes bifurcate, or, in other instances, two or more of them coalesce in front of the visceral portion, to form a larger one; crossing all of these, on the visceral region, are numerous, well-defined, regular, concentric wrinkles, producing a distinct reticulated appearance, while the whole surface of the ventral valve is sometimes provided with a few scattering, rather stout, erect spines, somewhat regularly arranged in quincunx. Sometimes nearly all the spines, excepting those on the lateral regions, apparently wanting.

Length of a medium-sized adult specimen, 1.06 inches; breadth, about 1.40 inches; convexity, about 0.70 inch.

This is the shell that has been in this country very generally identified with *P. costatus* of Sowerby, though its identity with the typical *P. costatus* may be questioned. At least I have never seen among thousands of specimens from numerous localities a single example of it presenting the large costæ and the extravagant forms sometimes assumed by that shell at foreign localities, such, for instance, as that figured on Pl. XV,

Fig. 13, of the Geology of Russia. The form under consideration is quite constant in its adult size and general characters. Its ribs vary more or less, being often irregular in size, some being rather distinctly larger than the others on the anterior slope, though this character is not well seen in the specimen figured, because the anterior edge of the valves is partly broken away, and the figures are not drawn in postures to show the anterior slope.

Some years since I sent specimens of this shell to Mr. Davidson, the well-known English authority on the *Brachiopoda*, and he wrote back that he did not think it could be *P. costatus*, but that he thought it more probably a small variety of *P. semireticulatus*. This may be true, and yet it is curious that we find hundreds of specimens of this shell, often directly associated with well-defined examples of *P. semireticulatus*, and still always readily distinguished at a glance by their smaller size and rather more unequal costæ on the anterior slope. It cannot be that these are the *young* of *P. semireticulatus*, because they give every evidence, in their gibbous form and produced front, of being adult shells.

Locality and position.—The specimens figured are from the division B of the Nebraska City section. We also found this form in what has been regarded as the same position (but as we believe at a higher horizon), two and one-half miles west of Nebraska City; at Wyoming, and again at the horizon of bed B at Bennett's Mill, and still lower at Rock Bluff, Cedar Bluff, Plattsmouth, Bellevue, Omaha, and, in short, throughout the Coal-Measures of all this region. It likewise ranges through the whole of the Coal-Measures of Illinois, Missouri, Iowa, &c., though I am not quite sure that it occurs in any of our Lower Carboniferous rocks. Professor Rogers has also figured apparently the same fossil from the Coal-Measures of Pennsylvania.

PRODUCTUS SEMIRETICULATUS, Martin, sp.

Pl. V, Fig 7 *a*, *b*.

Anomites semireticulatus, Martin, 1809, Petref. Derb., p. 7, Pl. XXXII, Figs. 1, 2, and Pl. XXXIII, Fig. 4.
——— *productus*, Martin, 1809, ib., p. 9, Pl. XXII, Figs. 1–3.
Productus scoticus, Sowerby, 1814, Min. Conch., Pl. LXIX, Fig. 3, and ib., Pl. 317, Figs. 2–4.
——— *antiquatus*, Sowerby, 1814, ib., p. 15, Figs. 1–5; Phillips, 1836, Geol. Yorks., p. 213, Pl. VII, Fig. 1.
———*semireticulatus*, de Koninck, Mon. Gen. Prod., Pl. VIII, Fig. 1, and Pl. IX, Fig. 1, and Pl. X, Fig. 1; also of Davidson, Salter, and others.
Producta Martini, Phillips, 1836, Geol. Yorks., Vol. II, p., 213, Pl. VII, Fig. 3; de Koninck, 1843, An. Foss. Carb. Belg., p. 160, Pl. VII, Fig. 2.
——— *pugilis*, Phillips, 1836, Geol. Yorks., Vol. II, p. 215, Pl. VIII, Fig. 6.
Leptæna antiquata, Fischer, 1837, Oryct. du Gouv. de Mosc., Pl. XXVI, Figs. 4-5.
——— *tubifera*, Fischer, 1837, ib., XXVI, Fig. 1; (not Deshayes).
Productus Inca, d'Orbigny, 1843, Paleont. Voyage dans Am. Merid., Vol. III, p. 51, Pl. IV, Figs. 1–3.
Producta flexistria, McCoy, 1844, Carb. Foss., Ireland, p. 109, Pl. XVII, Fig. 1.
Productus Calhounianus, Swallow, 1858, Trans. St. Louis Acad. Sci., 1, p. 181.*

This beautiful, and widely distributed species is too well known to require a detailed description or any comparisons. So far as I have been able to see, after examining numerous specimens from various places in our western Carboniferous rocks, at different horizons, I am not able to see any characters by which these American shells can be separated from the well-known European *P. semireticulatus*.

*The foregoing synonymy is adopted mainly from Mr. Davidson, Monogr. Brit. Carb. Brach.

I am satisfied that the form described by Professor Swallow, under the name *P. Calhounianus*, cannot be separated from *P. semireticulatus*; at least I know of no well-defined characters by which it can be distinguished, although I at one time rather inclined to think it might be distinct.

Locality and position.—Professor Geinitz mentions this species among the other fossils from division C of the Nebraska City section, and I remember finding an imperfect specimen of it in that horizon at that locality; but it was crushed to pieces in transporting the collections to Washington. It also occurs at apparently a higher horizon, 2¾ miles west of Nebraska City. The specimen figured on Pl. V is from division B of the Nebraska City section, where we found it quite abundant, as well as at Bennett's Mill, Wyoming, &c. This large variety also occurs at numerous lower horizons in the Upper Coal-Measures of Nebraska; such, for instance, as at Cedar Bluff on Weeping Water, and Rock Bluff, Bellevue, Omaha, and Plattsmouth; and again at higher horizons at Brownville, Peru, Aspinwall, &c., on the Missouri, as well as at various localities in the interior. It is also common in the Upper and Lower Coal-Measures of Iowa, Illinois, Kansas, Missouri, and south to New Mexico; and it is widely distributed in the Carboniferous rocks of Europe, South America, India, &c. It is likewise common at various horizons in our western Lower Carboniferous rocks.

PRODUCTUS LONGISPINUS, Sowerby ?

Pl. VI, Fig. 7, and Pl. VIII, Fig. 6, *a*, *b*, *c*.

Productus longispinus, Sowerby, 1814, Mineral Conch., Vol. I, p. 154, Pl. LXVIII, Fig. 1; de Koninck, 1843, An. Foss. Ter. Carb. Belg., p. 187, Pl. XII, Fig. 11, *a*, *b*, and Pl. XII *bis*, Fig. 2; Davidson, 1853, Introd. Brit. Foss. Brach., Pl. IX, Fig. 221; 1860, Monogr. Carb. Brach., Scotland, p. 39, Pl. II, Figs. 10 to 19.

——— *Flemingii*, Sowerby, 1814, Min. Conch., Vol. I, p. 154, Pl. LXVIII, Fig. 2; de Koninck, 1847, Monogr. Prod. and Chonetes, Pl. X, Fig. 2; Marcou, 1858, Geol. N. Am., p. 47, Pl. VI, Fig. 7.

——— *spinosus*, Sowerby, 1814, Min. Conch., Vol. I., p. 155, Pl. LXIX, Fig. 2.

——— *lobatus*, Sowerby, 1814, ib., Vol. IV, p. 16, Pl. 314, Figs. 2–6; Von Buch, 1841, Verhandl. der Königl. Akad. der Wissensch. zu Berlin, Theil, 1, p. 32, Pl. II, Fig. 17; de Vern., 1845, Geol. Russ., Vol. II, p. 266, Pl. XVI, Fig. 3, Pl. XVIII, Fig. 8.

——— *setosa*, Phillips, 1836, Geol. Yorks., Vol. II, p. 214, Pl. VIII, Fig. 9.

——— *capacii*, d'Orbigny, 1843, Voyage dans l'Amerique, Mer., Vol. III, p. 50, Pl. III, Figs. 24–26.

——— *tubarius*, de Keyserling, 1846, Petschora Land, p. 208, Pl. IV, Fig. 6.

——— *Wabashensis*, Norwood & Pratten, 1854, Jour. Acad. Nat. Sci., Philad., Vol. III, new. ser., p. 13, Pl. I, Fig. 6.

——— *splendens*, N. & P., 1854, ib., p. 11, Pl. 1, Fig. 5.

? ——— *muricatus*, N. & P. 1854, ib., p. 14, Fig. 8, *a*, *b*, *c*, *d*, *e*; (not of Phillips and others).

——— *Orbignyanus*, Geinitz, 1866, Carb. und Dyas in Neb., p. 56, Tab. IV, Figs. 8, 9, 10, 11; (not de Koninck, 1848).

——— *horridus* Geinitz, 1866, ib., Fig. 7; (not Sowerby, 1822).

Shell small, thin, wider than long; hinge line generally longer than the transverse diameter of the valves at any point farther forward, and terminating in more or less distinct, rather vaulted, and often slightly reflexed ears; anterior and anterior-lateral outlines approaching a semicircular curve, but the middle of the front is generally rather distinctly sinuous. Ventral valve gibbous, the greatest convexity being usually behind the middle, and the curve to the beak more rapid than to the front, provided with rather deep mesial sinus; posterior-lateral slopes descending nearly vertically to the ears; umbonal region moderately prominent, and usually projecting rather distinctly beyond the hinge, as seen in looking down upon the shell when lying with the dorsal valve beneath; beak small, strongly curved, but scarcely passing beyond the cardinal margin; surface ornamented with generally rather obscure, somewhat variable radiating costæ, which are often obsolete in the umbonal region, or in some examples over much of the valve farther forward, in other specimens quite distinct to the beak, sometimes bifurcating, and in other instances coalescing to form larger, faintly-defined ribs in front; fine indistinct marks of growth are also sometimes seen, and occasionally very obscure traces of small, concentric wrinkles may be observed near the beak; spines stout, erect, long, scattering, and arranged in quincunx. Ventral valve distinctly concave, or following nearly the curve of the other, and provided with a small mesial ridge corresponding with the sinus of the latter; surface marked as in the other valve, but apparently always without spines.

Length of a medium-sized adult specimen, 0.41 inch; breadth, 0.61 inch; convexity, 0.30 inch.

I follow Mr. Davidson in referring this little shell to *P. longispinus* of Sowerby, a conclusion arrived at by him after a comparison of authentic specimens sent from Illinois by Mr. Worthen, from some of the original localities of *P. splendens* and *P. Wabashensis*, N. & P. At first I was in some doubt whether or not our Nebraska shell is identical with the *P. Wabashensis*, on account of the costæ being represented so fine, distinct, and regular, and with so few spine bases, on Norwood & Pratten's figures, but on comparison with good specimens of that species from their original locality, I am entirely convinced that our shell is in all respects identical, the figures of *P. Wabashensis* alluded to being quite defective in representing the costæ, as stated above.

In regard to the identity of this shell with *P. Orbignyi*, I am compelled to differ from Professor Geinitz. I am also satisfied, as elsewhere stated, that the little shell figured by Professor Geinitz under the name *P. horridus* on his Pl. IV (Carb. und Dyas. in Nebraska), and copied on our Plate VIII, Fig. 6, *b*, *c*, is nothing but a young individual of the species under consideration. As already stated, this shell varies much in the distinctness of its costæ, which are usually rather obscure. It is but necessary to examine a few good specimens to see by their smooth, non-costate umbonal region, that they often attain a size even greater than that he has referred to *P. horridus*, without showing the slightest traces of radiating costæ. Indeed, some individuals of mature size show but faint indications of ribs even near the front margin, while the various individuals present every intermediate gradation in this character between these and the most distinctly ribbed specimens. In addition to this, the extreme improbability of there being in these rocks a large, conspicuous species like *P. horridus*, when no traces of such a shell have ever been seen among all the vast collections that have been obtained from them throughout the great area in which they occur in the West, would alone be a sufficient reason for rejecting the conclusion that such a mere mite as this is the young of that species. But the necessity for such an improbable conclusion is entirely removed by the fact that this specimen was found associated with a very common and abundant species, the young of which evidently agrees exactly with it.

Locality and position.—Nebraska City, division C (Geinitz); also in B at that place, Wyoming, Bennett's Mill, and lower at Rock Bluff, Cedar Bluff, Plattsmouth, Bellevue, and Omaha—in short, this species is found almost everywhere and at nearly all horizons in the Upper Coal-Measures of Nebraska, and in the upper, middle, and lower divisions of Iowa, Illinois, Kansas, Missouri, &c. We have no good specimens of it from division C at Nebraska City, and consequently reproduce Professor Geinitz's figures of individuals from that bed. We have good examples of it, however, from division B at Nebraska City, and the other localities mentioned above.

PRODUCTUS PRATTENIANUS, Norwood.

Pl. II, Fig. 5, *a*, *b*, *c*; Pl. V, Fig. 13; and Pl. VIII, Fig. 10, *a*, *b*.

Productus semireticulatus, Hall, 1852, Stansbury's Salt Lake Report, p. 411, Pl. III, Figs. 4 and 5; (not Martin, sp., 1809).
——— *Prattenianus*, Norwood, 1854, Jour. Acad. Nat. Sci. Philad., Vol. III, new ser., p. 17, Pl. I, Fig. 10. *a*, *b*, *c*, *d*.
? ——— *æquicostatus*, Shumard, 1855, Missouri Geol. Report, p. 201, Pl. C, Fig. 10.
——— *cora*, Marcou, 1855, Geol. N. Am. Pl. VI, Figs. 4 and 4 *a*; (not d'Orbigny).
——— *Flemingii*, Geinitz, 1866, Carb. und Dyas in Neb., p. 52, Tab. IV, Figs. 1, 2, 3, 4; (not Sowerby, 1814).
——— *Calhounianus*, Geinitz, 1866, *ib.*, p. 81; (not Swallow, 1858).
——— *Koninckianus*, ? Geinitz, 1866, *ib.*, p. 53, Tab. IV, Fig. 5, *a*, *b*; (not de Vern. 1845).

Shell attaining a medium size; breadth generally exceeding the length, especially when the ears are entire; cardinal margin usually somewhat longer than the transverse diameter of the valves at any point farther forward; anterior and antero-lateral outline regularly rounded. Ventral valve distinctly and rather evenly convex, and without any traces of a mesial sinus; umbonal region gibbous; beak incurved, but scarcely passing the hinge margin; ears large, rather compressed, and provided with a few large, strongly defined, concentric folds, which ascend a little upon the sloping sides of the umbo, and extend more or less along the posterior lateral margins, but never cross the beak, central region, nor front; surface ornamented with rather small, regular, rounded costæ or striæ, and armed with stout, erect, long spines, usually arranged over the whole valve in quincunx, while one or two rows along the hinge margin are more crowded, larger, and in part directed backwards, with an inward curve.* Dorsal valve concave, sometimes a little flattened in the visceral region, and following the curve of the other valve around the front and anterior lateral margins; ears with folds as in the other valve, and each separated from the concave central region by an oblique ridge or prominence; surface without spines, but with radiating striæ as in the ventral valve, and usually crossed by very obscure concentric wrinkles and a few embricating concentric marks of growth, particularly near the front and sides; cardinal process small, but slightly prominent, and bifid, while from its base a slender linear mesial ridge extends forward to, or a little beyond, the middle. Muscular and reniform impressions very obscure or obsolete in the specimens examined.

Length of a well-developed, medium-sized specimen, 1.40 inches; breadth, 1.47 inches; convexity, about 0.85 inch.

This species has been often referred to *P. cora*, d'Orbigny, from which, however, it differs (especially from the form referred to that species by European authors) in its greater convexity, more extended hinge, and much longer and stouter spines. It likewise differs from d'Orbigny's original figure of the South American type of that species, given in his Paleont. Voyage, dans l'Amerique Meridionale, Pl. 5, in several important characters. The fact, however, that M. de Koninck and others refer a European form to d'Orbigny's species, also differing widely from the figures of the South American shell alluded to, after having seen d'Orbigny's typical specimens, shows that he must have figured it incorrectly, and that we should possibly look more to the figures given by Mr. Davidson and M. de Koninck of the European form compared with

* In a few specimens they seem to be wanting over the greater part of the more convex region of the valve.

d'Orbigny's typical specimens, than to his own figures, in forming an idea of the true characters of *P. cora.*

Professor Geinitz has referred the species under consideration in part to *P. Flemingii*, of Sowerby, which Mr. Davidson says, after seeing Sowerby's type of that supposed species, was founded upon a bad specimen of the common *P. longispinus.* Some of the young individuals of *P. Prattenianus* look more or less like *P. longispinus*, but it is only necessary to have a good series of the Nebraska shell, of various sizes, for comparison, to be convinced that it is entirely distinct from *P. longispinus.* I am also compelled to differ from Professor Geinitz, in regard to the shell represented by Fig. 5, Pl. IV, of his Carb. und Dyas, being the *P. Koninckianus*, de Verneuil, a much smaller species, with a more prominent umbo, and a distinctly shorter hinge. To me, the shell figured under this name by Professor Geinitz seems to agree exactly with the usual adult characters of *P. Prattenianus*, but in order that others may have the means of forming their own conclusions on this point, I have given a copy of Professor Geinitz's figure (the ears of which I have restored in outline), and for comparison a copy of Mr. Davidson's figure of *P. Koninckianus*, from his Monogr. Brit. Carb. Brach., Pl. LIII, Fig. 7, natural size. Mr. Davidson's figure was drawn from a British specimen; but in the prominence of its beak and the shortness of its hinge, M. de Koninck's figure of a Belgian specimen of *P. Koninckianus* (under the name *P. cancrini*, but now generally regarded as the same as *P. Koninckianus*) differs in a more marked degree than Mr. Davidson's. The only figures I have seen of a Russian specimen of *P. Koninckianus* are those given by Count Keyserling, in his Petschora Land, Tab. IV, Fig. 4, *a, b, c.* These also show the umbo to be extremely prominent, and look in all respects quite unlike the Nebraska shell. It is also worthy of note that all of these authors both figure and describe the *P. Koninckianus* as a neat, pretty little species of the size of Mr. Davidson's figure, copied for comparison on our Plate V, Fig. 15, *a, b, c.*

Having only crushed and imperfect specimens of this shell from division C of the Nebraska City section, I have given on Pl. VIII, among the other fossils from that horizon, copies of two of the figures of this shell, published by Professor Geinitz in his Carb. und Dyas, Pl. IV.

Locality and position.—Divisions C and B of the Nebraska City section at that place, and in division B at Bennett's Mill; also at lower horizons at Cedar Bluff, Plattsmouth, Bellevue, Omaha, and, in short, at numerous other localities in the Upper Coal-Measures of Nebraska, Kansas, and Iowa, and in both Upper and Lower Coal-Measures of Illinois.

PRODUCTUS PERTENUIS, Meek.

Pl. I, Fig. 14, *a, b, c*, and Pl. VIII, 9 *a, b, c, d.*

Productus cancrini, Geinitz, 1866, Carb. und Dyas in Neb., p. 54, Tab. IV, Fig. 6, *a, b, c, d;* (not Murch. de Vern. and Keys., Geol, Russ., Vol. II, part Paleont., Pl. XVI, Fig. 8, *a, b, c*, and Pl. XVIII, Fig. 7).

Shell small, very thin, truncato-subhemispherical; sides and front regularly rounded; hinge line usually rather less than the greatest breadth of the valves. Ventral valve without any traces of a mesial sinus, moderately gibbous, the greatest convexity being slightly behind the middle, from which point it rounds off in all directions, but most abruptly to the beak and ears, which latter are flattened and subrectangular; beak small, slightly prominent, and but little incurved beyond the hinge line; surface with fine, regular, radiating striæ, crossed by small, rather distinct and regular, concentric wrinkles, which latter are generally most strongly defined on the ears; over the whole there are also regularly arranged in quincunx, very slender spines, 0.20 to 0.30 inch in length, rising from slight prominences or swellings of the radiating striæ. Dorsal valve distinctly concave, or following nearly the curvature of the other valve, its greatest concavity being in the central region, while its ears are nearly flat; surface with concentric wrinkles and radiating striæ as in the other valve, but apparently without spines, though a series of rather distinct pits are arranged over it in the same order as the spines of the other valve.

Length of one of the largest specimens seen, 0.46 inch; breadth, 0.57 inch; convexity, about 0.24 inch.

I have known this little shell since 1858, but had never been able to identify it with any of the known species, nor yet to feel sure that it was new, and consequently left it without a name in the Kansas collections, fearing it might be the young of some of the larger analogous forms. The comparisons, however, that I have recently had an opportunity to make, of a series of specimens from different localities, have satisfied me that it must be an adult shell, and as it cannot be properly identified with any of the described species, so far as yet known, I have ventured to regard it as new.

Professor Geinitz has referred it to *P. cancrini* of M. V. & K., which it certainly resembles quite closely in some of its characters. But its dorsal or smaller valve (ventral of some authors) is certainly *distinctly concave*, particularly in the visceral region, while this valve in *P. cancrini* is described as having its "disk entirely flat." It is an exceedingly thin shell, and the concavity of the smaller valve follows so closely the curve of the other, that the space occupied by the soft parts of the animal was very contracted.

As the only specimens in the collection under investigation from division C, of Nebraska City, are in a crushed condition, I have reproduced the figures 9, *a*, *b*, Pl. VIII, given by Professor Geinitz. Fig. 14, *a*, *b*, of Pl. I, however, are from a specimen obtained at another locality.

Locality and position.—Division C, of Nebraska City section, and from a shaft at that place, possibly below that horizon; also at Brownville, Nebraska, and a lower position at Atchison, Kansas. We likewise found it in 1858 in the Upper Coal-Measures on Grasshopper Creek, twelve miles west of Leavenworth, Kansas.

PRODUCTUS NEBRASCENSIS, Owen.

Pl. II, Fig. 2; Pl. IV, Fig. 6; and Pl. V, 11 *a*, *b*, *c*.

Productus Nebrascensis, Owen, 1852, Geol. Report, Wisconsin, Iowa, and Minnesota, p. 584, Pl. V, Fig. 3*; 1867, McChesney, Trans. Chicago Acad. Sci. 1, p. 24, Pl. I, Fig. 7.

——— *Rogersii*, Norwood & Pratten, 1854, Jour. Acad. N. S. Philad., Vol. III, new series, p. 9, Pl. I, Fig. 3, *a*, *b*, *c*; Hall, 1856, Vol. III, Pacific R. R. Report, p. 104, Pl. II, Figs. 14, 15.

——— *asper*, McChesney, 1860, Descr. New Palæozoic Fossils, p. 34; also illustrations same, 1865, Pl. I, Fig. 7 *a*, *b*.

Strophalosia horrescens, Geinitz, 1866, Carb. und Dyas in Neb., p. 81; (not Murch. de Vern. and Keyserling, 1845).

Shell of about medium size, approaching subhemispherical; length most usually a little less than the breadth; hinge line nearly or quite equaling the greatest transverse diameter; anterior outline nearly straight, or a little sinuous near the middle, rounding into the lateral

---

* This figure is *very* imperfect and gives no idea of the surface characters of the species.

margins, which are generally rather straight posteriorly, and ranging at an angle of from ninety to about one hundred degrees with the hinge; ears nearly rectangular, or a little rounded in outline, at their immediate extremities.

Ventral valve rather convex, most gibbous behind the middle, thence rounding regularly to the front and more abruptly to the beak, generally with a moderately distinct mesial sinus; posterior lateral slopes descending almost vertically to the ears; umbonal region gibbous, and with the strongly incurved beak projecting beyond the hinge line.

Dorsal valve somewhat flattened in the visceral region, but most concave near the beak and toward the anterior lateral regions, the concavity widening rapidly forward, so as to leave a kind of broad, obscure, oblique ridge between it and the flattened ears, and another in the middle; anterior and lateral margins following the curvature of the other valve; cardinal process prominent, bifid, and rather narrow; interior with mesial ridge, narrow, well defined, extending forward beyond the middle, and a little bifid at its connection with the base of the cardinal process; muscular and reniform impressions obscurely marked; interior surface with numerous pustular projections which are most prominent and pointed on a belt around near the anterior and lateral margins.

Surface of ventral valve with more or less defined, rather broad concentric undulations, and obscure striæ of growth, over the whole of which are arranged two sets of spines, connected at their bases with short interrupted ribs or elongated tubercles. One of these sets consists of small, short, appressed spines, and the other of stout, more erect, long ones. Surface of dorsal valve with small concentric ridges and striæ, with many little pits; spines nearly or quite all small, short and appressed.

Some confusion, in regard to the limits and relations of this species, has arisen from the differences presented by specimens, as broken from a hard limestone matrix, and the perfect shell as found weathered out of clay or soft shale. In the former case the spines, and nearly or quite all of the shell, are usually left in the matrix while the internal cast, when thus denuded, shows the concentric undulations and longitudinal ridges or elongated pustules, more strongly defined, with little indications of the spines. On the contrary, specimens from clay or shale are often found with the shell entire, and preserving the spines in a more or less perfect condition, so as to obscure, to some extent, the little interrupted ribs and concentric undulations.

The specimens first described by Dr. Owen, as I know from a careful examination of his types, kindly loaned to me by his brother, Professor Richard Owen, are from limestone and only show traces of the spines. Partly from this fact, and probably in part from an error of the engraver, Dr. Owen's published figure gives a *very* incorrect idea of the shell, though his description, and comparison with *P. Humboldti*, when taken in connection with the locality from which his specimen was obtained (Bellevue, Nebraska), would satisfy any one, familiar with the fossils of that region, that he must have had before him the form under consideration, even if his original types were lost.

The specimen upon which the species *P. Rogersii* was proposed, which I have also had an opportunity to examine, through the kindness of Professor McChesney, to whom it belongs, is also an internal cast showing the rib-like pustules and concentric undulations very clearly, but without any traces of the spines.

The specimens upon which Professor McChesney proposed the species *P. asper* are well preserved and retain much of the spines. I have not seen these identical specimens, however, but have had an opportunity to examine a large collection, consisting of several hundred individuals, belonging to Professor Powell, of Bloomington, Illinois, from the same locality and position at Lasalle, of that State. From these and numerous others, in all conditions, I have seen from various localities in the West, I have no hesitation in regarding *P. asper* as also synonym of *P. Nebrascensis.*

Professor Geinitz was certainly in error in referring this shell to *Strophalosia horrescens*, since it is positively not a *Strophalosia* at all, but a true *Productus*, as may be seen by the figures on plate V. It never has any traces of the cardinal area of the genus *Strophalosia*, as I know from a careful examination of hundreds of well-preserved specimens, its cardinal margin being linear, as may be seen by Fig. 11, *c*, of Plate V. By comparing this with Fig. 14 of the same plate, representing *Strophalosia horrescens*, from Professor Geinitz's work on the German Permian fossils (Dyas), the external difference between this genus, and *Strophalosia*, will be at once seen by the student, the latter genus having a cardinal area, (marked *a* in the figure.) The presence of an area alone, however, is not *always* a sufficient distinction, since there is, in some very rare instances, an abnormally developed area in true *Productus*. The total absence of cardinal teeth and sockets, however, in the latter genus clearly separates these types. That *P. Nebrascensis* is entirely destitute of any traces of hinge teeth is well known to all in this country who have examined the interior of this shell. Figure 11 *b*, of Plate V, represents the hinge and interior of a dorsal valve of this species, and shows it to be entirely without sockets for the reception of teeth.

P. Nebrascensis is evidently very similar to *P. scabriculus* of the Old World, even to the bifurcation of the internal mesial ridge of the dorsal valve. It differs, however, in having two very distinct sets of spines, the one small and appressed, and the other stout, erect, and long. (See Fig. 11 *d*, Pl. V.)

Locality and position.—Professor Geinitz mentions this shell from divisions C and B of the Nebraska City section. I have only seen fragments of it from division C at this locality, though we have it from division B at that place, Wyoming, Bennett's Mill, &c.; and from lower positions at Rock Bluff, Plattsmouth, Bellevue, Omaha, and numerous other localities in Nebraska as well as Iowa, Illinois, Missouri, Kansas, &c. In short it is a widely distributed Coal-Measure species, from Nebraska to New Mexico, and from the Rocky Mountains eastward to West Virginia.

PRODUCTUS SYMMETRICUS, McChesney.

Pl. V, Fig. 6, *a*, *b*; and Pl. VIII, Fig. 13.

Productus symmetricus, McChesney, 1860, Descriptions New Palæozoic Fossils, p. 35; 1865, illustrations of same, Pl. I, Fig. 9, *a*, *b*; 1866, Trans. Chicago Acad. Sci., 1, p. 25, Pl. I, Fig. 9.

Shell of medium size, suborbicular, or a little wider than long; hinge line somewhat less than the greatest breadth; sides rounding regularly to the front, which is rather broadly rounded in outline; ventral valve somewhat compressed, or only moderately convex, without any traces of a mesial sinus; ears compressed but not abruptly separated from the swell of the umbo, obtusely angular or a little rounded at the extremities; beak moderately large, incurved, but not curving much within the hinge margin. Dorsal valve rather evenly, and only moderately concave, cardinal process slender, prominent, curved, trifid, the middle division more prominent than the others, and emarginate at its extremity, the emargination being caused by a distinct mesial furrow that extends the entire length of the process; muscular scars somewhat convex, rather distinct in well-preserved specimens, and divided by a slender, simple mesial ridge, extending from the base of the cardinal process about three-fourths of the way to the front; whole interior, excepting the region of the muscular scars, roughened by little pustular projections which become more prominent near the anterior and anterior-lateral margins. Surface of both valves ornamented by small, rather obscure, and more or less regular concentric wrinkles, and covered by numerous small, short, rather appressed spines, which are larger on the ventral valve, where they are often connected with little, somewhat elongated tubercles.

Length of a well-developed specimen, 1.37 inches; breadth, 1.45 inches; convexity, 0.70 inch.

This shell will be readily distinguished from the last by its less convex ventral valve, without any mesial sinus, by its smaller concentric wrinkles, and particularly by having its spines consisting of a single series of small, rather depressed ones, instead of a large, stout, erect series, and another small, depressed series. Its cardinal process also differs in being trifid, and its internal ridge of the dorsal valve in being simple instead of bifid near the base of the cardinal process. The most striking differences, however, observable on comparing good specimens, consist in the differences of form mentioned, and the two sets of spines in the *Nebrascensis*.

Among European species this shell is perhaps most nearly like *P. scabriculus*, which it rather nearly resembles in most of its characters. It differs, however, in having no traces of a mesial sinus, in its finer and more distinct concentric wrinkles, and particularly in its distinctly trifid cardinal process, and the simple mesial internal ridge of its dorsal valve; that of *scabriculus* being like Owen's *Nebrascensis*, bifurcated near the base of the cardinal process.

It was probably imperfect specimens of this and the last-described species that Professor Geinitz and others have referred to *P. pustulosus* and *P. scabriculus*, from the outcrops at Bellevue and Plattsmouth; at least these two forms are more like those two foreign species, than any others known to me from the rocks of that region. They are both clearly distinct species from *P. pustulosus* and *P. scabriculus*, however.

Locality and position.—Professor Geinitz did not find this species in Professor Marcou's collections from division C, of the Nebraska City section. We found several fragments of it, however, in that bed at Nebraska, City, as well as at an apparently higher horizon 2½ miles west of there; and the distorted, but nearly entire individual, figured on Pl. VIII, Fig. 13, came from division C, at Nebraska City Landing. We also have imperfect specimens of it from division B, at that place and Bennett's Mill; also, from Peru, and from lower positions, three or four miles up Platte River, and from Bellevue, and numerous other localities in the Upper Coal-Measures of Nebraska and Kansas. It likewise occurs at the same horizon in Iowa, Missouri and Illinois, as well as in the Lower Coal-Measures of the last-mentioned State.

PRODUCTUS PUNCTATUS, Martin, sp.

Pl. II, Fig. 6, and Pl. IV, Fig. 5.

Anomites punctatus, Martin, 1809, Petref. Derb., Pl. XXXVII, Fig. 6; (not 7 and 8).
Trigonia rugosa, Parkinsen, 1811, Organic remains, Vol. III, Pl. XII, Fig. 11.
Anomites thecarius, Schloth., Nachtr. zur Petref., II, p. 63, Pl. XIV, Fig. 1.
Productus punctatus, J. Sowerby, 1822, Min. Conch., p. 22, Tab. 323; Von Buch, 1841, Abhandl. der K. Akad. der Wissensch. zu Berlin, Theil 1, Pl. II, Figs. 10, 11; de Kon., 1843, An. Foss. Carb. Belg., p. 196, Pl. X, Fig. 2; and 1847, Mong. Gen. Productus, Pl. XII, Fig. 2; de Verneuil, 1845, Russia and Ural Mts., Vol. II, p. 276, Pl. XVI, Fig. 11; Davidson, 1860, Scottish Carb. Brach., p. 42, Pl. IV, Fig. 20; and Monogr. Brit. Carb. Brach., p. 172, Pl. XLIV, Figs. 9–16.
——— *semipunctatus*, Shephard, 1838, Am. Jour. Sci., Vol. XXXIV, Fig. 9.
Producta punctata, Phillips, 1836, Geol. Yorks, p. 215, Vol. II, Pl. VIII, Fig. 10.
Leptæna sulcata, Fisher, 1837, Oryct. Mosc., Pl. XXIII, Fig. 2; (not Sowerby).
Productus concentricus, Potiez et Michaud., 1844, Galer. des Moll. du Mus. de Donái, Vol. II, p. 25, Pl. XLI, Fig. 1.
——— *tubulospinus*, McChesney, 1860, Descrip. New Pal. Foss., p. 37; also 1865, illustrations of same, Pl. I, Figs. 10 and 11.

Shell attaining a rather large size, thin, varying from rotundato-subquadrate to longitudinally subovate, being sometimes wider than long, and in other examples longer than wide, with all intermediate forms; hinge margin always shorter than the greatest breadth of the valves; anterior outline regularly rounded, or faintly sinuous in the middle. Ventral valve more or less gibbous, with a moderately distinct mesial sinus extending from near the beak to the front; beak incurved a little beyond the cardinal margin; ears rather compressed, but not distinctly defined from the swell of the umbo. Dorsal valve moderately concave with a small mesial elevation. Surface of both valves ornamented with numerous rather regular concentric ridges, increasing in size from the beaks towards the front, but becoming again smaller and more crowded in adult shells at the margin; in the ventral valve these ridges are a little prominent at the lower margin, separated from each other by smoother spaces, and support numerous small appressed spines, those of the upper row of which are larger and less crowded than the others; on internal casts, or partly exfoliated specimens, the spines are represented by small tubercles; surface of dorsal valve as in the other, excepting that the ridges are represented by little furrows.

Length of a fully developed, rather broad specimen, 2.66 inches; breadth of same, 2.50 inches; convexity about 1.10 inches.

This widely distributed species occurs in our Coal-Measures of the West, at numerous localities, as well as in the Lower Carboniferous rocks. At one time it was thought that the Coal-Measure specimens were specifically distinct, or ought to be so, from those in the lower rocks, but after a careful comparison of some fine specimens in the Illinois collection, showing the interior, cardinal process, &c., I was entirely unable to detect any specific differences from the well-known *P. punctatus*. The specimen figured by Professor McChesney (New Palæozoic Fossils, Pl. I, Fig. 11), has the cardinal process incomplete, and shows the muscular impressions imperfectly, without any traces of the reniform markings; but I have seen these, as stated above, in others in a better state of preservation from the Coal-Measures of Illinois, as they are represented by Mr. Davidson, in his beautiful figures of *P. punctatus*.

Locality and position.—Division B of Nebraska City section, at that place, Bennett's Mill, and Wyoming, and at, possibly, a higher horizon in a shaft near Nebraska City; also at Rock Bluff, Plattsmouth, and Bellevue; and numerous places in the Upper Coal-Measures of Illinois, Iowa, Missouri, Nebraska, Kansas, &c. It likewise occurs in the Lower Coal-Measures, and even in the Lower Carboniferous rocks of Missouri, Iowa, Illinois, &c.

Genus CHONETES, Fischer.

CHONETES VERNEUILIANA, N. & P.

Pl. I, Fig. 10 *a*, *b*.

Chonetes Verneuiliana, Norwood & Pratten, 1854, Jour. Acad. Nat. Sci. Philad., Vol. III, p. 26, Pl. II, Fig. 6, *a*, *b*, *c*; Shumard, 1855, Missouri Geol. Report, p. 216.

Shell rather small, varying from transversely sub-semicircular, to sub-oblong; hinge line more or less extended beyond the breadth of the valves at any other point; sometimes greatly produced. Ventral valve very convex, with a deep rounded mesial sinus, starting near the beak and deepening and widening rapidly to the anterior margin, to which it imparts a distinctly sinuos outline, thus dividing the gibbous part of the valve into two prominent, rounded lobes or diverging ridges, separated from the ear on each side by a broad rounded depression; ears more or less angular, sometimes extended and acutely pointed, slightly arching, and a little reflexed; beak rather prominent, and incurved; area moderately developed and common to both valves, but widest in the ventral; foramen wide; cardinal margin provided with four oblique spines on each side of the beak. Dorsal valve following rather nearly the curve of the other, and provided with a mesial ridge corresponding to the sinus of the other valve. Surface of each valve, ornamented with about 100 fine bifurcating, radiating striæ, and sometimes near the front, by a few marks of growth.

Length of a medium-sized specimen, 0.27 inch; breadth, 0.40 inch; convexity about 0.12 inch.

This species is remarkable for its convex ventral valve, with its deep mesial sinus, which imparts to it a peculiar bilobate appearance. I know of no species with which it is liable to be confounded.

Locality and position.—Plattsmouth, and various other localities in the Upper Coal-Measures of Nebraska, as well as Kansas, Iowa, Missouri, Illinois, &c. The remarkably extended specimen from which Fig. 10 *b*, of Plate I, was drawn, come from near the same horizon as the Plattsmouth outcrop, four miles up the Platte.

CHONETES GRANULIFERA, Owen.

Pl. IV, Fig. 9; Pl. VI, Fig. 10; Pl. VIII, 7.

[Cho]netes granulifera, Owen, 1855, Geol. Rep. Minn., Iowa, and Wisconsin, p. 583, Tab. V, Fig 12.

——— *mucronata*, Meek & Hayden, 1858, Proceed. Acad. Nat. Sci. Philad., p. 262; Paleont. Upp. Mo., p. 22, Pl. I., Fig 5, *a*, *b*, *c*, *d*, *e*; Geinitz, 1866, Carb. und Dyas, p. 58, Tab. IV, Figs. 12, 13, and 14; not *Chonetes mucronata* (=*Strophomena mucronata*, Conr., 1843).

Shell attaining a rather large size, semicircular in outline, having its greatest breadth on the hinge line, which often terminates in extended mucronate ears. Larger or ventral valve moderately convex, the greatest convexity being in the central region, or rather on each side of it, as there is usually a broad, shallow, mesial depression; ears and lateral regions compressed; front somewhat straightened along the middle; beak small, rather compressed, a little arched, and scarcely projecting beyond the cardinal margin, which is provided with from seven to about eleven oblique spines on each side of the beak; area rather narrow, ranging nearly parallel to the general plane of the valves, its fissure broad, partly closed above by an arching deltidium; hinge teeth well developed, compressed, and minutely striated; interior with impressions of cardinal muscles subovate, diverging, attenuate above;

adductor muscular scars small, narrow-subelliptical; mesial ridge prominent near the beak, much lower, and rarely extending forward to the central region; most of the interior occupied by granules, which are largest and most crowded on a narrow space around near the front and lateral margins; but around the immediate margin they are much smaller, and arranged in distinct radiating rows.

Dorsal or smaller valve, following nearly the curve of the other, the beak and central regions being concave, and the ears flat; area well developed, but narrower than in the other valve; bifid cardinal process and mesial prominence, nearly or quite closing the fissure of the other valve. From the base of this process, there are extending, on the inner side of the valve, five radiating ridges, two of which pass obliquely outward, forming the inner margins of the dental sockets, while a third mesial one extends at right angles to the hinge, a little more than half-way to the front; the other two are much shorter, oblique, and occupy intermediate positions between the middle and the two latter ones; granules of the interior as in the other valve.

Surface of both valves ornamented with a few subimbricating marks of growth, crossed by very fine, obscure, regularly and closely arranged radiating striæ, of which about one hundred and fifty may be counted around the free border of a large individual, where eight or nine of them may be counted in the space of one line.

Length of a large specimen, 0.62 inch; breadth on hinge line, 1.13 inches.

I am now, after seeing specimens from various localities along the Missouri River, nearly satisfied that this must be the same shell described by Dr. Owen, under the name *C. granulifera*. His figures and description are unfortunately so imperfect as to leave this question somewhat doubtful. He gives four figures, none of which are near so extended on the hinge, nor so mucronate at the lateral extremities as in well-developed specimens of the form for which we proposed the name *C. mucronata*, while they give the appearance of much coarser surface markings. Judging from the localities at which he obtained his typical specimens, however, and from numerous collections before me from the same and other localities, I can scarcely doubt that he must have founded his species on the same shell to which we afterwards gave the name *C. mucronata*.

Chonetes Smithii, of Norwood & Pratten, is a form with which this shell should be compared, though the form here described attains a much larger size, and is more compressed, as well as more extended on the hinge. The latter character, however, can scarcely be regarded as a specific distinction in this genus, as may be seen by the figures of *C. Verneuiliana*, given on Plate I.

Locality and position.—Division C, of the Nebraska City section; also in Division B, at the same place, and at Wyoming and Bennett's Mill; likewise at lower horizons at Plattsmouth, Bellevue, Omaha, and at numerous other localities in Nebraska, Kansas, Missouri, Iowa, &c., it being a very common shell in the Coal-Measures of that region. Specimens of a somewhat smaller size, but otherwise undistinguishable, also occur in the Coal-Measures of Illinois.

CHONETES GLABRA, Geinitz.

Pl. IV, Fig. 10; Pl. VIII, Fig. 8 *a*, *b*.

Chonetes glabra, Geinitz, 1866, Carb. und Dyas in Neb., p. 60, Pl. IV, Figs. 15–18.

Shell thin; transversely sub-semicircular, the length being more than half the breadth; hinge line a little longer than the greatest breadth of the valves, at any point farther forward; lateral extremities abruptly pointed and sometimes slightly recurved; anterior and anterior-lateral margins, forming a semi-circular curve in outline, excepting that the former is generally faintly sinuous in the middle; lateral margins curving abruptly outward just before intersecting the hinge extremities. Ventral or larger valve moderately convex, the most gibbous part being in the form of two broad, rounded, undefined prominences, which diverge from the beaks to the anterior lateral regions, leaving a broad, rounded, rather deep mesial sinus between them, extending nearly to the beak, but widening and deepening rather rapidly to the front; outside of these prominences, the posterior lateral regions are more or less compressed; beak small, compressed, slightly arched, and scarcely projecting beyond the cardinal margin; area narrow, inclined obliquely backward; its fissure small, nearly semi-circular and partly closed by the cardinal process of the other valve; cardinal margin armed on each side of the beak by four or five slender, moderately long, oblique spines, with sometimes remains of one or two much smaller rudimentary additional ones near the beak; cardinal teeth compressed, their longer diameter ranging nearly parallel to the hinge line—as seen under a strong magnifier, finely striated on the outside, at right angles to their length. Interior, excepting in the region of the muscular impressions, with numerous rather distinct granules, arranged in radiating rows—immediately within the fissure, provided with a short, rather prominent, compressed ridge ranging at right angles to the hinge; muscular impressions very obscure.

Dorsal or concave valve, following nearly the curve of the other; area of about the same size as in the dorsal valve, inclined forward from the hinge; cardinal process small, not very prominent, and as seen on the outer side somewhat trilobate, the middle lobe or ridge being divided by a linear sulcus; socket ridges very oblique; interior granulated as in the other valve. A very small, obscure linear ridge occupies the middle of the valve, without, however, extending up to the hinge; muscular impressions unknown.

Surface of both valves nearly smooth, but showing obscure concentric marks of growth. In some conditions of weathering, there is a faint appearance of radiating markings, but this is due rather to the structure of the shell, and not to proper external lines. When a single one of the thin valves is cleaned and examined by the aid of a good magnifier and a strong transmitted light, large, very scattering punctures or pores are seen arranged in quincunx, and passing obliquely through the shell. These appear to have been connected with minute tubular spines, arranged over the whole surface, during the life of the animal. At any rate, in one specimen, of which Fig. 8 *b*, Plate VIII, is a representation enlarged two diameters, there are numerous little projecting points having the appearance of minute spine bases projecting through a film of shale on the surface. They must, however, be extremely fragile and readily removed, as no traces of them are to be seen on any of the numerous other specimens in the collection.

Locality and position.—Division C of the Nebraska City section, where it is very abundant, and found in a fine state of preservation. It also occurs at that locality, though much less abundantly, in division B. It is likewise found common at a much lower horizon in the Coal-Measures at Atchison, Kansas. It is a remarkable fact that in a collection sent by Mr. Dall, from Russian America, there is a small rolled erratic mass of very hard siliceous rock, picked up on the shores of Youkon River, containing a *Chonetes* apparently agreeing with this in every respect.

Genus ORTHIS, Dalman.

ORTHIS CARBONARIA, Swallow.

Pl. I, Fig. 8, *a*, *b*, *c*.

Orthis carbonaria, Swallow, 1858, Trans. St. Louis Acad. Sci., 1, p. 218.
? ——— *Pecosii*, Marcou, 1858, Geol. N. Am., p. 48, Pl. VI, Fig. 14, *a*, *b*.
? ——— (sp. undt.), Meek, 1864, Paleontology of California, Vol. I, p. 10, Pl. II, Fig. 5, *a*, *b*, *c*.

Shell small, suborbicular, slightly wider than long, moderately convex in adult specimens; lateral margins rounded, or, in some examples, faintly straightened posteriorly; front more broadly rounded, but usually very slightly sinuous in the middle; valves nearly equally convex; hinge line very short, or only equaling about half the breadth of the valves. Ventral valve usually most convex in the umbonal region, sometimes a little flattened anteriorly, so as to give the shell slightly the form usually called "resupinate," though in gibbous specimens this character is nearly obsolete; beak moderately prominent, rather pointed and arched; area small, well defined, and arching with the beak; foramen narrow. Dorsal valve usually most convex between the middle and the beak, which is small, and nearly as prominent and arched as that of the other valve, generally with a shallow sinus extending from the middle to the front; area well developed, but smaller than in the other valve, arched and divided by a proportionally shorter foramen. Surface of both valves ornamented with concentric marks of growth and numerous fine radiating crowded striæ, which increase mainly by intercalation, and, as in many other species of the genus, show occasional perforations toward the front, apparently left by the removal of very small tubular spines.

Length of a well-developed gibbous specimen, rather above a medium size, 0.38 inch; breadth, 0.43 inch; convexity, 0.27 inch. Largest specimen about 0.44 inch in length, and 0.51 inch in breadth.

This little shell resembles so nearly, in form and surface characters, some varieties of *O. resupinata* and *O. Michelini*, that scarcely any one familiar with those shells, if shown a few of these, would hesitate to pronounce them young examples of one or the other, or both of those species. Still, from its uniform much smaller size, even at widely distant localities, and in beds of various lithological characters, where not associated with any similar larger shell, leads me to think it must be distinct, and that our specimens really represent the adult size. In addition to their smaller size, a careful comparison also shows that they certainly do present some appreciable and constant differences. For instance, *O. carbonaria* has its beaks more nearly equal than either *O. resupinata* or *O. Michelini*, and more prominent, particularly that of its dorsal valve, which is also more pointed. It is also generally proportionally more thickened within on each side of the rostral cavity in the ventral valve.

Locality and position.—The specimen figured was found, with others (some of which are less convex), at Rock Bluff, in Upper Coal-Measure beds, referred by Professor Marcou to the Lower Permian, (Dyas). We also have some crushed and distorted specimens of apparently the same shell from division B at Nebraska City, referred by Professors Marcou and Geinitz to the lower part of the Upper Permian. It likewise occurs at various lower horizons in the Upper Coal-Measures in Iowa, Nebraska, Kansas, Illinois, &c.; while Professor Swallow's typical specimens were from the Middle Coal-Measures at Lexington, Missouri.

Genus HEMIPRONITES, Pander.

HEMIPRONITES CRASSUS, M. and H.

Pl. V, Fig. 10, *a*, *b*, *c*; and Pl. VIII, Fig. 1.

Orthisina crassa, Meek and Hayden, 1858, Proceed. Acad. Nat. Sci., Philad., p. 260.
Orthis Lasallensis, McChesney, 1860, New Paleozoic Fossils, p. 32; and 1865, illustrations of same, Pl. I, Fig. 6, *a*, *b*.
——— *Richmondi*, McChesney, *ib.*, Fig. 5, *a*, *b*, *c*.
Hemipronites crassus, Meek and Hayden, 1864, Paleont. Upper Missouri, p. 26, Pl. I, Fig. 7, *a*, *b*, *c*, *d*.
——— *Lasallensis*, and *H. crassus*, McChesney, 1867, Trans. Chicago Acad. Sci., Vol. I, p. 28, Pl. I. Figs. 5 and 6.
Compare *Orthis Keokuk* and *O. robusta*, Hall, 1858, Iowa Geol. Report, Vol. I, Part 2, p. 640, Pl. XIX, Fig. 5, *a*, *b*, and p. 713, Pl. XXVIII, Fig. 5, *a*, *b*, *c*.

Shell varying from semi-circular to truncato-subcircular, or transversely suboblong, generally wider than long, varying from compressed to distinctly convex; hinge margin equaling, or shorter than, the greatest breadth of the valves, rectangular or sometimes more or rather less than rectangular at the extremities; anterior outline forming a more or less regular semi-circular curve. Dorsal valve always convex, sometimes very distinctly so, the greatest convexity being near the middle; beak not distinct from the cardinal margin. Ventral valve varying in convexity at the úmbo, sometimes very prominent, and occasionally distorted there; less convex, flattened, or not unfrequently a little concave around near the front; area varying in height in proportion to the elevation of the beak, and either flat or with the beak a little arched, usually rather distinctly striated; its closed fissure varying in the proportions of height and breadth, with the greater or less elevation of the beak; interior always provided with a prominent mesial septum extending from the beak forward to near the middle of the valve. Surface of both valves marked by numerous strong, raised radiating striæ, of unequal size, there being generally one or several smaller ones between each two of the larger crossing the whole are also numerous fine, regular concentric striæ, more or less defined both between and upon the radiating striæ, to which latter they impart a neatly crenate appearance.

After proposing the name *H. crassus* for this species, I was led to think it most probably only a variety of the well-known widely distributed European, *H. crenistria*, which it certainly very closely resembles in external characters. On sending specimens showing the interior of the valves from the original locality, however, to Mr. Davidson, he wrote back that he could not think it properly belongs to that species, because among all the specimens of that form he had examined he had never seen one showing the peculiar mesial septum always so strongly developed in the interior of the ventral valve of this shell. I have no doubt in regard to its identity with *Orthis Lasallensis* and *O. Richmondi*, of McChesney, and I suspect the same species has been described by Professor Hall in the Iowa Report under the names *Orthis Keokuk* and *O. robusta*, particularly the former, which seems to have the same mesial septum within the ventral valve, and agrees nearly in other characters excepting in its larger size.

Locality and position.—Divisions B and C of the Nebraska City section, and from apparently above the latter horizon $1\frac{3}{4}$ and $2\frac{3}{4}$ miles west of there; and again from bed B at Bennett's Mill and Wyoming; also at lower horizons in the Upper Coal-Measures at Cedar Bluff, Rock Bluff, Plattsmouth, Bellevue, and Omaha. It likewise occurs at Peru, Rulo, and Brownville, as well as at numerous other localities and positions in the

Coal-Measures of Nebraska, Kansas, Iowa, Missouri, and Illinois; and I have recently identified it among specimens collected by Professor J. J. Stevenson from the horizon of the Chester limestone in West Virginia.

Genus MEEKELLA, White and St. John.

MEEKELLA STRIATO-COSTATA, Cox, sp.

Pl. V, Fig. 12, *a*, *b*, *c*.

Plicatula striato-costata, Cox, 1857, Owen's Geol. Report, Ky., Vol. III, p. 568, Pl. VIII, Fig. 7.
Orthis striato-costata, Geinitz, 1866, Carb. und Dyas in Neb., p. 48, Tab. III, Figs. 22–24.
Meekella striato-costata, White and St. John, 1867, Trans. Chicago Acad. Sci., Vol. 1, p. 120 and 121, Figs. 4, 5, and 6.
Compare *Streptorhynchus pectiniformis*, Davidson, 1863, Liege, Brach., Pal. l'Inde, Pl. X, Fig. 17.

Fig. 5.

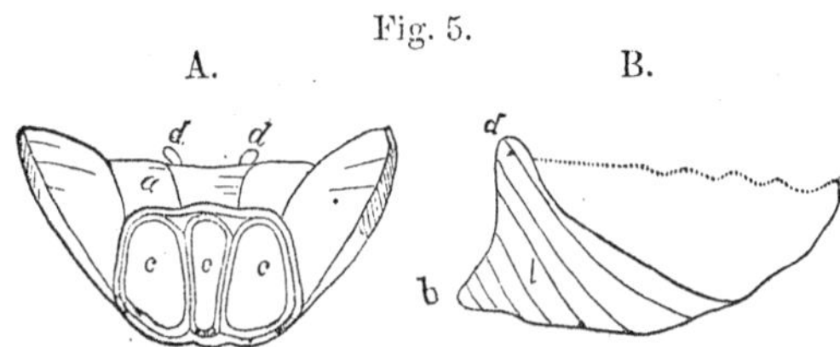

Meekella striato-costata.

A. Showing a transverse section of the beak of the ventral valve, (*a*) being the remaining portion of the cardinal area, (*d d*,) the dental processes, and (*c c c*) the tree-chambers into which the interior is divided by the two intervening septa.
B. Showing a longitudinal section of the same valve, (*l*) being the side and outline of one of the septa or dental laminæ, (*b*) the point of the beak, and (*d*) the dental process. (After Dr. White's and Professor St. John's figures.

Fig. 6.

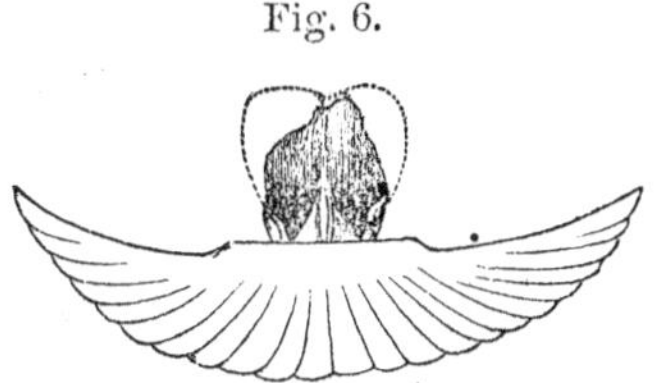

Meekella striato-costata.

A posterior view of dorsal valve, showing above the remarkable cardinal process, with its broken alate part restored in outline; from a specimen borrowed from Dr. White.

Shell trigonal-subglobose, becoming very convex with age, generally a little longer than wide; hinge line very much shorter than the breadth of the valves. Dorsal valve convex, the greatest prominence near the umbo, thence rounding over to the front, being usually somewhat flattened over the central and anterior regions, but without any mesial sinus; beak strongly incurved and with its most prominent part sometimes projecting a little beyond the hinge line, but in others flattened, and with its immediate apex nearly always terminating at the margin of the hinge; posterior lateral margins laterally compressed and converging toward the umbo at nearly a right angle; surface ornamented by about ten to thirteen large radiating, more or less angular, simple or rarely bifurcating plications, which are themselves (as well as the furrows between) marked by fine but distinct radiating striæ, which, toward the front, instead of continuing parallel to the furrows and plications, converge forward on each side of the latter so as to intersect along the crests of the same at acute angles; crossing all of these, there are usually near the front and lateral margins, a few strong zigzag marks of growth. Ventral valve more convex than the other, the greatest convexity being at or near the beak, which is elevated and usually more or less distorted, being sometimes twisted to one side, and in other examples straight or somewhat arched backward; cardinal area narrow transversely, but proportionally high, being often distinctly higher than wide but well-defined, and usually finely striated transversely and vertically, either flat or more or less arched backward; false deltidium closing the fissure, narrow and provided with a slender, rounded, prominent mesial ridge extending to the apex of the beak; surface as in the other valve.

Length of a medium-sized well-developed adult specimen, 1.06 inches; convexity of same, 0.97 inch; breadth, 1.12 inches.

Individuals of this shell vary much in size and form, and especially in the elevation of the beak and area of the ventral valve. As in the allied genus *Hemipronites* (=*Streptorhynchus*), the beak of its ventral valve is often considerably distorted, and occasionally with its immediate apex truncated as if from adhesion by the substance of the young shell. In most of its external characters it is *very* closely allied to an East Indian form described by Mr. Davidson under the name *Streptorhynchus pectiniformis.* (See Fig. 16 of our Pl. V.) Professor Geinitz and Mr. Davidson (the latter after a direct comparison of specimens) think there are no well-grounded specific differences by which these shells can be distinguished; and I confess, after comparing a specimen of the Indian form sent by Mr. Davidson to Dr. White, and along with it a fine series of the western shell from Iowa, kindly loaned to me for comparison by the latter gentleman, that one would scarcely suspect any specific difference to exist between these shells if found together at the same locality and associated in the same beds. Without undertaking to decide this question with but a single specimen of the Indian form for comparison, I would merely remark that both on Mr. Davidson's figure and on the specimen before me of the latter, the radiating striæ are perceptibly coarser than on our shell, and not converging to the crests of the plications, as may be seen by comparing the figures of the two on Plate V. In addition to this, a transverse section across the beak of the ventral valve of the Indian specimen quite unexpectedly reveals no traces of the strongly developed dental laminæ of the American shell. These, however, may have been broken out by some accident; otherwise there would seem to be even a generic difference.

Dr. White and Professor St. John have, as I think, very properly separated the form under consideration generically from *Hemipronites* (=*Streptorhynchus*), on account of its dental supports in the ventral valve instead of merely converging to the apex of the beak under the area, having the form of two strongly-developed laminæ attached to the bottom of the valve and produced forward with slight divergence nearly to the middle of the same. The cardinal process of the other valve is also much more prominent than that of *Hemipronites*. and very differently formed, as may be seen by the annexed cut. These characters, and the differences in the general physiognomy of the two types, seem to me to distinguish them generically. There are also some large scattering punctures about the beak of the ventral valve of the form under consideration that I have never seen in *Hemipronites*. It, therefore, appears to bear the same relations to the latter that *Syntrilasma* does to *Orthis*.

Locality and position.—This shell was first described by Professor Cox from the Coal-Measures of Kentucky. It also occurs at various horizons in the Middle and Upper Coal-Measures of Iowa, Missouri, and Illinois. The specimen we have figured is from division B, at Nebraska City, and we found it at lower horizons at Bellevue and Plattsmouth, and at higher positions at Otoe City, Aspinwall, &c., in Nebraska. I have never seen it from any locality east of Illinois, nor anywhere in the Lower Coal-Measures.

Genus SYNTRILASMA, M. & W.

SYNTRILASMA HEMIPLICATA, Hall, sp.

Pl. VI, Fig. 1, *a*, *b*; and Pl. VIII, Fig. 12 *a*, *b*.

Spirifer hemiplicatus, Hall, 1852, Stansbury's Salt Lake Report, p. 409, Pl. IV, Fig. 3, *a*, *b*,
Syntrilasma hemiplicata, Meek & Worthen, 1866, Report Geol. Survey, Ill., Vol. I, p. 323. Fig. 36, and p. 324, Fig. 37.
Rhynchonella angulata, Geinitz, 1866, Carb. und Dyas, p. 37, Tab. III, Figs. 1–4; (not *R.* [*Anomia*] *angulata*, Linnæus, 1767).

Fig. 7.

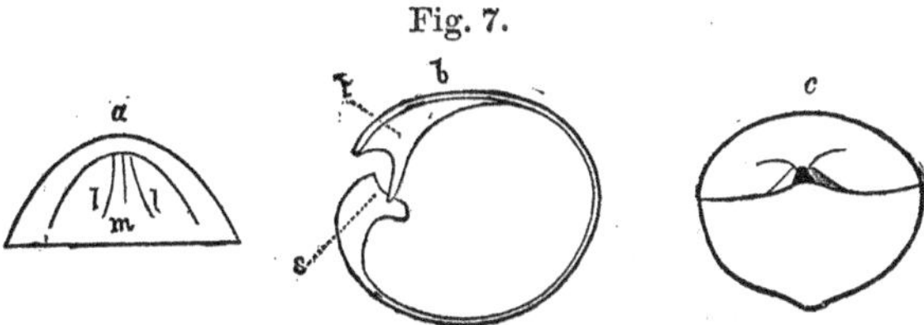

Syntrilasma hemiplicata.

a. A transverse section of a ventral valve, showing the three laminæ (*l*, *m*, *l*) of the interior.
b. A longitudinal section of the two valves united, showing the side of one of the dental laminæ (*l*) and the side of one of the socket-plates, (*s*).
c. A cardinal view, showing the small area and the triangular foramen.

Fig. 8.

a *b*

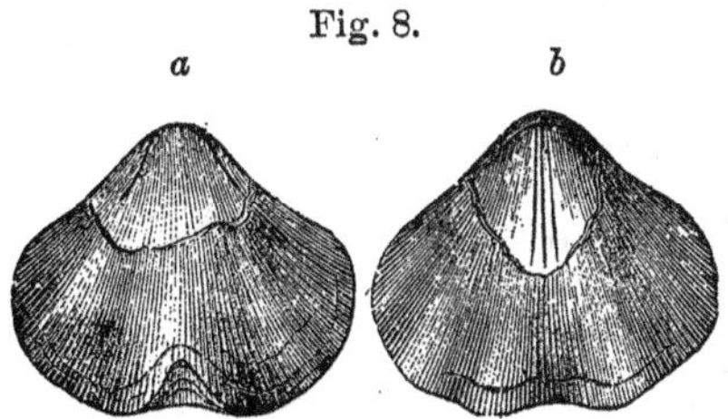

Syntrilasma hemiplicata.

a. Shows a dorsal view with a portion of the shell removed so as to expose in the cast the diverging socket plates.
b. A similar view of the ventral valve, showing the three closely approximated parallel plates within.

Shell in young examples only moderately convex, and having all the external appearances of a true *Orthis*; in adult specimens, often globose, or even more convex than long or wide. Hinge line very short, or not more than one-third the greatest breadth of the valves, and, owing to the gibbosity of the shell, imparting little or no angularity to the outline of the lateral slopes. Dorsal valve more convex than the other, and very strongly arched, particularly in mature shells; umbonal region gibbous, and often, in adult examples, projecting somewhat beyond the beak of the other valve; beak strongly incurved, so as to bring its apex under the beak, and nearly against the area of the other valve; area rather narrow and distinctly incurved with the beak.

Ventral valve convex; beak only moderately prominent and arched, or more or less incurved; area triangular, small, about one-third as high as wide, and moderately well defined; its triangular foramen scarcely as wide as high. Surface of both valves ornamented with rather fine, regular, crowded radiating striæ, and a few very large, rounded, or more or less angular, radiating plications, which latter are never defined on the umbones, but occupy the anterior half, and become more prominent toward the front, where they often terminate in deeply interlocking angular marginal projections. Of these plications there are two—rarely three—on each side of the slightly larger and more prominent one forming the mesial fold on the dorsal valve, while on the ventral valve there are three—rarely four—on each side of the mesial sinus; a few zigzag marks of growth also traverse the anterior and lateral margins of the valves, parallel to their deeply-notched edges.

Length of a medium-sized adult specimen, 0.72 inch; breadth, 0.74 inch; convexity, 0.75 inch. Radiating striæ numbering twelve to fifteen in 0.10 inch.

This interesting and peculiar shell was, by an oversight, referred by Professor Geinitz to *Rhynchonella angulata*, Linn. sp. Although it resembles that species somewhat in general external appearances, it is really very widely removed from it by its generic and even family characters, being a distinctly punctate shell, with a well-defined, though small, cardinal area, divided by an open triangular fissure, and having in the other valve a cardinal process exactly as we see in *Orthis*.* It differs generically, however, from the latter, in having the dental laminæ very closely approximated, prominent, and, with a third mesial plate, continued nearly parallel to each other far forward.

Specifically, it is very closely allied to a South American Carboniferous shell referred by Mr. Salter in the Quart. Jour. Geol. Soc. Lond., XVII, p. 64, Pl. IV, Fig. 3, to *Terebratula Andii*, d'Orbigny. I have also, through the politeness of Colonel Romanowski, of the mining engineers' department of Russia, some imperfect specimens from the Mountain limestone of Rjasan, Russia, with the name *Spirifer Lamarckii*, Fischer, attached, that I believe to belong to the species under consideration. I have not seen the original figures and description of Fischer's species, but if correctly represented by Murchison, de Verneuil and v. Keyserling in the Geology of Russia, I should think it distinct, specifically, from the form sent by Colonel Romanowski. At any rate, I can scarcely doubt that Fischer's species ought to be called *Syntrilasma Lamarckii*, as it resembles our typical species apparently too nearly to belong to any other genus.

Locality and position.—Divisions B and C of the Nebraska City section; also at the former horizon at Wyoming, Bennett's Mill, &c., and at numerous lower horizons in the Coal-Measures of Illinois, Iowa, Nebraska, Kansas, Missouri, &c.

---

* Mr. Davidson, to whom I sent specimens of this shell, writes that he thinks it a good genus.

Genus RHYNCHONELLA, Fischer.

RHYNCHONELLA OSAGENSIS, Swallow.

Pl. I, Fig. 9, *a*, *b*, and Pl. VI, Fig. 2, *a*, *b*.

Rhynchonella (*Camarophoria*) *Osagensis*, Swallow, 1858, Trans. St. Louis Acad., S. 1, p. 219.
Terebratula Uta, Marcou, 1858, Geol. N. Am., p. 51, Pl. VI, Fig. 12, *a*, *b*, *c*.
? *Rhynchonella*, sp., Salter, 1861, Jour. Geol. Soc. London, Vol. XVII, p. 64, Fig. 5, *a*. *b*.
Camarophoria globulina, Geinitz, 1866, Carb. und Dyas in Neb., p. 38, Tab. III, Fig. 5; (not *C. globulina*, Phillips, sp., 1834).

Shell small, more or less variable in form, often subtrigonal, generally wider than long, more or less gibbous; front truncated, or sometimes sinuous in outline; anterior lateral margins rounded in outline; posterior lateral margins convex, or nearly straight and converging to the beaks at an angle of from about 90° to 120°. Dorsal valve more convex than the other, greatest convexity near the middle or between it and the front, which has a broad, rather deep, marginal sinus, for the reception of a corresponding projection of the front of the other valve; mesial fold somewhat flattened, but slightly prominent, and rarely traceable back of the middle of the valve; generally composed of three, but sometimes of four—rarely more—plications; sides rounding down rapidly on each side of the mesial fold, and each occupied by about three or four simple plications; beak curving strongly beneath that of the other valve; interior with a faint linear mesial ridge, on each side of which is a raised curved line inclosing an ovate space, occupied by the adductor muscular impressions. Ventral valve distinctly less convex than the other, with a broad, shallow, short sinus occupied by about two or three short plications; anterior lateral margins on each side of the sinus, with from two to four simple plications; beak moderately prominent, and more or less arched, rather pointed; foramen small.

Length of a medium sized specimen, 0.40 inch; breadth, 0.45 inch; convexity, 0.32 inch.

Like most other species of the genus, this varies considerably in form; some individuals being proportionally wider and more gibbous than others. It also varies in the number of its plications, from six or seven to about twelve, or rarely thirteen or fourteen to each valve. In a majority of instances there are only two plications in the mesial sinus and three on the fold. Some specimens have more, though I have not seen any I could confidently refer to this species, with as many as six plications in the sinus, as stated by Professor Swallow to be sometimes the case.

As remarked by Professor Swallow, some individuals of this species resemble *Camarophoria Schlotheimi*, or at least certain forms of that species, quite closely. Professor Geinitz has even referred it to *Camarophoria globulina*, Phillips, a shell by some believed to be only a variety of *C. Schlotheimi*; but as I have elsewhere stated, it is certainly distinct from those forms, because it is not a *Camarophoria* at all, but true *Rhynchonella*. This I have ascertained both by internal casts and by sections made across the beak of the ventral valve, of specimens from various localities.

As we obtained no good specimens from division C, at Nebraska City, I have copied on Pl. VIII the figure given by Professor Geinitz. Fig. 2 *a*, *b*, Pl. VI, represents another specimen from division B of the same section, two and a half miles west of Nebraska City; while Fig. 9, *a*, *b*, Pl. I, represent an internal cast from Rock Bluff, on the Missouri, at a lower horizon in the Coal-Measures. From these figures, particularly Fig. 9 *b*, it may be at once seen that this shell is a true *Rhynchonella*, and not a *Camarophoria*, as it shows no traces whatever of the rostral chamber characteristic of that genus. I have been unable to find a figure of *Camarophoria globulina*, showing the interior, but have copied for comparison from Professor King's work on the Permian Fossils of England, a figure of an internal cast of the closely-allied form *C. Schlotheimi* (Fig. 15, Pl. I), from which students will at once see by comparison with Fig. 9 *b*, of the same Plate, the generic differences between these shells; (*a*) of Fig. 15, being the cast of the rostral chamber, which is separated by a wall from the other part of the interior; while in Fig. 9 *b*, the cast of the rostral cavity is a mere protuberance, without any traces of the deep slits left by the walls of a chamber.

Locality and position.—This species occurs in divisions C and B, at Nebraska City. It ranges through a great thickness of our Coal-Measures, and is found at numerous localities in Iowa, Missouri, Nebraska, Kansas, &c. I have also seen specimens of it from the Upper, Middle, and Lower Coal-Measures of Illinois; and it is probably the same shell that was figured by Mr. Salter in 1861, among Professor Forbes's collections from the Andes, South America.

Genus ATHYRIS, McCoy.

ATHYRIS SUBTILITA, Hall, sp.

Pl. I, Fig. 12; Pl. V, Fig. 8, and Pl. VIII, Fig. 4.

? *Terebratula argentea*, Shepard, 1838, Am. Jour. Sci., XXXIV, p. 152.
? *Spirifer Roissyi* (*Terebratula Peruviana*, on plate), d'Orbigny, 1847, Palæont. del'Amer. Merid., p. 46, Pl. III, Fig. 17–19.
Terebratula subtilita, Hall, 1852, Stansbury's Grt. Salt Lake Report, p. 409, Pl. IV (by error in text II), Fig. 1 *a*, *b*, and 2 *a*, *b*; Swallow, 1855, Missouri Geol. Report, p. 79, 80, 81, &c.; Shumard, ib. p. 216; Schiel, 1855, Pacif. R. R. Rep., II, p. 108, Pl. 1, Fig. 2; Hall, 1856, Pacif. R. R. Rep., Vol. III, p. 101, Pl. 2, Fig. 4.
———? *subtilita*, Davidson, 1857, Monogr. Brit. Carb. Brach., p. 18, Pl. 1, Fig. 21, 22.
Spirigera subtilita, Meek and Hayden, 1859, Proceed. Acad. Nat. Sci., Philad., p. 20, &c.
Athyris subtilita, Newberry, 1861, Ives's Colorado Report, p. 126; Davidson, 1863, Fossils of S. India, Pl. IX, Fig. 7.
——— *subtilita*, Salter, 1861, Quart. Jour. Geol. Soci., Lond., XVII, Pl. IV, Fig. 4 *a*, *b*.

Shell ovoid, being usually widest a little in advance of the middle, and nearly always somewhat longer than wide, moderately convex, becoming rather gibbous with age. Ventral valve, usually a little more convex than the other, its greatest convexity being generally behind the middle; beak prominent, rounded and distinctly incurved upon that of the other valve; foramen round, of moderate size, and truncating the immediate apex of the beak; mesial sinus, absent or very shallow, in young or compressed individuals, but well defined, and round, flattened, or angular in adult gibbous specimens, in which it rapidly increases in size, from near the middle to the front, where it produces a more or less prominent marginal projection, fitting into a corresponding sinuosity in the margin of the opposite valve. Dorsal valve moderately convex, the greatest convexity in small or compressed specimens often near the middle or between it and the umbo, but in large, gibbous individuals, with a well defined, prominent mesial fold, sometimes near the front; beak rather distinctly incurved under that of the opposite valve. Surface of both valves nearly smooth, or with mere lines of growth, in young shells, but in large or mature specimens with well defined, imbricating marks of growth on the anterior half; exfoliated surfaces also show, under a magnifier, traces of obscure radiating striæ.

Length of a well-developed, medium-sized specimen, 0.95 inch, breadth, 0.81 inch; convexity, 0.56 inch. Very large specimens sometimes measure as much as 1.54 inches in length. At some localities, however, the average length of evidently adult examples is only about 0.60 inch.

As may be seen by the above measurements, this shell varies considerably in form, as well as in size. Young individuals, or those that have had their growth arrested by unfavorable circumstances, are always more compressed and proportionally rounder in outline, while they show no traces of a mesial fold or sinus, and have less distinct marks of growth. Notwithstanding these differences of form, proportions, &c., there is something in the general physiognomy of this shell that enables one, once familiar with its appearance, to recognize it without difficulty.

It is quite probable that we may have to adopt the name *Athyris* or *Spirigera** *argentea*, for this species, since Shepard, in 1838, figured under the name *Terebratula argentea*, a shell from the Upper Coal-Measures of La Salle, Illinois, agreeing more nearly with this than with any other known in these rocks, while this species is very common there. If his type can be found, and should prove to be another form, then it will become a question whether we will not have to call it *Athyris* or *Spirigera Peruviana*, because d'Orbigny figured a South American shell in 1847, under the name *Terebratula Peruviana*, almost exactly like this, and believed by Mr. Salter and others to be the same.

Locality and position.—We did not find this species in division C of the Nebraska City section, and have therefore copied Professor Geinitz's figure of a specimen from that horizon. We found it, however, at apparently a higher position, one and three-fourths and two and three-fourths miles west of there, and at the horizon of B at Bennett's Mill, Wyoming, &c. It is also common at lower positions at Cedar and Rock Bluffs, Plattsmouth, Bellevue, and Omaha; and it occurs at almost all horizons in the Coal-Measures, of Illinois, Missouri, Iowa, West Virginia, and Ohio; also, at Pecos Villages, New Mexico; and less abundantly in the Coal-Measures and up through the Permo-Carboniferous beds into the Permian, in Kansas. In England and India it has been identified in Lower Carboniferous rocks; but in this country, unless *A. subquadrata* of Hall, is only a variety of this species, it has not been found below the Coal-Measures.

Genus RETZIA, King.

RETZIA PUNCTULIFERA, Shumard.

Pl. I, Fig. 13; and Pl. V, Fig. 7.

Retzia punctulifera, Shumard, 1858, Trans. St. Louis, Acad. Sci., Vol. I, p. 220.
Terebratula Mormonii, Marcou, 1858, Geol. North Am., p. 51, Pl. VI, Fig. 11.
Retzia Mormonii, Meek and Hayden, 1859, Proceed. Acad. Nat. Sci. Philad., p. 27; Geinitz, 1866, Carb. und Dyas in Neb., p. 39, Tab. III, Fig. 6.
Retzia subglobosa, McChesney, 1860, Descrip. New Pal. Fossils, p. 45; and Pl. I, Fig. 1, illustrations of same, 1865.

Shell small, ovate; in mature specimens, gibbous; hinge line short, or scarcely extended enough to show distinctly the little ears at the extremities. Ventral valve more convex than the other, the greatest convexity being between the middle and the umbo, which is prominent, rounded, more or less strongly arched, and provided with a moderately large circular foramen; area well defined, triangular, and arching with the beak. Dorsal valve most convex near the middle; beak extending

* It is an unsettled question, whether the name *Spirigera* or *Athyris* should be retained for this genus. *Athyris* has priority of date, but is very objectionable, because it implies a plain contradiction of fact, the shell being provided with a distinct foramen.

a little beyond the hinge margin, and distinctly incurved. Surface of each valve ornamented by fourteen or fifteen (very rarely sixteen to seventeen) simple, rather prominent, radiating costæ, one or two of which are sometimes slightly more depressed than the others, near the front of the ventral valve, so as to cause some appearance of an obscure mesial sinus, but without producing any corresponding mesial elevation on the other valve, or visibly interrupting the general straightness of the uniting margins of the two valves; lines of growth obscure; punctures visible under a good pocket lens, and very regularly disposed.

Length of a well-developed, rather gibbous specimen, 0.35 inch; breadth, 0.30 inch; convexity, 0.25 inch.

If the shells generally included in the genus *Retzia*, but differing from Professor King's type of that genus (*R. Adrieni*) in the possession of a small cardinal area, are, as has been proposed,* to be considered generically distinct, the name of this species would become *Eumetria punctulifera*, as it has the small cardinal area characterizing that group.

Specifically, this form is related to *Retzia radialis*, Phillips, sp., and an occasional specimen may be picked out very closely resembling certain varieties of that species. It differs, however, in having the umbo of its ventral valve more prominent, thus giving a more ovate form to the entire outline than we see in *R. radialis*. Our shell is also much less variable in the number of its costæ, of which there are nearly always fourteen to each valve. I have never seen a specimen with more than sixteen or seventeen, while *R. radialis*, although sometimes not having more than fourteen or fifteen, *generally* has about twenty. A more important difference, however, is to be observed in the nature of the obscure mesial sinus sometimes observed in the ventral valve of our shell, which is not produced by a wider and deeper sulcus between two of the costæ, but by the depression of one or two of the costæ themselves; nor does this faint sinus produce any traces of a corresponding ridge in the other valve, as in *R. radialis*.

Professor Geinitz cites my *R. compressa*, from the Carboniferous rocks of California, as a synonym of the form under consideration. In this, however, he is certainly in error, that species being a much more compressed shell, with only about half the number of costæ, which are also proportionately more prominent. To me, the *R. punctulifera* seems more nearly allied to *R. radialis*, Phillips, than *R. compressa* is to *R. punctulifera*; so that if we were to include the California shell as a variety of *R. punctulifera*, both of the latter might, with at least as much propriety, be arranged as synonyms of Phillips's species.

I am in doubt whether Dr. Shumard's name, *R. punctulifera*, or Professor Marcou's name, *Terebratula Mormonii*, for this shell, has priority, both having been published in 1858. If Professor Marcou's name was published earlier in the year than Dr. Shumard's, it would of course have to be retained. If the two names, however, were published at exactly the same date, or so near it as to leave the question of priority in doubt, the name *punctulifera* should be retained, as it was proposed along with a much better description, and with a correct knowledge of the affinities of the shell.

Locality and position.—The specimen figured on Pl. I was, with many others, found at Rock Bluff, on the Missouri, referred by Professor Marcou to the Lower Dyas, and by Professor Geinitz to the Coal-Measures, to which latter horizon this outcrop certainly belongs. We also find it at Plattsmouth; in the division B, at Nebraska City; at the same horizon at numerous other localities, as well as at various positions in the Coal-

* See Report Regents University, N. Y., on State Cab. N. H., 1864, p. 64.

Measures of Nebraska, Kansas, Illinois, and the adjoining States; in short, it is a widely-distributed Coal-Measure species.

Genus SPIRIFER, Sowerby.

SPIRIFER CAMERATUS, Morton.

Pl. VI, Fig. 12; and Pl. VIII, Fig. 15.

Spirifer cameratus, Morton, 1836, Am. Jour. Sci., Vol. XXIX, p. 150, Pl. II, Fig. 3; Hall, 1856, Pacific Railroad Report, Vol. III, p. 102, Pl. II, Figs. 9, 12, and 13; and 1858, Iowa Geol. Report, 1, Part II, p. 709, Pl. XXVII, Fig. 2 *a*, *b*.
———*meusebachanus*, Roemer, 1852, Kreid. Von Texas, p. 88, Pl. XI, Fig. 7.
———*triplicatus*, Hall, 1852, Stansbury's Report Salt Lake Exp., p. 410, Pl. II, Fig. 5, (by error Pl. 4.)
———*fasciger?* (v. Keyserling), Owen, 1852, Report Wisconsin, Iowa, and Minn., Pl. 5, Fig. 4.
Spirifer ——— H. D. Rogers, 1858, Geological Report Pennsylvania, Vol. II, p. 833, Fig. 694.
Compare *S. fasciger*, v. Keyserling, 1847, Petschora-land, p. 231, Pl. 3, Fig. 3.

Shell attaining a medium size, transversely semicircular or subtrigonal, more or less gibbous; cardinal margin nearly always equaling the greatest breadth of the shell, sometimes extended into pointed ears. Dorsal valve rather less convex than the other; beak but little prominent, and scarcely extending beyond the hinge—together with the narrow area, rather distinctly incurved; mesial fold prominent, subangular or narrowly rounded, commencing narrow at the beak and widening and deepening rapidly to the front. Ventral valve with beak rather prominent and incurved; area distinctly defined, of moderate height, with subparallel margins, and extending nearly or quite to the lateral extremities of the hinge, more or less straightened and arched; foramen forming a nearly equilateral triangle; mesial sinus corresponding to the fold of the other valve, extending to the beak. Surface ornamented with distinct, irregular, or unequal, more or less bifurcating, radiating costæ, both on the mesial fold and within the sinus, as well as on each side of the same, those on the sides showing a more or less marked tendency to group themselves into fasciculi, which are each composed, at the front, of from three to seven or eight ribs, only the middle one of which usually extends quite to the beaks; marks of growth moderately distinct near the free margins, but never forming raised lamellæ.

Length of a well-developed adult specimen, 1.35 inches; breadth, 2 inches; convexity, 1 inch.

This is one of our most common and widely distributed Coal-Measure fossils, and, like other species of the genus, it varies considerably both in form and some of its other characters. Some individuals are quite gibbous with a comparatively short hinge, and rather obtuse lateral extremities; while others are less convex, have the hinge line considerably extended, and the lateral extremities acutely angular. The fasciculated character of the costæ is also, to some extent, a variable character, though it is generally more or less marked, and often very strongly defined. Some of the extreme varieties look quite distinct enough to be considered different species, if it were not for the numerous intermediate gradations. One of these extremes, with an extended hinge line terminating in pointed extremities, and provided with rather fine, imperfectly fasciculated costæ, Professor Swallow has proposed to call *S. cameratus*, var. *Kansasensis*.

It is a remarkable fact, that this latter variety is almost exactly like certain varieties of the well-known *S. striatus*. Indeed, some of the best authorities of Europe have regarded even the typical *S. cameratus* as merely a variety of *S. striatus*. Yet in this country, although we have a form (*S. Logani*, Hall) agreeing well with the large typical *S. striatus*, the *S. cameratus* is in all its varieties strictly a Coal-Measure shell, and can be distinguished at a glance from all of our Lower Carboniferous species.

Professor Geinitz cites Mr. Davidson's *S. Moosakhailensis*, from Punjaub, India, as a synonym of *S. cameratus*, but in this he is certainly in error, as that species is characterized by distinct, regularly disposed, raised lamellæ of growth, over the entire surface; a character never seen on the American shell. It is far more probable, however, as supposed by Dr. Owen, that *S. fasciger* of Keyserling is a synonym of this shell.

Locality and position.—This species ranges through the whole Coal-Measures of this country, and up into the series called Permo-carboniferous by Dr. Hayden and myself in Kansas. It is more abundant, however, in the Upper than the Lower Coal-Measures. It occurs in division C of the Nebraska City section, and has been found by Dr. White in Iowa, and by Dr. Hayden and myself in Kansas, in beds believed to hold the same position. It is common in division B at Nebraska City and vicinity, and at various lower positions in the Coal-Measures of that and the adjoining States. It also occurs in the Coal-Measures of Eastern Ohio, West Virginia, and Pennsylvania, and ranges south into Texas and New Mexico.

SPIRIFER [MARTINIA] PLANOCONVEXUS, Shumard.

Pl. IV, Fig. 4, *a*, *b*; Pl. VIII, Fig. 2, *a*, *b*.

Spirifer planoconvexus, Shumard, 1855, Report Missouri Geological Survey, p. 202; Meek and Hayden, 1864, Palæont. Upp. Missouri, Part 1, pp. 20 and 21, Figs. *a*, *b*, *c*, *d*, *e*; Geinitz, 1866, Carb. und Dyas in Neb., p. 42, Tab. III, Figs. 10, 11, 12, 13, 14, 15, 16, 17, 18.

Ambocœlia gemmula, McChesney, 1860, Descr. Pal..Foss., p. 41; also 1865, Fig. 3 *a*, *b*, *c*, Pl. I, illustrations of same.

Martinia planoconvexa, McChesney, 1869, Trans. Chicago Acad. Sci., Vol. I, p. 34, Pl. I, Fig. 3.

Compare *Spirifer Urii*, Fleming, 1828, British Animals, p. 376.

Shell very small, plano-convex, or very rarely even a little concavo-convex, sometimes wider than long, in other examples slightly longer than wide; hinge margin always shorter than the greatest transverse diameter of the valves, and rounded at the extremities; lateral margins and front regularly rounded; surface apparently smooth, excepting a few very obscure concentric marks of growth, but when examined by the aid of a magnifier, showing remains of the bases of minute hair-like spines. Dorsal or smaller valve truncato-suborbicular in outline, generally nearly flat, with a faint longitudinal depression in front, sometimes slightly convex near the beak, and concave around the anterior and lateral margins; beak scarcely distinct from the cardinal margin; area narrow, but well developed, or about half as large as in the other valve; socket plates a little projecting, like diverging teeth on each side of the small fissure. Ventral valve very gibbous, particularly in the umbonal region, sometimes with obscure traces of a narrow longitudinal depression along the middle, but without a proper mesial sinus; beak very prominent, and strongly arched back over the hinge; area subtriangular, being moderately high under the beak, but narrowing rapidly, with moderately defined concave margins, to the extremities of the hinge, and arching with the beak; fissure rather narrow, or higher than wide, apparently rounded above under the beak, and spreading at the hinge.

Length of a medium-sized specimen, 0.30 inch; breadth, 0.31 inch; convexity, 0.17 inch.

This abundant little shell agrees so closely with *Spirifer Urii* of Fleming, that I have been somewhat inclined, as elsewhere stated, to think it may really be a variety of that species. The only differences I have been able to see, consist in its more flattened, or, in some instances, partly concave dorsal valve, and the general (though not constant) absence of any traces of a mesial sinus in its ventral valve. The fissure of its ventral valve seems also to be narrower, and, so far as known, not partly closed by a false deltidium.

Locality and position.—Very abundant in division C of the Nebraska City section; also common in division B at that place; at Bennett's Mill, Wyoming, and at apparently a higher horizon at Morton's and Werth's places, one and three-fourths to two and three-fourths miles west of Nebraska City. It is likewise common in still lower positions at Cedar Bluff, on Weeping Water; Rock Bluff, Plattsmouth, &c., on the Missouri; and less abundantly at Bellevue and Omaha. We also found it at Brownville, Otoe City, Rulo, &c.; indeed, it is one of the most common fossils of our western Coal-Measures in Iowa, Missouri, Nebraska, Kansas, &c.; and I have seen it among some collections sent by Dr. Stevens from the Coal-Measures near Pittsburg, Pennsylvania. I have likewise identified it among specimens sent by Professor J. J. Stevenson from the Lower Coal-Measures of West Virginia.

Genus SPIRIFERINA, d' Orbigny.

SPIRIFERINA KENTUCKENSIS, Shumard.

Pl. VI, Fig. 3 *a*, *b*, *c*, *d*; and Pl. VIII, Fig. 11 *a*, *b*.

Spirifer octoplicatus, Hall, 1852, Stansbury's Salt Lake Report, p. 409, Pl. XI, Fig. 4, *a*, *b*; (not Sowerby ?, 1827).
——— *Kentuckensis*, Shumard, 1855, Missouri Geological Report, p. 203; Hall, 1856, Pacific Railroad Report, Vol. III, p. 103, Pl. II, Fig 10–11.
——— *laminosus*, Geinitz, 1866, Carb. und Dyas in Neb., p. 45, Tab. III, Fig. 19 *a*, *b*, *c*, *d*; (not *S. laminosus*, McCoy).

Shell rather small, varying from subglobose, to semicircular, or even subfusiform, always wider than long; breadth sometimes twice or even three times the length; hinge line always equaling the greatest breadth of the valves, occasionally greatly extended, and terminating in slender mucronate ears; anterior and lateral margins generally forming a nearly semi-circular curve.

Ventral valve somewhat more convex than the other, the greatest convexity being between the beak and the middle; beak moderately prominent, and rather distinctly arched or incurved; area arched, usually of moderate height, well defined, and extending nearly or quite to the lateral extremities, highest in gibbous specimens, in which it is narrow near the extremities, while it increases rapidly in height, with concave lateral margins, toward the beak; foramen, generally higher than wide, with a marginal furrow on each side, and, so far as known, not closed by a deltidium; mesial sinus narrow, rather deep, sometimes with a small obscure rib along its middle, but more frequently without it; plications on each side of the sinus from about five to eight or nine, rather narrow, simple, prominent, and a little rounded; mesial septum of interior moderately prominent.

Dorsal valve with greatest convexity near the middle; beak scarcely projecting beyond the hinge margin, more or less incurved; area very narrow, and incurved with the beak; mesial fold narrow, not very prominent, nor greatly larger than the first plication on each side, most generally rounded, but not infrequently with an obscure sulcus along the middle, near the front; lateral plications as in the other valve.

Entire surface of both valves ornamented with numerous closely crowded, very regularly arranged, subimbricating lamellæ of growth, strongly arched in passing over the costæ; over the whole may also be seen, by the aid of a magnifier, numerous granules, apparently connected with the punctures passing through the shell, which are comparatively large and distant, though regularly arranged.

Breadth of a well developed, rather gibbous specimen, with the hinge line not greatly extended, 0.55 inch; length, 0.38 inch; convexity, 0.33 inch.

This species varies extremely in form, some specimens being almost subglobose, with the hinge line not extended beyond the general breadth of the valves near the middle; while other individuals have the lateral extremities greatly extended and pointed, even in some cases more so than in the specimen represented by our figure. For these most extended forms, Professor Swallow has proposed in the St. Louis Academy's Transactions the name *Spirifer propatulus*, as a variety of the *S. Kentuckensis.* After a careful comparison and study of an extensive series of specimens, it seems to me that there are all degrees of intermediate gradations between these extremes of form.

Some of the less drawn out, or more nearly globose varieties, appear to me to be so nearly like *Spiriferina octoplicata*, Sowerby, that it would certainly be very difficult to point out any reliable characters (yet known) by which they can be separated. The fact, however, that these are so inseparably connected with the others, greatly extended transversely, and in this respect so very different from *S. octoplicata*, as figured and described in European works, is at least a sufficient reason for doubting the specific identity of the American and European forms.

Mr. Davidson, however, wrote to me that gibbous specimens sent to him from Nebraska, would scarcely be thought distinct from *S. octoplicata* if found in British Carboniferous rocks.

The only specimens I have seen from division C of the Nebraska City section being too imperfect to figure, I have reproduced Professor Geinitz's figures of specimens from that division. We found nearly all the varieties of this shell, however, at that locality in the next division below, and the figures 3 *a* and 3 *d*, Pl. VI, are given to illustrate the extremes of form. Fig. 3 *c*, same plate, shows the interior of a ventral valve from the latter locality and position with its mesial septum.

Prof. Geinitz was certainly in error in referring this shell to *S. laminosus*, McCoy, which is a much larger species, with so high an area that Professor McCoy even thought it a *Cyrtia.*

Locality and position.—This species is found in divisions C and B, at Nebraska City; and at numerous localities and various position in the upper Coal-Measures of Kentucky, Illinois, Missouri, Iowa, Nebraska, Kansas, &c., south to Texas and New Mexico; also in the Lower Coal-Measures of Illinois and Iowa.

Genus TEREBRATULA, Llhwyd.

TEREBRATULA BOVIDENS, Morton.

Pl. I, Fig. 7 *a*, *b*, *c*, *d*, and Pl. II, Fig. 4.

Terebratula bovidens, Morton, 1836, Am. Jour. Sci., Vol. XXIX, p. 150.
——— *bovidens?*, Hall, 1858, Iowa Geol. Report, Vol. I, Part II, p. 711; McChesney, 1869, Trans. Chicago Acad. Sci., Vol. I, p. 37, Pl. I, Fig. 2.
——— *millepunctata*, Hall, 1856, Pacific Railroad Report, Vol. III, p. 101, Pl. II, Fig. 1 and 2.
——— *geniculosa*, McChesney, 1861, Descriptions New, Palæozoic Foss., p. 82; also 1865, illustrations same, Pl. I, Fig. 2 *a*, *b*, *c*.
Compare *Terebratula arcuata*, Swallow, 1862, Trans. St. Louis Acad. Sci., II, p. 83.

Shell ovate, rounded and rather compressed at the anterior and anterior lateral margins, and most convex a little behind the middle; valves nearly equally convex; ventral valve strongly arcuate longitudinally, and presenting a regularly increasing curve, from the front to the beak, which is moderately prominent, and very strongly and closely curved over and upon that of the other valve; foramen a little oval, and not truncating the immediate apex of the beak, but situated directly outside of it; mesial sinus rather wide, and rounded at the front, but narrowing and becoming less deep further back, until it dies out near the curve of the umbo, which is sometimes slightly flattened. Dorsal valve often nearly straight, or but slightly convex, along the middle, from the beak to the front, where its margin is usually somewhat raised for the reception of the slightly produced margin of the other valve at the termination of the sinus; sides sloping from the middle to the lateral margins along nearly the entire length of the valve; beak terminating directly under that of the other valve, without any distinct curvature. Surface nearly smooth, or only showing moderately distinct marks of growth; and, by the aid of a magnifier, exhibiting very distinctly the moderately large regularly arranged punctures.

Length of a medium-sized individual, 0.64 inch; breadth of ditto, 0.50 inch; convexity, about 0.28 inch.

This shell often exhibits almost *exactly* the external form and appearance of *T. elongata*, Schlot., as figured by Professor Geinitz, in his work on the German Permian fossils (Dyas), Pl. XV, particularly the form represented by his figures 14 and 15. This resemblance is so striking that there can be no doubt whatever that he would have unhesitatingly referred the Nebraska shell to that species, had Professor Marcou found it.

Mr. Davidson and several other good authorities on the *Brachiopoda* regard the Permian *T. elongata* and *T. sufflata* as only varieties, or recurrent forms of the Carboniferous *T. hastata*. However this may be, I can only say that our shell is *very constant* in its characters, never presenting but slight variations from the forms figured on Pl. I, which will be seen to resemble most nearly the *T. elongata* of the European Permian rocks, as figured by Professors King and Geinitz. Were it not for the fact that the foramen of that shell truncates (according to Professor King) the immediate extremity of the beak, instead of perforating it a little outside of its apex, I confess it would be extremely difficult, if not impossible, to separate our shell, upon any known external characters from certain forms of *T. elongata*. Yet I am assured by Dr. White that he has ground off Iowa specimens, in no way distinguishable from *T. bovidens*, so as to show the loop through the translucent calcareous matter filling the interior, and that it is elongated as in *Waldheimia*, and not short as in true *Terebratula*. If the appearance of a long loop, seen by Dr. White, is not due to some other cause, of course our shell cannot be properly referred to any variety of *T. elongata*, or other allied European form of true *Terebratula.*

Professor Swallow has described, under the name *T. arcuata*, from the Lower Carboniferous rocks at Chester, Illinois, a shell agreeing apparently exactly, in all external characters, with this, but it is desirable that their internal characters should be compared before arriving at any positive conclusions respecting their identity.

Locality and position.—The specimen represented on Pl. II, Fig. 4, was found at Plattsmouth, and figures 1 *a*, *b*, *c*, on Plate I, are from a specimen found three miles up Platte River, on the north side, in a bed holding a position a little lower than those at Plattsmouth. Fig. 7 *d*, represents a larger specimen from the Coal-Measures of Indian Creek, Kansas. Professor McChesney figured it from the Upper Coal-Measures of La Salle, Illinois; and Professor Hall from the same horizon in New Mexico. Dr. Morton first figured and described it from the Coal-Measures of Ohio. It also occurs at various localities in the Upper Coal-Measures of Iowa, Missouri, and Kansas, as well as in the Lower Coal-Measures of Illinois, though it is not usually abundant at any one locality.

LAMELLIBRANCHIATA.

Genus LIMA, Bruguiere.

LIMA RETIFERA, Shumard.

Pl. IX, Fig. 5.

Lima retifera, Shumard, 1858, Trans. St. Louis Acad. Sci., 1, p. 214; Geinitz ?, 1866, Carb. und Dyas in Neb., p. 36, Tab. II, Figs. 20 and 21.

Shell obliquely subovate, moderately convex, apparently not gaping in front; hinge line short, or between one-half and one-third the antero-posterior diameter of the valves; base forming a nearly regular semicircular curve; anterior side extended obliquely forward, rather narrowly rounded below, and straight or slightly concave in outline, with a rather long oblique slope to the hinge above; posterior side distinctly shorter than the other, and rounding from near the ear into the base; ears subequal, the front margin of the anterior one forming an obtuse angle with the hinge line, rather distinctly flattened from the swell of the umbo, and somewhat extended along the anterior margin below; posterior ear a little more convex than the other, with its upper margin incurved, and its lower margin separated from the umbo by a faint oblique furrow, sometimes faintly sinuous behind, and nearly rectangular at its extremity; umbones rather convex or moderately compressed, extending very little above the cardinal margin, and placed near the middle of the same; surface ornamented by about twenty-five slightly irregular, angular, radiating costæ, about equaling the spaces between, and occasionally bifurcating on the umbones and lateral margins of the body part of the valves, where they become obsolete; crossing all of these, as well as on the ears and lateral margins, are numerous fine concentric striæ.

Height of a well-developed specimen, 0.50 inch; antero-posterior diameter of same, 0.62 inch; convexity, about 0.40 inch; length of hinge, 0.27 inch.

A fine series of specimens of this shell now before me shows it to agree well with Dr. Shumard's description, and I have no doubt in regard to its identity with his species. It should be borne in mind, however, in making comparisons with his description, that he has inadvertently described the longer or anterior side as the anal side, and the posterior as the buccal side. I fully agree with Dr. Shumard in regarding it as having, at least, all the external characters of a true *Lima*. I am aware this fact may be appealed to as an evidence that these beds should be included in the Permian; but it should not be forgotten in this connection that the species was originally described by Dr. Shumard from acknowledged Coal-Measure beds in the valley of Verdigris River, Kansas, where it was found directly associated with the well-known Carboniferous types *Fusulina cylindrica* and *Productus Nebrascensis*. The only remaining portion of the shell of this species, in all the specimens I have seen, is a thin layer preserving the surface markings so distinctly as to leave the impression that it is the outer layer; yet, in examining it by the aid of a high magnifier and a strong transmitted light, it seems to present a prismatic structure.

Locality and position.—Division C of the Nebraska City section, and in the Coal-Measures on Verdigris River, Kansas. It also ranges through the whole of the Coal-Measures of Illinois.

Genus ENTOLIUM, Meek.

ENTOLIUM AVICULATUM, Swallow, sp.

Pl. IX, Fig. 11 *a*, *b*, *c*, *d*, *e*, *f*.

Pecten aviculatus, Swallow, 1858, Trans. St. Louis Acad. Sci., Vol. I, p. 215.

Shell compressed lenticular, very thin, nearly or quite equivalve, suborbicular, or broad subovate in outline exclusive of the ears, the antero-posterior diameter being often a little less than that at right angles to the same; sides and base more or less regularly rounded; lateral margins above the middle apparently a little gaping, straight, and converging to the beaks at an angle of 115° to 125°; cardinal margin very short, or less than one-third the transverse diameter of the valves, and in the left valve generally concave, or more or less sloping in outline, from the extremities of the ears to the beaks; straight or nearly so in the right valve; ears small, flat, very nearly equal, obtusely angular at the extremities, and separated from the body of the valves by an impressed line, not defined by any proper sinus in either valve, though the broad obtuse notch separating the anterior one from the straight, sloping adjacent margin is slightly more defined than the other; beaks small, rather compressed, equal, and not projecting beyond the cardinal margin. Each valve with two shallow undefined impressions diverging from the beak nearly to the anterior and posterior margins; that on the posterior side being longer than the other.* Surface with very fine close concentric striæ scarcely visible without the aid of a magnifier; crossing these are also sometimes seen traces of extremely minute radiating striæ, curving gracefully outward toward the lateral margins.†

Antero-posterior diameter of a specimen a little under medium size, 0.85 inch; height, 0.89 inch; length of hinge line, 0.27 inch. Specimens are sometimes found of nearly double these dimensions.

* Owing to the thinness of the shell these impressions appear as ridges on the inside of the valves.

† In most of the specimens these radiating striæ are entirely obsolete, even as seen under a magnifier; and it is generally only on specimens that have been slightly weathered that they are most distinctly seen, while even on these they seem to be more due to some peculiarity of the shell *structure*, than proper *surface* sculpturing, the shell showing a disposition to crack along these curved lines. Both these and the concentric striæ are almost invisible to the unassisted eye.

This shell evidently belongs to a group for which I proposed, in the California Report, the name *Entolium*, with *Pecten demissus*, Phillips (not Fleming), as illustrated in Quenstedt's Der Jura, Pl. 48, Fig. 6, as the type. At the time of proposing this name I was under the impression that the valves of these shells were closed on each side, but the species here under consideration seems to have been gaping on the sides above the middle. This being the case, I am not sure the group is more than subgenerically distinct from *Pseud-amussium*, Brug., 1789.* It differs from *Amussium* mainly in having no internal costæ, and in having the valves more nearly equal, with, sometimes, minute radiating striæ, and no traces of a sinus under the anterior ear in either valve. The species known to me have the cardinal margin of the left valve angulated in outline by the elevation of the extremity of the ears; while that of the right valve seems to be straight, and articulatd in a little transverse groove of the other valve, not always defined, however (see Fig. 14 *g*, *g*, Pl. IX, copied from Quenstedt's figure of *Pecten demissus*, taken from an impression of the hinge left in the matrix). The cartilage pit is as in other allied types of the *Pectinidæ*, while diverging from it are two elongated tooth-like ridges (*t*, *t* of Quenstedt's figure). These, however, do not seem to have been properly teeth, fitting into sockets, but appear to have been a little raised in both valves, and occupy a position between the ears and the broad diverging impressions, descending obliquely from the beaks.

On some of the internal casts of this shell I have noticed some very singular fine sculpturing, rather difficult to account for. It closely resembles the zigzag markings seen on the surface of the group of *Nucula* for which H. and A. Adams proposed the name *Acila;* but is exceedingly fine and obscure, being formed by numerous very regularly and closely arranged, distinctly zigzag lines, traversing the valves with a general direction parallel to the curve of the marks of growth. Fig. 11 *d*, of Plate IX, shows these markings as seen by the aid of a magnifier. The most singular fact in regard to these zigzag lines is that they have no connection whatever with the surface striæ, since they are only seen on internal casts, often retaining portions of the shell, with the usual minute concentric lines, and showing no trace externally of these inner markings. They seem also to be more frequently seen on casts of the left valve, though on casts of many of the left, and apparently all of those of the right, they are entirely wanting. In regard to their complete absence from the interior of right valves, there may be some room for doubts, as it is only under the most favorable circumstances, where casts have been formed of exceedingly fine clays, that they are to be seen. They are possibly, as it were, a sort of effort at internal markings, of a very different kind, but in some respects analogous to the internal costæ of *Amussium.*

Professor Winchell, in 1865, proposed the name *Pernopecten* for a type (*Aviculopecten limaformis*, White and Whitfield) from near the base of the Lower Carboniferous, that agrees exactly with this, in all external characters, but differs in having, in addition to the central cartilage pit, a row of minute pits or crenulations along the whole hinge line. From *Camptonectes* of Agassiz, the shell under consideration differs in having

* Klein first used the name *Pseudo-Amusium*, in 1753, and H. and A. Adams have proposed to adopt it for a group including some forms more or less like our shell, and others quite distinct. If the name is to be retained, however, Bruguiere should, I would think, have to be regarded as the author of the genus, as Klein was not a binomial author, while Bruguiere is the first binomial author that used it after the introduction of the binomial nomenclature, and with a diagnosis. He limited it to smooth species with only fine radiating striæ.

no sinus under the anterior ear of either valve, and in wanting the distinct radiating curved surface striæ.

It is certainly congeneric with *Pernopecten Shumardi* of Winchell (which seems to be the same shell previously described by Dr. Shumard in the Missouri report under the name *Avicula Cooperensis**), and I must confess that I cannot clearly see how it differs even specifically, although the two shells came from widely different horizons. The species *Cooperensis*, or *Shumardi*, came from the lowest division of the Carboniferous in Missouri and Iowa, and occurs in the same horizon in Ohio. I have seen its hinge, however, and it has not the crenated character of the type of *Pernopecten*, but agrees with that of the shell under consideration, as the species does in all other known characters.

Compared with European species, our shell will be seen to be very nearly allied to *Pecten Sowerbyi* of McCoy (Carb. Fossils of Ireland, p. 100, Pl. XIV, Fig. 1), from the Carboniferous rocks of Ireland, which can scarcely belong to any other group than this, though it would appear to differ specifically, in having stronger concentric markings, without any traces of the minute obscure radiating striæ sometimes seen on this, as well as in having its ears more pointed and elevated, as may be seen by Fig. 13 *a* of our Plate IX, reproduced from McCoy's original figure. It is worthy of note that McCoy has noticed (British Palæozoic Fossils, p. 475) that exfoliated laminæ of his species exhibit minute zigzag divaricating, scratch-like markings, like those observed in that here under consideration.

Locality and position.—This species is common in division C of the Nebraska City section. It also occurs in division B, at Bennett's Mill, three miles northwest of Nebraska City, and at lower positions in the Coal-Measures at Plattsmouth, and at several localities near Rock Bluff on the Missouri. Dr. White has found it at various localities in the Upper Coal-Measures of Iowa, and it occurs in the same horizon in Kansas, from which position it was first described, by Professor Swallow. It likewise occurs in both the Upper and Lower Coal-Measures of Illinois.

Genus AVICULOPECTEN, McCoy.

AVICULOPECTEN OCCIDENTALIS, Shumard, sp.

Pl. IX, Fig. 10.

Pecten occidentalis, Shumard, 1855, Missouri Report, p. 207, Pl. C, Fig. 18; (not Winchell).
——— *Cleavelandicus*, Swallow, 1858, Trans. St. Louis Acad. Sci., Vol. I, p. 184.
Aviculopecten ——— ?, Meek and Hayden, 1864, Paleont Upper Mo., p. 50, Pl. II, Fig. 10.
——— *occidentalis* ?, Meek and Worthen, 1866, Geol. Report, Illinois, Vol. II, p. 331, Pl. XXVII, Fig. 4 and 5.
Pecten Missouriensis ?, Geinitz, 1866, Carb. und Dyas in Neb., p. 35, Tab. II, Fig. 18; (not Shumard, 1855).

Shell distinctly inequivalve, not oblique; subovate exclusive of the ears; lateral and basal margins regularly rounded; hinge margin nearly or quite equaling the greatest breadth of the valves; cardinal plate of moderate breadth. Left valve convex, with ears subequal; anterior one with distinct radiating costæ, more convex, shorter, and more obtuse than the posterior, as well as more defined from the swell of the umbonal

* It should be mentioned here that the engraver exaggerated the very obscure ribs sometimes (but rarely) seen on Shumard's species, in the figure given in the Missouri report, and that the out line of the right ear of the same figure is not exactly correct. The species is nearly always destitute of any traces of ribs, as I know from examining numerous specimens from the original locality. These agree exactly with Prof. Winchell's *P. Shumardi*.

slope; posterior ear flattened and more angular at the extremity than the other, sometimes without radiating costæ, but in other instances having them more or less developed, each separated from the margin below by a rounded, rather broad, more or less deep, sinus. Right valve nearly flat, and having the general outline of the other, excepting that its beak is scarcely distinct from the cardinal margin, and its anterior ear much narrower, and defined by a deep, sharply angular sinus. Surface of left valve ornamented with rather depressed or flattened irregular radiating costæ, of which only about twelve or fourteen of the largest reach the beak, the others dying out at various distances between the margins and the umbo, in proportion to size, the larger of the intercalated ones being longer than the smaller; crossing all of these are numerous fine concentric striæ, some of which on the ears, particularly on the anterior one, often form little vaulted scales; in well-preserved specimens these vaulted projections are strongly developed on one of the posterior costæ of the body part of the valve. Surface of right valve with generally only very obscure radiating costæ, and fine crowded lines of growth.

Height of a rather large specimen, 1.65 inches; breadth, from 1.40 to 1.55 inches; convexity, about 0.28 inch.

The foregoing description is taken from well-preserved specimens, showing the surface markings better than any examples I had previously seen. As usually found in the condition of casts, the vaulted scales mentioned on the wings, and a few of the lateral costæ, are entirely absent. This character is also variable in specimens showing the surface well preserved, it being scarcely possible to find any two individuals with it equally well developed, while in many instances it is nearly or entirely wanting.

I have copied Professor Geinitz's figure, because we found no example of this shell in Professor Marcou's bed C, at Nebraska City. If this figure is accurately drawn, the specimen from which it was made may possibly be a different species, since it represents the anterior ear of the left valve more angular than I have ever seen it in this shell. This, however, is probably a slight error in the drawing in restoring a portion of the ear partly broken away or hidden in the matrix, since among the great numbers of specimens I have seen from these rocks, at numerous localities in that and the adjoining States, there is no otherwise similar shell agreeing with this figure in this respect. His specimen being a cast, of course does not show the vaulted scale-like surface character mentioned in the foregoing description. Professor Geinitz was certainly mistaken, however, in referring the form he has figured to *A. Missouriensis* of Shumard, which is a smaller species, unknown above the St. Louis Limestone of the Lower Carboniferous series.

Locality and position.—*Aviculopecten occidentalis* has an extensive geographical distribution, as well as a considerable vertical range. It occurs at numerous localities in the Upper Coal-Measures of Illinois, Missouri, Iowa, Eastern Nebraska, Kansas, and Kentucky, as well as in the Lower Coal-Measures of the first mentioned State; and has been found near the Black Hills. It occurs in the Coal-Measure rocks above the Platte, in Iowa and Nebraska, referred by Mr. Marcou to the Mountain Limestone; in those at the mouth of Platte River referred by him to the Lower Dyas; and at Rock Bluff, Bennett's Mill, Wyoming, Nebraska City, &c., in beds included by him and Professor Geinitz to the Upper Dyas. In Kansas it ranges through the whole Upper Coal-Measures, and Permo-carboniferous, into the Permian. I have never seen it at any western locality in any of the Lower Carboniferous or older rocks; but from Nova Scotia I have seen casts of a form that could not be distinguished from it, obtained from beds generally regarded as Lower Carboniferous.

AVICULOPECTEN NEGLECTUS, Geinitz, sp.

Pl. IX., Fig. 1 *a*, *b*.

Pecten neglectus, Geinitz, 1866, Carb. und Dyas in Neb., p. 33, Tab. II, Fig. 17.
Aviculopecten neglectus, Meek, 1867, Am. Jour. Sci. & Arts, Vol. XLIV, sec. ser., p. 183.

Shell very small, broad subovate exclusive of the ears, very thin, rather compressed; sides and base more or less regularly rounded; cardinal margin shorter than the breadth of the valves. Left valve (according to Professor Geinitz's figure) with ears nearly equal, the anterior one separated from the margin below by a broad, very shallow sinus, and forming less than a right angle at its extremity; posterior ear extending farther down the margin than the other, very faintly sinuous behind, and forming an angle of about 100° at the extremity. Right valve with anterior ear narrow and rather acutely angular, defined by a deep, narrow sinus, extending back about half its length; posterior ear of about the same length, but of greater vertical breadth than the other, rather pointed at the extremity, and defined by a moderately deep, broadly rounded sinus, and a subangular umbonal slope. Surface of the body part of both valves apparently only marked by fine concentric striæ; ears with a few radiating costæ, crossed by fine striæ and a few coarser marks of growth.

Height and breadth each, 0.26 inch; length of hinge, 0.21 inch.

This little species is rather remarkable in having the body part of the valves with apparently only fine concentric striæ, while the ears are ornamented with a few comparatively distinct radiating costæ. I know of no species with which it is liable to be confounded.

The specimen figured by Professor Geinitz is a left valve, while those I have seen are all right valves. The latter are mainly casts, but one of them retains portions of the shell. On raising a small piece of this with the point of a knife, and placing it under the microscope, where it could be examined by a strong transmitted light, it was found to present distinct indications of a prismatic structure, apparently not due to crystallization. As this, the only remaining portion of the shell, is exceedingly thin, and consists of a single apparently prismatic layer, I have little doubt that the inner laminated portion of the shell has been dissolved away, as seems to have been the case in other species in these rocks.

Locality and position.—Division C of the section at Nebraska City. It also occurs both in the Upper and Lower Coal-Measures of Illinois.

AVICULOPECTEN CARBONIFERUS, Stevens, sp.

Pl. IV., Fig. 8, and Pl. IX, 4 *a*, *b*.

Pecten carboniferus, Stevens, 1858, Am. Jour. Sci. and Arts, Vol. XXV, p. 261.
——— *Broadheadii*, Swallow, 1862, Transactions St. Louis Academy Sci., Vol. II, p. 97.
——— *Hawni*, Geinitz, 1866, Carb. und Dyas in Neb., p. 36, Tab. II, Fig. 19 *a*, *b*.

Shell rather small, slightly oblique, moderately convex, length and breadth nearly equal; hinge line nearly or quite straight, and somewhat less than the greatest breadth of the valves, provided with a marginal ridge in both valves; basal margin regularly rounded. Left valve more convex than the other; posterior ear rather well defined from the swell of the umbo, somewhat extended and terminating in an acute point, separated from the margin below by a deep rounded sinus; anterior ear about two-thirds as long as the other, and rather more distinct from the umbo and more obtuse, but still rather acutely angular, defined by a moderately distinct subangular sinus. Right valve nearly flat, or distinctly less convex than the other; its anterior ear narrow, and defined by a deep, rather sharp sinus; posterior ear of the same size and form as in the left valve. Surface ornamented in the left valve with about fifteen or sixteen regular, distinct, angular, radiating plications, separated by furrows of the same size, each one of which terminates at the free border in a little spine-like projection with curved-up margins; lines of growth fine on the body of the valve, but becoming more distinct and irregular on the ears, where there are rarely any defined radiating costæ. At a few distantly separated intervals there are prominent imbricating laminæ of growth, showing the same digitate margins as the free borders of the shell. In the right valve the surface markings are somewhat like those of the other valve, but much more obscure, excepting on the anterior wing, where there are a few more distinct radiating costæ.

Height of the largest specimen seen, 0.73 inch to extremity of projecting marginal spines; breadth, 0.75 inch; length of hinge, 0.57 inch.

In all of sixteen or eighteen individual specimens I have seen, of the left valve, the sinuous posterior margin under the wing is smoothly rounded in outline. In one individual, however, agreeing in all other respects with the others, this part of the margin is evidently also provided with a few smaller projecting points than those at the terminations of the furrows between the costæ on the body of the valve. There were no radiating furrows or costæ, however, on the ear corresponding to those little projections.

The flattened right valves of this species I have seen, are not in a condition to show very clearly whether its free margin is digitate, like that of the other, though it seems not to be: at any rate not so distinctly so.

On first seeing Professor Geinitz's figure of this shell, I believed it to be the same species described from the Upper Coal-Measures of Northwestern Missouri by Professor Swallow, under the name *Pecten Broadheadii*. Some time after, I wrote to Mr. Broadhead, who discovered the typical specimens described by Professor Swallow, and requested him to send me a specimen of the *P. Broadheadii* from the original locality, which he kindly did, and it proved to be exactly the same shell figured by Professor Geinitz. Still later, I notice among the descriptions of some fossils from the Coal-Measures of Illinois, published by Dr. Stevens in 1858, one of a *Pecten* that seemed to agree quite well with the *P. Broadheadii*, and on examining a good series of specimens from the same locality, I found the same shell quite common among them; and Dr. Stevens has also assured me that Professor Geinitz's figure, of which I sent him a tracing, certainly represents his species.

The only specimens of this species I have seen consist entirely of what seems to be the thin outer layer of the shell, in which there appears to be a prismatic structure, as seen by the aid of the microscope and a strong transmitted light. They show no flattened cardinal plate, but a furrow along the inner side of the hinge margin of each valve. The cardinal plate or area was doubtless composed, as in other cases, of the inner laminated portion of the shell, that has been destroyed during the fossilizing process; if not, it would seem to be a new genus.

Locality and position.—Division C of the Nebraska City section. It was likewise found in Mr. Morton's shaft, a short distance west of Nebraska City. Dr. White has also found it in the upper part of the

Middle Coal-Measures, and near the middle of the Upper Coal-Measures, in Iowa; and it occurs in the Upper and Lower Coal-Measures of Illinois at numerous places; also, in the Lower Coal-Measures of West Virginia.

AVICULOPECTEN WHITEI, Meek.

Pl. IV, Fig. 11 *a*, *b*, *c*.

? *Avicula*—, Professor H. D. Rogers, 1858, Geological Report of Pennsylvania, Vol. II, p. 833, Fig. 689.

Shell truncato-suborbicular, thin, compressed, nearly or quite equivalve; length and breadth nearly equal; basal and anterior and posterior basal margins regularly rounded; posterior margin not sinuous above, but rather straight and intersecting the hinge at an angle of about 60° to 65°; hinge line nearly or quite straight, and generally a little less than the greatest breadth of the valves; posterior ear of both valves flat and somewhat alate, but not defined by any marginal sinus, nor separated from the umbones by any depression or sulcus; anterior ear of left valve small, nearly rectangular, or somewhat rounded, and defined by a shallow, subangular sinus; same in the right valve, nearly rectangular at the extremity, and defined by a rather deep, sharply-angular sinus, from which a shallow furrow extends obliquely up to the beak. Surface of both valves ornamented with rather irregular, obscurely-defined, depressed, and more or less flexuous radiating costæ, which are obsolete on the anterior ear and compressed posterior alation; lines of growth rather well defined, particularly on the ears and around the free borders.

Height (of one of the largest specimens), 0.67 inch; breadth of same, 0.76 inch.

This species is in some respects related to *A. rectilaterarius* of Cox, and *A. papyraceus* of Sowerby, but differs from them both in its surface markings, as well as in some other details. In general form it resembles *Aviculopecten* (*Meleagrina*) *echinatus*, McCoy (Carb. Fos. Ireland, p. 79, Pl. XIII, Fig. 18), but in that species the deep byssal sinus was in the left instead of the right valve, and its surface markings were different.

All the specimens of this species I have seen from Nebraska consist only of the extremely thin, outer apparently prismatic layer of the shell, and are compressed nearly flat in the shaly matrix, though they show the form and surface markings perfectly. They retain no traces of the cardinal plate, but that was probably composed of the laminated portion of the shell that is wanting in these specimens. I have seen others, from the Upper Coal-Measures of Iowa, apparently retaining both the inner and the outer layers of the shell, and hence thicker than those from Nebraska, though none of them were in a condition to show the hinge. I have some doubts whether it is a true *Aviculopecten*.

The specific name is given in honor of Professor C. A. White, the efficient State geologist of Iowa.

Locality and position.—Nebraska City, from a shaft sunk apparently to near the horizon of the outcrops seen at the landing there; also from near the same or a higher horizon at Brownville, on the Missouri; and from Middle Nodaway River, Iowa, in Upper Coal-Measures. It occurs in both the Upper and the Lower Coal-Measures of Illinois; and I suspect that it is the same shell figured by Professor Rogers from the Coal-Measures of Pennsylvania in his report cited; though it may be only an allied species.

AVICULOPECTEN COXANUS, M. & W.

Pl. IX, Fig. 2 *a*, *b*.

Aviculopecten Coxanus, Meek & Worthen, Proceed. Acad. Nat. Sci., Philad., Oct., 1860., p. 453; 1866, Illinois Geol. Report, vol. 2, p. 326, Pl. 26, Fig. 6 *a*, *b*.

Shell very small; thin, compressed, slightly oblique; broad subovate, exclusive of the ears; basal margin, rounded; anterior margin more or less rounded, rather straight and oblique above; posterior margin more prominent than the anterior, often sub-angular at the point where the postero-basal margin rounds up to meet the obliquely-sloping edge above. Hinge generally a little less than the greatest breadth of the valves below. Left valve with anterior ear of moderate size, flat, triangular, with the extremity generally a little less than a right angle, sometimes very slightly rounded, separated from the margin below by an abruptly rounded or sub-angular sinus; posterior ear slightly larger and much more acutely angular than the other, but shorter than the most prominent part of the margin below, from which it is separated by a moderately deep rather broadly rounded sinus; beak small, compressed, scarcely projecting beyond the cardinal margin, and placed a little in advance of the middle of the hinge; surface ornamented with linear, simple, often more or less flexuous costæ, which alternate in size, the smaller ones dying out at various distances between the free margins and the umbo —crossing all of these are numerous, extremely fine, regular, closely arranged concentric striæ, which, like the costæ, are more or less distinctly defined on the ears, as well as on the body of the valve. Right valve unknown.

Height of left valve, 0.37 inch; breadth of do. 0.39 inch; convexity about 0.05 inch.

The little shell I have here referred to *A. Coxanus* agrees well in size, general form and ornamentation, with the typical specimens of that species, excepting that it is a little more oblique than any examples I have seen from the Illinois locality. Yet I am not prepared to regard this as a specific difference, since the specimens vary more or less in this character.

These little shells are somewhat related to *Aviculopecten rectilaterarius* of Cox, and *Aviculopecten* (*Pecten*) *papyraceus*, Sowerby; but differ in being much smaller, and in having the posterior ear much more acute, in consequence of the deeper marginal sinus separating it from the posterior border below.

In all the specimens I have seen, the shell, or at any rate the only remaining portion of it, is extremely thin, and shows under the microscope the distinct appearances of prismatic structure. It is highly probable, however, that this is only the outer layer of the shell, and this may also account for the fact that the hinge shows scarcely any traces of the usual flattened and furrowed cardinal plate or area of the genus *Aviculopecten*; the cardinal margins being, as we now see the shell, apparently linear.

Locality and position.—Nebraska City, bed C; also in the Coal-Measures in Adams County, Illinois. It is worthy of note, that at the latter locality it is in the Lower Coal-Measures; while at Nebraska City it is found in the Upper. It also occurs in the Upper Coal-Measures of Illinois.

Genus AVICULOPINNA, Meek.

AVICULOPINNA AMERICANA, Meek.

Pl. IX, Fig. 12 *a*, *b*, *c*, *d*.

Avicula pinnæformis, Geinitz, 1866, Carb. und Dyas in Neb., p. 31, tab. II, Fig. 13; (not *Avicula pinnæformis*, Geinitz, 1857).
Aviculopinna Americana, Meek, 1867, Am. Jour. Sci. and Arts, Vol. XLIV, p. 282.

Shell small, compressed, with the slender elongated form of some of the Carboniferous species of *Pinna*; cardinal and ventral margins generally nearly straight (the latter being the more convex in outline) and converging gradually from behind to the rather obtusely pointed anterior extremity; posterior side truncated, rounding into the base, and intersecting the posterior extremity of the hinge very nearly at right angles—a little sinuous just below the extremity of the hinge. Cardinal margin so slightly convex in outline as to appear quite straight, very nearly equaling the greatest length of the valves, and provided with a well-defined marginal ridge, which narrows to a mere line, or dies out before reaching the beaks, and widens very gradually to the posterior extremity. Beaks nearly or quite obsolete, extremely oblique, and very slightly behind the very narrow, obtusely pointed, anterior extremity. Surface with two or three broad, nearly obsolete radiating ridges on the posterior dorsal region, and ornamented by numerous slender, very regularly disposed, and abruptly elevated lines or lamellæ, much narrower than the spaces between, and curving gracefully parallel to the posterior border; while on the basal half of the valves they are closely approximate and curve forward.

Length of the largest specimen seen, 1.35 inches; height of same, 0.43 inch; convexity, about 0.08.

Professor Geinitz identified this little shell with *Avicula pinnæformis*, Geinitz (=*Pinna prisca* of Munster). On comparing a good series of of specimens, however, with his figures and description of that shell given in his Dyas, of typical German examples, I was at once, without a shadow of doubt, led to regard the Nebraska shell as an entirely distinct species. Later comparisons with specimens of the same from German localities have also fully confirmed this conclusion. In the first place, it scarcely ever attains one-fourth the size of the foreign species, from which it differs entirely in the character of its surface-marking, in having its lamellæ of growth very regularly and abruptly elevated and arranged with great regularity, while they never form wrinkles along the lower side; nor are they and the spaces between crossed by any traces of the faint radiating lines, sometimes seen on the German species. It is also a proportionally much more compressed shell. Again our shell has its beaks *always* nearer the anterior extremity than *A. pinnæformis*, so much so, indeed, that it requires close looking, even in examining good specimens, to satisfy one that they are not really terminal. In fact, it was not until after examining a number of specimens that I ascertained from some of the internal casts that there is an almost minute lobe-like portion of the anterior extremity, projecting slightly beyond the beaks. Other important differences are the presence of a well-defined ridge along the dorsal margin of the Nebraska species, and its slightly sinuous posterior margin, by which it is made to intersect the cardinal edge at a right angle; while neither Professor Geinitz's figures, nor the specimens I have seen of the European form, show any traces of such dorsal marginal ridge, and they have the posterior margin, and marks of growth curving forward above so as to intersect the cardinal margin at a decidedly obtuse angle.

The impropriety of referring such a shell to the genus *Avicula* must be evident to every conchologist. At first I was inclined to place it in the genus *Pinna*, but on finding that its beaks, although *very* near the anterior extremity, are yet not exactly terminal, I became satisfied that it could not be properly referred to that genus, and referred it provisionally to the genus *Aviculopinna*, though later comparisons have led me to suspect that it may possibly be found to belong to McCoy's genus *Pteronites*, and have to take the name *Pteronites Americana*. It has the prismatic shell structure of the allied *Pinna* and *Avicula* groups.

Locality and position.—Division C, of the Nebraska City section. Dr. White has also found it associated with many of the same species of other fossils near the middle of the Upper Coal-Measures of Western and Central Iowa. I have not yet met with it at any locality further eastward, though it will probably be found in Illinois.

Genus PINNA, Linnæus.

PINNA PERACUTA, Shumard.

Pl. VI, Fig. 11, *a*, *b*.

Pinna peracuta, Shumard, 1858, Trans. St. Louis Acad. Sci. 1, p. 19.
——— *Adamsi*, McChesney, 1860, New Palæozoic Fossils, p. 74.

Shell thin, very narrow, elongated, and tapering gradually and regularly from the larger to the smaller extremity; convex or almost subcylindrical, excepting toward the posterior extremity, which is compressed and obliquely rounded, or subtruncated. Hinge margin very long, and almost perfectly straight—carinated in consequence of the sudden erection of the dorsal edges of the valves; ventral margin equally as straight as the dorsal, and ranging at an angle of about 12° with the latter. Surface nearly smooth, or only showing very obscure lines of growth.

I have never seen a complete specimen of this species, and hence cannot give accurate measurements. Some specimens, however, indicate a length of nearly 12 inches.

I know of no other American species with which this is liable to be confounded. Its smooth surface will at once distinguish it from the other described, narrow, elongated Carboniferous species of this country. I am much inclined to think an imperfect shell figured by Professor de Koninck (Anim. Foss. Pl. 5, Fig. 3), under the name *Solen siliquoides*, is a true *Pinna*, and related to this, though specifically distinct. Our shell, however, seems to be most nearly allied to *P. spatula*, McCoy, as figured in his Palæozoic Fossils, Pl. 3 E, Fig. 9-10, and may possibly be the same, though it seems to be straighter on the margins than McCoy's species.

Locality and position.—The specimen figured is from Bennett's Mill, three miles northwest of Nebraska City; it also occurs at Wyoming, on the Missouri, and at Nebraska City, in beds B, referred by Professor Geinitz and Professor Marcou to the lower part of the Upper Dyas. It likewise occurs at a lower position at Plattsmouth, Rock Bluff, Bellevue, and, in short, at numerous localities through the Coal-Measures of Iowa, Nebraska, Kansas, Missouri, Illinois, &c., &c.

Genus AVICULA (Klein), Brug.

AVICULA LONGA, Geinitz, sp.

Pl. IX, Fig. 8.

Gervillia longa, Geinitz, 1866, Carb. und Dyas in Neb., p. 32, Tab. II, Fig. 15.

Shell nearly or quite equivalve; body part obliquely elongated and more or less arcuate; posterior end narrow and abruptly rounded; base nearly straight and parallel to the cardinal margin behind, but ascending obliquely forward from near the middle of the valves; anterior side oblique, and broadly and faintly sinuous under the ear. Hinge line about three-fourths the length of the valves, and provided with a marginal ridge, produced behind into a very narrow, elongated ear, considerably shorter than the oblique body portion of the valves, from which it is separated by a deep sinus which narrows to an abruptly rounded or subangular extremity close under the ear; anterior ear shorter and much broader than the other, in the left valve convex, with its extremity pointed, and faintly sinuous just below the point—separated from the swell of the umbo by an oblique sulcus extending from the anterior side of the same to the back part of the broad, shallow marginal sinus defining the ear. Beaks of both valves convex, very oblique, placed one-fourth to one-fifth the length of the hinge back of the anterior extremity; in the right valve, rising little above the hinge, but in the left somewhat more prominent, according to Professor Geinitz's figures.

Length of medium-sized specimen, measuring obliquely from the extremity of the anterior ear to the posterior end of the body part of the valves, 0.61 inch; height, measuring at right angles to the hinge, 0.33 inch; length of hinge, about 0.23 inch.

Having seen the hinge-plate of Illinois specimens of this little shell, I am positively sure that it has not the peculiar cartilage pits of the genus *Bakevellia* or *Gervillia*, though some of the casts I have seen show that it has the hinge teeth of *Avicula*, to which genus I cannot hesitate to refer it.

I have not seen a left valve—all those before me from Nebraska City being right valves. In all of these there is, as represented in the figure given, a little short sulcus or indentation, extending nearly at right angles to the hinge, just in front of the beak. In Professor Geinitz's figure of a left valve, there would seem to be several of these little indentations. They are not cartilage pits, however, but are formed by pressure upon anterior teeth within, so as to show on the outside of the valves, and are not impressions in a cardinal area. I am not acquainted with any species with which this is liable to be confounded.

Locality and position.—The specimen figured is from division C of the Nebraska City section. While passing through Iowa with Dr. White, we found good specimens of it near St. Charles, Madison County, in the upper beds of the Middle Coal-Measures, directly associated with *Aviculopecten carboniferus* (=*Pecten Hawni* Geinitz), and various Lower Coal-Measure types. I also found it, at a considerable lower position than the Nebraska City beds, in the Coal-Measures at Riverside, three miles below Atchison, Kansas. It likewise occurs in the Upper, Middle, and Lower Coal-Measures of Illinois.

AVICULA? SULCATA, Geinitz.

Pl. IX, Fig. 9.

Gervillia (Avicula) sulcata, Geinitz, 1866, Carb. und Dyas in Neb., p. 33, Tab. II, Fig. 16.

I have seen no good specimens of this species, and consequently reproduce the figure given by Professor Geinitz. From his figure (of a left valve) it seems to be a subrhombic oblique shell, with a short, compressed, triangular anterior ear, defined by a faint, wide marginal sinus; and a larger compressed, somewhat alate, posterior ear, with a marginal ridge, showing a tendency to be produced into a narrow appendage behind, separated from the margin below by a rather deep rounded sinus. The posterior basal extremity is rather narrowly rounded, but not much produced; while the outline of the base is broadly semielliptical, and the umbo convex, and rising somewhat above the hinge line, which is apparently shorter than the greatest length of the valve. Surface ornamented behind by fine lines of growth, and before by two or three sulci, extending from the anterior side of the beak to the antero-basal margin, leaving ridges between, which are more or less crenated by the crossing of the marks of growth; just behind the posterior one of these sulci, the margin of the latter is ornamented with regular, rather strongly defined wrinkles, or little folds, some of which are prolonged backward parallel to the lines of growth.

This must be a very handsome species, and is more probably an *Avicula* than a *Bakevellia;* but, as its hinge and cardinal area are unknown, we have not the means of settling this question. I know of no similar species with which it might be confounded.*

Locality and position.—Bed C of the Nebraska City section.

Genus PSEUDOMONOTIS, Beyrich.

PSEUDOMONOTIS, sp.

Pl. II, Fig. 11.

This fragment agrees so exactly in its surface-marking (consisting of a larger series of flexuous costæ, with vaulted, scale-like projections, and smaller, irregular, intermediate ribs) with shells of the genus *Pseudomonotis*, that I can scarcely entertain a doubt in regard to its belonging to that group. It is not sufficient, however, for specific identification, though it would not be difficult for those whose method of making paleontology easy leads them to include, under the single species *speluncaria*, all the known forms of this group, to see that species in it. So far as can be seen, however, it agrees well with some real or supposed varieties of the species *speluncaria*, and it may really belong to that species: and yet, a perfect specimen *might* show it to be quite different.

It may be proper to explain here that the genus *Pseudomonotis* was proposed by Beyrich in 1862 for the group typified by the species *Avicula speluncaria*, Munster. This shell had been for some time previous referred to Bronn's genus *Monotis*, from the type of which (*M. salinaria*) it differs very materially in its inequivalve character, and especially in having a deep, sharply defined byssal notch in the anterior margin of the right valve. On comparing one of these shells with specimens of

*NOTE.—Professor Geinitz has figured another form from Kansas on the same plate with the foregoing under the name *Gervillia parva*, M. & H., which I at one time believed to be a variety of our *Bakevellia parva;* but from such comparisons as I have been able to make of casts (in which condition only it is known to me) during the last three or four years, I have been led to the suspicion that it may be a distinct species.

Bronn's type in 1864, I was satisfied that they were not congeneric, and proposed the name *Gryphorhynchus* for the *speluncaria* group, not being at that time aware that Beyrich had previously proposed to separate it, upon exactly the same grounds, under the name *Pseudomonotis*. As the latter name has priority of date, of course it will have to take precedence.

Locality and position.—The fragment figured was found loose at the base of Cedar Bluff, on Weeping Water, twelve miles northwest of Nebraska City, composed of the same beds referred by Professor Marcou to the Lower Dyas, and by Professor Geinitz to the Coal-Measures, at Rock Bluff on the Missouri. As elsewhere explained, species of this genus occur in Kansas, Iowa, and Illinois, far down in the Coal-Measures, below the horizon of the Nebraska City beds.*

PSEUDOMONOTIS RADIALIS, Phillips ? ?, sp.

Pl. IX, Fig. 3.

? *Pecten radialis*, Phillips, 1834, Encyc. Meth., Vol. IV, Pl. III, Fig. 5.
? *Monotis radialis*, King, 1848, Catalogue, p. 9; 1850, Monogr. Perm. Foss., England, p. 157, Pl. XIII, Figs. 22, 23.
? *Avicula speluncaria*, Geinitz, 1866, Carb. und Dyas in Neb., p. 78; (not Münster).

Left valve small, oblong, higher than wide, moderately convex; basal margin regularly convex; posterior margin nearly straight along the middle, but rounding into the base below, and a little inclined forward above, so as to intersect the hinge at an obtuse angle, and form a slight, undefined, compressed posterior alation; anterior side rounding into the base below, and rather distinctly sinuous above, just beneath a short, round, lobe-like anterior ear; umbo not oblique, moderately convex, and rising very little above the cardinal margin, which is rather shorter than the breadth of the valve. Surface ornamented with fine, unequal, flexuous, radiating ribs, or striæ, more or less roughened, apparently by the crossing of little concentric markings, and a few larger wrinkles of growth; some of the costæ apparently terminating in little short spine-like projections at the base. (Right valve unknown.)

Height, 0.56 inch; antero-posterior diameter, 0.47 inch; convexity (left valve), about 0.13 inch.

I merely refer this form provisionally to *P. radialis*, more from a reluctance to attempt to name and describe a new species from a single valve in a genus like this, than from any strong impression that it is really idential with that form. It has much the general appearance of Professor King's Fig. 22, cited above, though its costæ are much finer and more crowded, and its anterior margin differs in being sinuous below the ear. These differences may or may not be specific, or the shell may even belong to another genus, as it is impossible to determine these questions without specimens, showing both valves, for a comparison.

Locality and position.—Division C of the Nebraska City section. It must be very rare, as only a single specimen was found.

Genus MYALINA, de Koninck.

MYALINA [?] SWALLOVI, McChesney.

Pl. IX, Fig. 7 *a*, *b*.

Myalina Swallovi, McChesney, 1860, Descriptions New Palæozoic Fossils, p. 57; and 1865, Pl. II, Fig. 6, illustrations of same; Meek & Worthen, Illinois Palæont. Report, p. 341, Pl. 27, Fig. 1.
? *Aucella Hausmanni*, Geinitz, 1866, Carb. und Dyas, in Neb., p. 25, Tab. II, Fig. 8; (not Goldfuss, 1834, sp).

Shell rather small, nearly or quite equivalve, modioliform or mytiloid, convex, or even subangular, along the umbonal slopes from the beaks to

* Imperfect casts of perhaps the same species occur at Bennett's Mill.

the anterior basal margin; posterior and postero-dorsal regions, cuneate; cardinal margin nearly straight, and about half the length of the shell,—passing almost imperceptibly, or without any angularity, into the posterior margin, which rounds down with a semicircular curve, to the narrowly rounded basal extremity; antero-basal margin ascending obliquely forward, more or less sinuous near the middle, or sometimes a little above, usually swelling out into a kind of lobe or protuberance above the middle in front of the umbonal slopes, as in *Modiola*. This prominence sometimes extends a little beyond the beaks, and varies more or less in breadth. Beaks small, very oblique, not projecting beyond the cardinal margin, and located so near the anterior extremity as often to appear very nearly terminal. Surface rather smooth, but showing fine concentric lines, which in well-preserved specimens are sometimes crossed by very fine, obscure traces of radiating striæ, that curve upward on the posterior dorsal region.

The specimens figured are of nearly natural size, but smaller than the average size of the species.

This species varies somewhat in the size of the protuberance under the beaks; in some individuals it is well developed, and gives the shell much the form of *Modiola;* while in others it is smaller, so as to present more the appearance of a *Mytilus*. The beaks, however, are rarely quite terminal. The radiating striæ mentioned in the description are very obscure, and may be easily overlooked, as they appear to be rather due to the texture of the shell than to true surface striæ, and in some cases they seem to be entirely obsolete. The cardinal plate, as seen in authentic specimens from Illinois, is quite narrow, and shows only obscure traces of one or two cartilage furrows. As I have also been unable to see any traces of a prismatic structure in the shell, there may be some reason for doubting whether it is a true *Myalina*.

The specimen figured by Professor Geinitz under the name *Aucella Hausmanni*, may possibly be a distinct species from this, but I have little doubt in regard to its being really an internal cast of this shell. That it is not an *Aucella*, however, I have no doubt whatever, that being a Jurassic genus, unknown in the Coal-Measures or Permian rocks, and presenting radical differences. Of a large collection now before me, from the same locality and position, the shell I have figured is the only one resembling that figured by Professor Geinitz. It never has any traces of the little anterior ear, and byssal emargination of *Aucella*.

Locality and position.—Bed C of the Nebraska City section. It is also widely distributed in the Upper and Lower Coal-Measures of Illinois, Kentucky, Iowa, Nebraska, Kansas, and Missouri.

MYALINA SUBQUADRATA, Shumard.

Pl. IV, Fig. 12; and Pl. IX, Fig. 6.

Myalina subquadrata, Shumard, 1855, Geol. Report Missouri Survey, p. 207, Pl. C, Fig. 17; Geinitz, 1866, Carb. und Dyas in Neb., p. 27, Tab. III, Fig. 25 and 26.
Compare *M. deltoidea*, Gabb, Proceed. Acad. N. S. Philad., Nov. 1859.

Shell large and thick, oblong or subquadrate, the height being greater than the antero-posterior diameter; right valve nearly flat; left convex, both somewhat compressed and alate above and behind the umbonal prominence. Hinge line nearly straight, about equaling the greatest breadth of the valves, and ranging at right angles to the vertical axis; basal margin regularly rounded; posterior margin nearly vertical, rounding into the base below, a little sinuous above the middle, and intersecting the hinge above at very nearly right angles; anterior margin thickened within, rounding into base, thence rising nearly vertically, with a broadly rounded concavity mainly above the middle. Beaks terminal and directed forward. Cardinal plate or area usually rather broad, with cartilage furrows distinctly defined. Surface of left valve marked with fine concentric striæ, and stronger imbricating lamellæ of growth. These markings are much less distinct on right valves.

A fully developed specimen of this species, from the original locality on the Missouri near the mouth of Nemaha River (from which locality Dr. Shumard's typical specimens were obtained), measures about three inches in height from the base to the hinge margin, and 1.90 inches in breadth.

This is one of the largest species of the genus known. We did not succeed in finding examples of it in division C at Nebraska City, excepting in the condition of mere fragments, and hence I have copied Professor Geinitz's figure of a specimen from that horizon on Pl. IX. Fig. 11, Pl. IV, however, represents an impression, with portions of the thin outer fibrous layer, of a fully mature specimen, as seen in the matrix, taken from Mr. Morton's shaft, one and three-quarter miles west of the Nebraska City landing. It is widely distributed in the Upper Coal-Measures of Iowa, Nebraska, Kansas, Illinois, &c.

The large species figured by Dr. Hayden and the writer on p. 33, of the Palæontology of the Upper Missouri as *M. subquadrata*, is, as we then suspected, a distinct species, differing in having the posterior margin rounding forward into the hinge above, and not sinuous or meeting the hinge at right angles, as in *M. subquadrata*. It will have to take the name *M. ampla*, suggested by us for it in case it should be found distinct. It and *M. deltoidea* of Gabb, are the only forms resembling *M. subquadrata* with which I am acquainted; and the latter may be only a variety of Shumard's species.

The outer layer of *M. subquadrata* is so coarsely prismatic that its structure can be readily seen, when well preserved, by the aid of a common pocket lens.

Locality and position.—The specimen figured by Professor Geinitz, is from division C of the Nebraska City section. We also have it from division B of that place, and Bennett's Mill; and from Bellevue, in lower positions of the Coal-Measures; likewise from apparently a higher position a short distance west of Nebraska City.

Genus NUCULA, Lamarck.

NUCULA BEYRICHI, v. Schauroth?

Pl. X, Fig. 18.

? *Astarte Geinitziana*, Liebe, 1853, Leonhard u. Bronn, Jahrb., p. 773; (without description).

? *Nucula Beyrichi*, v. Schauroth, 1854, Zeitschr. d. Deutsch. geol. Gess., VI, p. 551, Tab. 21, Fig. 4; Geinitz, 1861, Dyas, p. 67, Tab. XIII, Fig. 22–24; also 1866, in Carb. und Dyas in Neb., p. 21, Tab. 1, Fig. 36–37.

Comp. *Nucula parva*, McChesney, 1860, Descr. New Palæozoic Foss., p. 54; and illustrations same, 1865, Pl. 2, Fig. 8 *a*, *b*, *c*.

Shell very small, longitudinally subovate, moderately convex, widest posteriorly;* anterior end somewhat narrowly rounded; base forming a semiovate curve, the most prominent part being near the shorter end; posterior side very short, comparatively wide, and subtruncated; beaks near the posterior extremity; hinge-line nearly rectangular at the

* In true *Nucula*, the longer side is the anterior.

beaks; denticles comparatively large, about seven on the longer side, and five or six on the shorter; surface marked with moderately distinct regular concentric striæ.

Length, 0.16 inch; height, 0.10 inch.

It is with very great doubt that I have referred this little shell to *N. Beyrichi*, as I only know it from a single specimen, consisting mainly of an internal cast. There is also room for some doubt whether it is the same form referred by Professor Geinitz to *N. Beyrichi*, from the same locality and position. Unless Professor Geinitz's specimens were distorted or incorrectly drawn, I should think his a different species, both from that described above and from the true *N. Beyrichi*. The form described above agrees in size, and tolerably nearly in outline with German examples of von Schauroth's species now before me, or at least with specimens sent to the Smithsonian Museum with that name attached. It differs from von Schauroth's description in having only some seven, or possibly eight, hinge-teeth behind the beak, and five or six in front, instead of twelve of the former and seven of the latter.

In order that others may have some means of forming their own conclusions in regard to the relations of these shells, I have given an enlarged figure of the form described above, and copies of two of Professor Geinitz's figures (see our Pl. X, Fig. 19 *a*, *b*) of the Nebraska shell, referred by him to *N. Beyrichi*. On the same plate (Fig. 25 *a*, *b*, *c*), I have also given copies of von Schauroth's original figures of his species, and of Professor Geinitz's figures of German examples of the same (Fig. 24 *a*, *b*); likewise, figures, natural size and enlarged, drawn directly from a foreign specimen sent to the Smithsonian Institution, labeled *N. Beyrichi*. (See Fig. 23 *a*, *b*.)

Locality and position.—Division C of the Nebraska City section. An undistinguishable form also occurs in the Coal-Measures of Illinois; while another form described from Illinois by Professor McChesney, under the name *N. parva*, is also closely related, though apparently distinct.

NUCULA VENTRICOSA, Hall.

Pl. X, Fig. 17 *a*, *b*, *c*.

Nucula ventricosa, Hall, 1858, Iowa Report, I, Part II, Pl. 716, P. 29, Fig. 5 *a*, *b*.
Compare *N. tumida*, Phillips, 1835, Geol. Yorksh., p. 20, Pl. V, Fig. 15.

Shell small, thick, subovate, very convex; the greatest convexity slightly in advance of the middle of the valves; posterior (shorter) end obliquely truncated from the beaks to its narrowly rounded or subangular connection with the base, rather deeply excavated just behind the beaks; anterior (longer) end rather narrowly rounded, its most prominent part being near or slightly above the middle; dorsal outline declining gently, with moderate convexity from the beak to the anterior extremity; basal margin forming a nearly semiovate curve, being a little more prominent before than behind the middle; beaks convex, rather prominent, and placed about half-way between the middle and the most projecting part of the postero-ventral extremity. Surface with (at least near the base) fine, regular, concentric striæ.

Length, 0.42 inch; height, 0.28 inch.

The only specimen of this little shell in the collection is incrusted by calcareous matter, so as to obscure the surface striæ, excepting near the lower margin; hence the striæ, on other parts of the figures given, are restored. It is therefore barely possible that the more convex portions of the valves may be smooth, or nearly so.

It is possible that the last-described form may be only the internal cast of this. If so, I could have no doubts in regard to the species being entirely distinct from *N. Beyrichi*, since its form is altogether different.

Although this shell differs slightly in outline from the particular specimen of *N. ventricosa*, figured in the Iowa Report, a comparison with a series of specimens of that species (which varies more or less in outline) leaves little room to doubt the identity of our shell with *N. ventricosa*. It is also proper to remark here that after comparing good specimens of *N. ventricosa* from the Coal-Measures of Illinois with examples of a shell sent to Mr. Worthen by Mr. Thomas Davidson, of Brighton, from Carluke, Scotland, with the name *N. tumida*, Phillips, attached, no satisfactory differences were observed.

Locality and position.—Division C of the Nebraska City section: also, at Rock-Bluff, in bed 6, of the section at that place. This species is common all through the Coal-Measures of Illinois; and I have identified it among Lower Coal-Measure species from West Virginia, collected by Professor J. J. Stevenson, of Morgantown, in that State.

Genus YOLDIA, Möller.

YOLDIA SUBSCITULA, M. & H. ?

Pl. X, Fig. 10.

Leda subscitula, Meek & Hayden, 1858, Trans. Albany Inst., Vol. IV.*
Yoldia ? subscitula, M. & H., 1864, Palæont. Upp. Mo., Part 1, p. 60, Pl. II, Fig. 4, *a. b.*
Nucula (*Leda*) *subscitula ?* Geinitz, 1866, Carb. und Dyas in Neb., p. 22, Tab. 1, Fig. 35.

Shell longitudinally subovate or subelliptic, compressed, the greatest convexity a little in advance of the middle, about twice as long as high; anterior extremity wider than the other, but rather narrowly rounded, the most prominent point being usually slightly above the middle; outline of base forming a broad semiovate curve, being more prominent anteriorly than behind; posterior side narrowed, its margin rounding up gradually from the base, so as to meet the dorsal margin nearly at right angles, sometimes faintly truncate at the immediate extremity; posterior dorsal margin compressed or cuneate, and declining gradually, with a nearly straight, or slightly concave outline; anterior dorsal margin not cuneate, sloping forward gradually, and a litle convex in outline; beaks rather depressed and subcentral, or very little in advance of the middle; umbonal slopes without any defined ridge or angle. Surface smooth, or only showing traces of very minute concentric striæ.

Length, 0.77 inch; height, 0.37 inch; convexity, about 0.14 inch.

It is with considerable doubt that I have concluded to refer this shell to *Y. subscitula*, M. & H., because the specimens from Nebraska City are distinctly more compressed than the type upon which that species was founded, their convexity being uniformly not more than half as great proportionally. The five or six individuals in the collection are constant in this character, and yet show no evidences whatever of accidental compression. Otherwise the two forms are *very* similar in their general outline, but we know nothing of the internal and hinge characters of the form under consideration. I strongly suspect, however, that it will be found to be a distinct species, in which case I would propose to call it *Yoldia propinqua*, from its near resemblance to *Y. subscitula*. Of course

* This paper was issued in the form of extras on 4th March, 1858, some time in advance of the volume.

it is only placed provisionally in the genus *Yoldia*, its internal characters being unknown. It agrees exactly, however, in all external characters with that genus, and has a crenated hinge.

Professor Geinitz's figure of this shell is slightly defective (that is, supposing it to represent the same species, of which there is scarcely any reason to doubt, as it agrees in other respects, and was taken from a specimen from exactly the same locality and position) in having the posterior margin rounding into the hinge instead of meeting it at a more or less obtuse angle; this may have resulted from a slight imperfection of his specimen.

I have described, from the Lower Coal-Measures of West Virginia, in the Third Annual Report of the Regents of the University of that State for 1870, under the name *Y. Stevensoni*, another very closely allied species, only distinguished by having rather distinct, regularly disposed lines and furrows of growth.

Locality and position.—Division C of the Nebraska City section, Nebraska City.

Genus NUCULANA, Link.

NUCULANA BELLISTRIATA, *var.* ATTENUATA.

Pl. X, Fig. 11 *a*, *b*.

Leda bellistriata, Stevens, 1858, Am. Jour. Sci. and Arts, Vol. XXV, p. 261; Hall, 1858, Iowa Report, Vol. 1, Part II, p. 717, Pl. XXIX, Fig. 6.

Nucula Kazanensis, Geinitz, 1866, Carb. und Dyas in Nebr., p. 20, Tab. 1, Figs. 33 and 34; (not *Nucula Kazanensis*, de Verneuil ?, 1845).

Shell longitudinally subovate, moderately convex, or rather gibbous in the umbonal and anterior regions, as well as along the posterior umbonal slopes; basal margin semiovate; anterior margin more or less narrowly rounded; posterior very attenuate, and, at the extremity, subangular; umbonal ridge well defined; cardinal margin behind the beaks slightly concave in outline, carinate, and sloping backward, with a rather distinct, broad concavity or impression between it and the umbonal ridge, in front of the beaks, sloping more abruptly with a convex outline; beaks moderately prominent, and located generally about two-fifths the entire length of the valves behind the anterior margin. Surface ornamented with very fine and regular, concentric striæ, which become obsolete on the posterior umbonal ridges.

Length, 0.42 inch; height, 0.20 inch; convexity, 0.10 inch.

The specimens of this little shell yet obtained from Nebraska are smaller and more pointed behind than in the average adult size of *N. bellistriata* from the typical localities in Illinois, as well as rather more finely striated, but they agree so closely in form and general appearance as to render their identity with that shell almost certain. I have not seen any specimens from the Nebraska localities showing the hinge, but Dr. Stevens describes the Illinois specimens as having "about twenty-five teeth, five of which are smaller than the others, and clustered under the beaks."

Whether Professor Geinitz is right in referring this species to *N. Kazanensis* of Verneuil, is a question I have not the means of determining, only knowing that shell from figures of moulds of the exterior left in the matrix. If I may form an opinion from these figures, I should say that it is very similar to our shell, but scarcely more so than to Illinois specimens, from the typical localities in the Coal-Measures of that State. If the smaller size of the Nebraska specimens should be an objection to their being specifically identical with *N. bellistriata*, it should be remembered that this objection will equally apply to their reference to the Russian species, which is very nearly of the same size as the typical Illinois specimens of *N. bellistriata.*

Supposing the published figures of *N. paruncului*,v. Keyserling, and *N. Kazanensis*, de Verneuil, to be even nearly correct, I must differ widely from Professor Geinitz in regard to their representing the same species. Indeed, it is rarely the case that two species of this genus are more unlike, even if we compare recent and Carboniferous species. The *N. paruncului*, if correctly represented, is *certainly* distinct from the Nebraska and Illinois shell, which, although varying somewhat, never assumes the form of Count von Keyserling's species.

Locality and position.—Division C of the Nebraska City section; also at a lower position in the Coal-Measures at Leavenworth, Kansas, and various localities in Iowa, Illinois, &c. I have likewise identified *precisely* the same smaller and more attenuate variety of this shell among specimens sent by Professor Stevenson from the Lower Coal-Measures of West Virginia.

Genus MACRODON, Lycett.

MACRODON TENUISTRIATA, M. & W.

Plate X, Fig. 20 *a*, *b*.

Macrodon tenuistriata, Meek & Worthen, 1867, Proceed. Chicago Acad. Sci., I, p. 17.
Arca striata, Geinitz, 1866, Carb. und Dyas in Neb., p. 20, Tab. 1, Fig. 32; (not *Mytilites striatus*, Schloth., 1817,=*Arca striata*, V. Schauroth, 1856).

Shell small, rhombic-oblong, rather distinctly convex, along the umbonal slopes, and near the front, a little more than twice as long as high; basal and cardinal margins parallel; the former nearly straight, or somewhat sinuous near the middle; cardinal margin straight, not quite equaling the greatest antero-posterior diameter; anterior side rounding up from below so as to meet the hinge nearly at right angles; posterior basal margin narrowly rounded; posterior margin obliquely truncated, often a little sinuous above; dorsal region behind the umbonal slope compressed; beaks convex, a little flattened, incurved, and rising somewhat above the hinge margin, located about half-way between the middle and the front; flanks broadly impressed or concave from the umbonal regions obliquely backward to the faintly sinuous part of the base; cardinal area unknown; posterior linear teeth about three; surface ornamented with distinct marks of growth crossed by radiating markings, which on the compressed posterior dorsal region form rather well-defined radiating lines; anteriorly, however, these diminish in size so as to become very minute or scarcely visible, crowded, obsolescent striæ.

Length, 0.75 inch; height, 0.31 inch; convexity, about 0.30 inch.

Professor Geinitz has identified this with the well-known *M. striatus*(= *Mytilites striatus* of Schlotheim). It may be at once distinguished from that species, however, by its costæ becoming mere minute obsolescent striæ on the middle and anterior portions of the valves. In order that the student may have the means of making the comparison for himself, I have also given, on the some plate, Fig. 27, an exact copy of Professor Geinitz's figure of the *M. striatus*, from the Permian rocks of Germany, given in his "Dyas," Taf. XIII, Fig. 33 *a*, from which our shell differs in so many points that I think there can be but one opinion among paleontologists generally, in regard to these shells being clearly distinct species. I do not wish to disguise the fact, however, that *M. striatus* (or at least forms referred to it) varies in the size of the radiating costæ, some of them having the ribs not more than half as large as those of the example copied from Professor Geinitz. An examination of good European specimens, as well as of all the published figures, shows that they are always as coarse (and generally even more so) on the anterior part of the valves as behind, and never become mere minute lines on any part of the valves, while there are other differences not to be overlooked.

The surface markings mentioned, together with its much less ventricose form, will also distinguish our shell from *M. tumidus*(=*Bysso-arca tumida*) of Sowerby. From *Macrodon carbonarius*(=*Arca carbonaria*), Cox, it will be at once distinguished by its finer radiating striæ, and its beaks being placed less nearly over the anterior margin.

Locality and position.—Division C of the Nebraska City section. It also occurs in the Upper Coal-Measures of Springfield, Illinois, and at other localities of that State in the Lower Coal-Measures. Dr. White has likewise found it in the Upper Coal-Measures of Western Iowa.

Genus SCHIZODUS, King.

SCHIZODUS CURTUS, M. & W. ?

Pl. X, Fig. 13 *a*, *b*, *c*, (*d*?) *e*.

Schizodus curtus, Meek & Worthen, 1866, Proceed. Chicago Acad. Sci., 1, p. 18.
? ———*Rossicus*, Swallow, 1858, Trans. St. Louis Acad. Sci., 1, p. 193.
———*Rossicus*, Geinitz, 1866, Carb. und Dyas in Neb., p. 18, Tab. 1, Fig. 28; (not *S. Rossicus*, de Verneuil ?)
Comp. *Sch. rotundatus*, Brown sp., 1841, as figured in King's Perm. Foss., England, Pl. XV, Fig. 30.

Shell small, suborbicular, rather compressed, thin; anterior side obliquely subtruncated, with a convex outline above, and rounded into the base below; base deeply rounded anteriorly, and ascending with a slightly straightened outline, or even sometimes very faintly sinuous, behind; posterior side narrower than the front, nearly vertically truncated, so as to form almost a right angle with the base at the termination of the umbonal ridge, and more or less rounding into the cardinal margin above; posterior dorsal region behind the umbonal ridge compressed and cuneate; cardinal margin sloping more or less behind the beaks; beaks elevated, incurved, and placed very slightly in advance of the middle; umbonal slope rather distinctly angular from the beaks to the posterior basal extremity; flanks just in advance of this ridge, sometimes faintly concave; surface marked with very fine concentric striæ.

Length of the largest specimen of the form from which the above description was drawn up, 0.43 inch; height, 0.38 inch; convexity, about 0.16 inch.

The foregoing description was made out from a fine series of specimens from Nebraska City, but this shell agrees so nearly with *S. curtus* from the Coal-Measures of Illinois (see Fig. 13, *e*, *f*, *g*, of same plate, given for comparison from an Illinois specimen from the original locality), that nearly every word of the description would apply equally well to that shell; the only differences being in the slightly straighter and more sloping posterior margin, and greater convexity of the Illinois shell. The latter character, however, is almost certainly due to the accidental compression of the Nebraska specimens, all of which, when carefully examined, show minute cracks on each side of the beaks, evidently produced by accidental pressure. The Illinois specimens being perfectly preserved, with the substance of the shell remaining, show the surface striæ rather more distinctly than those from Nebraska, which are casts of the exterior. Without intending to express a positive opinion in regard to the relations of these two forms, I have concluded to refer the Nebraska City specimens provisionally to *S. curtus*.

In regard to both of these forms being distinct from *S. Rossicus*, however, *provided the figures given in the Geology of Russia are exact and represent normal forms of that shell*, I think there is much less room for doubt. On comparison with these, Figs. 21 *a* and 21 *b*, which I have reproduced for that purpose, particularly Fig. 21 *a*, which is the typical form of *S. Rossicus*,* it will be seen that the Nebraska shell has the beaks much more prominent, and the antero-ventral region much more deeply rounded, while its posterior margin is more nearly vertically truncated. Its posterior basal margin is also more angular in outline. Now, in all these characters the Nebraska shell is *very* constant. It is true there is another associated form, of which all of the few specimens I have seen are larger. This (see Fig. 13 *d*) I think almost certainly a distinct species, having seen no intermediate forms.

The shells here described, including both the Nebraska and Illinois specimens, are certainly *very* closely allied to *S. rotundatus* of Brown, and if identical with any European shell I should think would have to be referred to that species.

The drawings I have given are made with great care, and all the points of difference mentioned above and represented in the figures, both from the specimens and the figures reproduced for comparison, may be relied upon.

Locality and position.—The specimens represented by Fig. 13 *a*, *b*, *c*, and *d*, are from division C of the Nebraska City section. That represented by Fig. 13 *e*, *f*, *g*, is from the Wabash cut-off, Illinois Upper Coal-Measures. I also have somewhat larger specimens, agreeing nearly with the latter, from the Coal-Measures near Atchison, Kansas. The same form is likewise found in the Lower Coal-Measures of Illinois. According to Eichwald, however, *S. Rossicus* occurs both in the Carboniferous and Permian rocks of Russia.

SCHIZODUS WHEELERI, Swallow, sp.

Pl. X, Fig. 1 *a*, *b*, *c*, *d*, (and *e*, *f*?).

Cypricardia ? Wheeleri, Swallow, 1862, Trans. St. Louis Acad. Vol. I, p. 96.
Schizodus obscurus, Geinitz, 1866, Carb. und Dyas in Nebraska, p. 20, Tab. I, Figs. 30 and 31; (not Sowerby, 1821).

Shell attaining a medium size, longitudinally subovate, moderately convex; anterior side wider than the other, and regularly rounded; posterior side narrowed, and obliquely truncated; basal outline rather prominently rounded anteriorly, and straightened, or slightly sinuous between the middle, and sharply rounded or subangular posterior basal extremity; dorsal margin straight, and sloping from the beaks to the truncated posterior edge; beaks rather depressed (for a species of this genus), incurved, and placed about half-way between the middle and the front, or perhaps nearer the middle; posterior umbonal slope rather prominent, or usually forming a rather obtuse ridge near the posterior basal extremity; surface with merely fine lines and obscure marks of growth.

Professor Swallow's typical specimen of *Cypricardia ? Wheeleri*, according to a tracing he permitted me to make from his drawing, measures 1.15 inches in length, and 0.81 inch in height.

---

* The other figure, 21 *b*, is said not to be exactly correct, and besides may belong to another species.

It is possible the shell described under the above name by Professor Swallow may be distinct from those figured by Professor Geinitz, as the hinge and interior of the two cannot be compared; yet the tracing of the *S. Wheeleri*, as may be seen by Fig. 1 *b* of Plate X, is so nearly like Professor Geinitz's figures (one of which I have also copied on the same plate (Fig. 1 *a*), that I can scarcely doubt their identity.

Figures 1 *c*, *d*, of Plate X, represent the exterior and interior, hinge, &c. of a beautiful specimen, apparently of the same shell, from the Upper Coal-Measures of Adams County, Iowa, kindly loaned to me by Dr. White, the Iowa State geologist. Figures 1 *e* and 1 *f* of the same plate, represent an internal cast of a larger specimen, also loaned by Dr. White, from the Upper Coal-Measures of Union County, Iowa. Although the latter is larger, and differs slightly from the outline of Professor Geinitz's figures, if we make allowance for its larger size and for the fact of its being an internal cast, I see no very satisfactory reason for regarding it as a distinct species, though it may be so. At any rate, all of these four shells—that is, the typical *S. Wheeleri*, the one figured by Professor Geinitz, and the two from Iowa—differ more decidedly from the published figures of *S. obscurus*, to which Professor Geinitz refers the Nebraska City specimens, than they do from each other. For instance, it will be observed that they all have the umbones less prominent and less ventricose than is represented in the published figures of *S. obscurus;* while they all, both in internal casts and testiferous specimens, present a straightness of the posterior dorsal slope, and a distinctly truncated outline of the anal margin, never yet represented, so far as I have been able to see, in *S. obscurus*. In most of these points of difference, it will be observed that these shells agree more nearly with a form figured by Professor King, under the name *S. Schlotheimi*, Geinitz, on Plate XV, Fig. 32 of his Monograph of the English Permian fossils. Professor Geinitz, however, thinks this specimen figured by Professor King does not belong to his *S. Schlotheimi*. It certainly looks very unlike any of those figured by Professor Geinitz under that name in his work on the German Permian fossils, although the figures there given under the name *S. Schlotheimi* represent at least three forms differing much more widely from each other than the Nebraska and Iowa shells differ from each other, or from the typical *S. Wheeleri* of Swallow.

Until these shells can be better known, I have preferred to refer them, provisionally, to Professor Swallow's species, and at the same time to give figures of the several types, so that others can have the means of forming their own conclusions on these points.

Locality and position.—Professor Swallow's type of *S. Wheeleri* is from the Upper Coal-Measures of Caldwell County, Missouri. The specimens figured by Professor Geinitz are from division C of the Nebraska City section; and those I have figured from Iowa came from Adams and Union Counties, where they were found by Dr. White near the middle of the Upper Coal-Measures, associated with many of the same fossils found in division C of Nebraska City. All of these forms also occur in the Coal-Measures of Illinois.

SCHIZODUS, undt.

Pl. X, Fig. 2.

Of this form I have seen but a single specimen, consisting of an external cast of one valve. In its general outline it is decidedly more like *S. obscurus* than that figured by Professor Geinitz under that name from the same bed at Nebraska City; and yet it seems to differ from the latter, as well as from *S. obscurus*, in being more compressed. It will also, on comparison, be at once seen to differ from the other Nebraska City specimens, as well as from those from Iowa, in having its umbones more elevated and its anterior ventral margin more deeply rounded in outline. Without other specimens for comparison, I would not be willing to refer it to *S. obscurus*, nor yet to name it as a new species.

It is also worthy of note that the form under consideration is quite as nearly like one of the specimens figured by Professor Geinitz on Plate XIII of his work on the Permian fossils of Germany, under the name *S. Schlotheimi*, though quite unlike the others. I mean his Figure 12 of the plate referred to. Not having access to his original description of the *S. Schlotheimi*, I do not know which of the forms he has figured on the plate above cited is the type of that species.

Locality and position.—Division C of the Nebraska City section.

Genus MODIOLA, Lamarck.

MODIOLA? SUBELLIPTICA, Meek.

Pl. X, Fig. 5.

Clidophorus (Pleurophorus) occidentalis, Geinitz, 1866, Carb. und Dyas in Neb., p. 23, Tab. II, Fig. 6; (not *Pleurophorus occidentalis*, M. and H., 1858).

Pleurophorus subellipticus, Meek, 1867, Am. Jour. Sci. and Arts, Vol. XLIV, new series, p. 181.

Shell narrow, subelliptical, rather convex, extremely thin, usually a little more than twice as long as high; basal margin nearly straight, or sometimes very slightly convex or sinuous near the middle, rounding up at each extremity; anterior margin narrowly rounded; posterior extremity more compressed, and more broadly round, sometimes a little oblique above; cardinal margin somewhat straightened along the middle but rounding imperceptibly into the anterior and posterior extremities; beaks much depressed, or scarcely distinct from the cardinal margin, moderately convex and placed very near the anterior margin, but not terminal; umbonal slopes forming a very obscure narrow ridge, which extends, with a slight curve from each umbo, to the posterior basal margin. Surface marked with moderately distinct lines of growth, which on the posterior dorsal rigion above the umbonal ridge, are crossed by very minute or microscopic radiating and rather distinctly divaricating striæ.

Length of largest specimen seen, 1.03 inches; height of same, 0.45 inch; convexity, about 0.25 inch.

Professor Geinitz referred this species to *Pleurophorus occidentalis*, M. and W., but it is a widely distinct shell, differing greatly in having its beaks scarcely distinct from the cardinal margin, instead of prominent and nearly terminal. It also differs in the outline of its dorsal margin, which is concave in *P. occidentalis;* while the latter is a much thicker shell. Indeed, a careful examination of a good series of specimens, obtained from Nebraska City by Dr. Hayden's survey, since the publication of my review, in the American Journal of Sciences, of Professor Geinitz's work on the Nebraska fossils, has clearly satisfied me that this shell is not a *Pleurophorus* at all, as its internal casts show no traces of the posterior hinge teeth or internal ridges of that genus. Its general form and divaricating minute radiating striæ, depressed beaks, &c., seem to indicate affinities to some of the sections of the genus *Modiola*, though it may possibly be found to fall into the genus *Cardiomorpha*, and have to take the name *C. subelliptica*, or into *Modiomorpha* and have to be called *M. subelliptica.*

Locality and position.—Division C of the Nebraska City section; also at a lower position in the Coal-Measure at Riverside, three miles below Atchison, Kansas.

Genus PLEUROPHORUS, King.

PLEUROPHORUS OBLONGUS, Meek.

Pl. X, Fig. 4 *a*, *b*, *c*.

Pleurophorus Pallasi, Geinitz (pars), 1866, Carb. und Dyas in Neb., Tab. II, Fig. 4; (not *Modiola Pallasi*, de Vern., 1845).

Shell small, longitudinally oblong, about twice as long as high, moderately convex, particularly along the umbonal slopes from the beaks to the posterior basal margin, but without any defined angle or ridge there; cardinal margin nearly straight, and subparallel to the base, about equaling two-thirds the entire length of the valves; basal margin more or less distinctly sinuous near the middle, at the termination of a broad, oblique impression or concavity extending from the anterior side of the beaks under the umbonal slopes to the lower margin; anterior margin narrowly rounded below; posterior side much wider, rounded, or sometimes obliquely subtruncated above; beaks convex, very oblique, obtuse, located one-seventh to one-eighth the length of the valves behind the anterior extremity; surface with apparently only fine concentric marks of growth; muscular impressions faintly marked; ridge behind the anterior one small; posterior lateral tooth slender and elongated.

Length of largest specimen seen, 0.44 inch; height, 0.24 inch; convexity, about 0.14 inch.

Although this shell resembles in external characters some of the short varieties of the so-called *Pleurophorus Pallasi*, from the Russian Permian rocks, I cannot agree with Professor Geinitz in referring it to that species, because some of the internal casts before me show that the Nebraska shell has a distinct, elongated, linear posterior-lateral tooth, as we see in true *Pleurophorus;* while the Russian species is both figured and described as being "completely edentulous," and is consequently not a *Pleurophorus*, but more probably a *Cardiomorpha.* In addition to this, our shell has a much straighter and proportionally longer hinge, is not, so far as yet known, so variable in form, and has its anterior muscular impression defined by a weaker and less oblique ridge.

I am also compelled to differ with Professor Geinitz in regard to the identity of this and the form represented by his Fig. 3 of the same plate; and as I have elsewhere stated, more decidedly, in regard to the latter being identical with the Russian species *Pallasi*, from which I believe it to differ generically.

Locality and position.—Division C of the Nebraska City section.

PLEUROPHORUS OCCIDENTALIS, M. & H.?

Pl. X, Fig. 12.

Pleurophorous occidentalis, Meek & Hayden, 1858, Trans. Albany Inst., IV; 1864, Palæont. Upper Missouri, p. 35, Pl. I, Fig. 11 *a*.*

Clidophorous Pallasi, Geinitz (pars), 1866, Carb. und Dyas in Neb., p. 23, Pl. II, Fig. 3; (not *Mytilus Pallasi*, de Vern., 1845).

Internal cast small, narrow-oblong, moderately convex along the umbonal slopes; basal and dorsal margins nearly straight and sub-parallel,

* This figure is very imperfectly lithographed, the outline of the base being made too convex, and the anterior basal margin ascends too obliquely.

or converging slightly forward; posterior side wider than the anterior, rather compressed, obliquely sub-truncated above and narrowly rounded below; anterior side very short and rounded; beaks very oblique, located almost directly over the anterior extremity; posterior dorsal region, with three oblique radiating ridges, extending from behind the beaks to the anal margin, the lower one forming the umbonal ridge; below the latter traces of two or three much smaller radiating linear marks are seen; impression of posterior lateral hinge tooth distinct along the cardinal margin; internal ridge bounding the anterior muscular scar, well defined and ranging nearly vertically.

Length, about 0.46 inch; height, 0.24 inch.

It is not without some doubts that I have referred this form, which I only know from Professor Geinitz's figure of an internal cast (copied on Pl. X, Fig. 12), to *P. occidentalis*, M. & H., though strongly inclined to believe it the same. It is much more like that species than the figure of it given in our Palæontology of the Upper Missouri would lead one to suppose, the error in the engraving giving the shell too much of a pointed appearance anteriorly. When allowance is made for this fault in the engraving, and it is borne in mind that Professor Geinitz's figure represents an internal cast, and that the thickened shell in the region of the umbones must have made the beaks look much more prominent than in the cast, it will be understood that the form under consideration must have closely resembled *P. occidentalis*.

Whatever may be its relations to that shell, however, it seems to me very clear that it must be widely distinct, as I have elsewhere shown, from the so-called *Mytilus Pallasi*, de Verneuil, which, as already stated, is both figured and described as a "completely edentulous" shell; while that under consideration, as shown by the cast, has the long posterior lateral tooth, as well as the general physiognomy of true *Pleurophorous*. In short, as suggested by Professor King, the so-called *Mytilus Pallasi*, has the hinge characters of *Cardiomorpha;* but whether a *Cardiomorpha* or not, it can scarcely be possible that it belongs to the same genus as the shell under consideration.

I should be much more inclined to think our shell identical with *Pleurophorus costatus*, Brown, than with the *M. Pallasi;* for on comparison with Professor King's figures of that species (Perm. Foss., England, Tab. XV, Figs. 13 and 14), they will be seen *very* closely to resemble it.

Locality and position.—The typical specimens of *P. occidentalis* were from the Upper Coal-Measures in Otoe City, Nebraska, opposite the northern boundary of Missouri. The specimen figured by Professor Geinitz, and copied on Plate X, was found in division B of the Nebraska City section. I have also seen the same shell from the Upper Coal-Measures of Illinois, at Grayville.

Genus EDMONDIA, de Koninck.

EDMONDIA REFLEXA, Meek.

Pl. X, Fig. 6, *a*, *b;* and Pl. IV, Fig. 7.?

Shell sub-elliptical, the length being about twice the height, moderately convex in the umbonal and central regions, but without any defined anterior or posterior umbonal ridges; basal margin forming a broad, sem-ielliptic curve; anterior and posterior extremities subequally rounded; dorsal margin somewhat straightened along the middle, but rounding into the extremities, without a defined escutcheon or lunule;

beaks much depressed, obtuse, and rising very little above the hinge margin, placed near half-way between the middle and the front; surface with fine lines and small undulations of growth. (Hinge and interior unknown.)

Length, 0.60 inch; height, 0.30 inch; convexity, about 0.19 inch.

Internal casts of this species show an impression near the hinge, behind the beaks, apparently made by a cartilage lamina, such as we see in the genus *Edmondia*, to which I have provisionally referred it. At first I was inclined to think the depressed character of the beaks might be due to some accident; but finding it to be constant in several examples, showing no evidences of distortion, I can but regard it as a natural character of the shell, though this would not *necessarily* prove it to belong to this genus.

In all the specimens yet obtained the anterior margin above the middle seems to be slightly reflexed, as if the valves had been a little gaping there.

Locality and position.—Division C, and Morton's shaft, Nebraska City

EDMONDIA ? GLABRA, Meek.

Pl. X, Fig. 7 *a*, *b*.

Shell subelliptical, rather compressed, or moderately convex; length about once and a half the height; extremities rounded, the posterior a little wider than the other; basal margin semielliptic in outline; umbones rather obtuse, rising moderately above the dorsal margin, and placed less than half-way forward from the middle toward the front; cardinal margins sloping very slightly from the beaks backward, and rounding imperceptibly into the posterior margin, declining more abruptly anteriorly, with a slight excavation in front of the beaks, but without a defined lunule; both valves without anterior or posterior umbonal ridge or escutcheon. Surface with faint lines and very obscure traces of stronger concentric marks of growth.

Length, 0.67 inch; height, 0.45 inch; convexity, about 0.20 inch. A single imperfect specimen from the same locality and position, probably of this species, shows obscure rounded concentric undulations, and measures 1.60 inches in length, and about 1.16 inches in height.

All the specimens I have seen being casts of the exterior, giving no clew to the hinge and internal characters, it is only provisionally I have referred this species to *Edmondia*. I know of no species with which it may be confounded, unless it may be some of the forms briefly characterized, by Professor Swallow, under the names *Cypricardia*, *Cardinia*, &c. As near as I can determine from his diagnoses, however, it seems to be distinct from all of these.

Locality and position.—Division C of the Nebraska City section.

EDMONDIA ? NEBRASCENSIS, Geinitz, sp.

Pl. X, Fig. 8 *a*, *b*.

Astarte Nebrascensis, Geinitz, 1866, Carb. und Dyas in Neb., p. 16, Tab. 1, Fig. 25.
? *Astarte*, sp., ib., Fig. 27.

Shell subovate, compressed, more or less rounded at the extremities; length nearly once and a half the height; basal margin broadly semielliptic or semiovate in outline; dorsal margin sloping from the beaks, but more abruptly in front than behind, rounding into the extremities; beaks moderately prominent, and located somewhat in advance of the middle. Surface marked by broad, rounded, rather regular concentric furrows, separated by sharp, moderately prominent concentric linear ridges, which sometimes show under a magnifier indications of being minutely crenate; impressions or furrows between the ridges, showing concentric striæ, which, by the aid of a lens, in a cross light, appear to be crossed by fine, nearly obsolete radiating markings.

Length of largest specimen seen, 1.35 inches; height, 0.95 inch; convexity, about 0.30 inch.

It will be observed that our figure 8 *b* does not exactly agree in outline with that given by Professor Geinitz, and represents a larger shell. Hence I am in some doubt whether or not it is the same. As it came from the same locality and position, however, and agrees more nearly with his figure than any of the other specimens of this size obtained there, I have concluded to refer it provisionally to that species.

From the same bed, at the same locality, a number of smaller, more rounded specimens with more nearly central beaks were obtained, apparently agreeing almost exactly, in their furrows and ridges, with that described above. Fig. 8 *a* of Pl. X represents one of these forms. It was probably from one of these that Professor Geinitz's Fig. 27, of his Tab. 1, was drawn, though that figure represents the beaks more exactly central, and the hinge margin straighter, than any of our specimens. It is possible that these may all belong to the one species, *Nebrascensis;* though I suspect them to be distinct, not merely from the differences of size and form, but because I have not seen satisfactory evidences of their concentric ridges being minutely crenate, as in the larger, more elongated shell. Should these smaller, shorter individuals prove distinct, they may take the name *Edmondia? Geinitzii.*

In regard to the generic relations of these shells, it is proper to remark that Professor Geinitz only referred them provisionally to the genus *Astarte*, and says he thinks they may belong either to *Edmondia* or *Cardiomorpha*. In this opinion I fully concur with him, as it is manifest that they are not true *Astarte*, because they have the dorsal margin behind the beaks, erect, and not inflected or excavated, and want the deeply impressed, sharply defined lunule of *Astarte*. They, moreover, seem to have been much thinner shells than we usually see in that genus. The absence of a defined lunule and inflected dorsal margin also separates these shells from *Astartella*, some of the species of which they closely resemble in form and surface-markings.

These shells seem to be related specifically to some of the forms described by Professor Swallow, under the names *Cardinia*, *Edmondia*, *Cypricardia*, &c., and may possibly be identical with some of them, but so far as I have been able to determine from descriptions alone, they seem to be distinct.

Locality and position.—Division C of the Nebraska City section.

EDMONDIA SUBTRUNCATA, Meek.

Pl. II, Fig. 7.

Shell longitudinally oblong, being between one-third and one-fourth longer than high, moderately convex, the greatest convexity being near the middle; posterior margin subtruncated, though convex in outline and rounding abruptly into the basal and cardinal margins; base but slightly convex in outline along the middle, and rounding up rather more gradually anteriorly than behind; anterior margin short and narrowly rounded below, abruptly and obliquely truncated above; dorsal margin nearly straight and scarcely declining behind the beaks; beaks rather depressed, convex, and placed nearer the anterior end than the middle.

Surface of cast marked with moderately distinct irregular undulations.

Length, 1.33 inches; height, 0.95 inch; convexity, about 0.50 inch.

It is possible that this may be the form figured by Professor Geinitz on Plate I of his work on the Nebraska fossils, as *Astarte gibbosa* of McCoy, though it is proportionally longer, being more oblong in outline, and less ventricose. Neither of these shells belong to the genus *Astarte*, however; they are probably *Edmondia* or *Cardiomorpha*. The identity of either of them with McCoy's species is exceedingly doubtful, however.

Locality and position.—Rock Bluff, Nebraska, Upper Coal-Measures; we also have it from a still lower position in the Coal-Measures at Atchison, Kansas, and, I believe I have seen it in the Illinois Lower Coal-Measures, though of this I am not quite sure.

EDMONDIA ASPINWALLENSIS, Meek.

Pl. IV, Fig. 2 *a*, *b*, *c*.

Edmondia Aspinwallensis, Meek, 1871, Dr. Hayden's Report of Geol. Survey Wyoming, p. 299.

Shell longitudinally subovate, moderately convex, the greatest convexity being a little in advance of and above the middle; base nearly semielliptic in outline; posterior side rather narrowly rounded, or sometimes very faintly subtruncate obliquely above; dorsal margin nearly straight just behind the beaks, but very gradually declining with a slightly convex outline posteriorly; anterior side quite short and declining very abruptly from the beaks above, and rounded below; beaks rather depressed, incurved, and located nearer the anterior end than the middle. Surface of cast with moderately distinct, irregular concentric undulations; showing behind the beaks distinct impressions of the cartilage fulcra.

Length, 1.45 inches; height, 1.03 inches; convexity, about 0.68 inch.

This species resembles the last somewhat, but differs in being more ovate, instead of nearly oblong, owing to its narrower posterior extremity, and more elevated beaks. Its undulations are also more obscure, and its beaks more pointed.

Locality and position.—Aspinwall on the Missouri, in apparently a somewhat higher horizon than any of the beds exposed at Nebraska City. It also occurs at various horizons in the Coal-Measures of Illinois, and in the Lower Coal-Measures of West Virginia.

Genus CHÆNOMYA, M. & H.

CHÆNOMYA LEAVENWORTHENSIS, M. & H.

Pl. II, Fig. 9.

Allorisma Leavenworthensis, Meek & Hayden, Dec'r, 1858, Proceed. Acad. Nat. Sci. Philad., p. 263.

Chænomya Leavenworthensis, Meek, 1864, Palæont. Upper Missouri, Part 1, p. 43, Pl. II, Fig. 1, *a*, *b*, *c*.

Of this curious shell, we have but a single fragment, consisting of the posterior, extremely widely gaping portion of the two valves united. The species is so peculiar, however, and so entirely unlike any other form known in these rocks, that it can scarcely be possible to confound it with any other shell. It shows the lines of growth, and small concentric wrinkles, abruptly deflected parallel to the distinctly truncated and widely gaping posterior margin, very clearly; and even the radiating rows of minute granules are well preserved on the lower half.

Locality and position.—The specimen figured is from near the middle of the Rock Bluff section of the Upper Coal-Measures. The species was originally described, from a dark argillaceous limestone of the Coal-Measures, at nearly the level of the Missouri River, at Leavenworth, Kansas. It also occurs in the Coal-Measures of Iowa, Missouri, and Illinois, being sometimes found below the middle of the series in the latter State.

CHÆNOMYA MINEHAHA, Swallow, sp.

Pl. —, Fig. 13 *a b*.

Allorisma ? Minehaha, Swallow, 1858, Trans. St. Louis. Acad. Sci., Vol. I, p. 193.
Chænomya Minehaha, Meek & Hayden, 1858, Palæont. Upp. Mo., p. 43.

Shell rhomboidal in outline, very convex, distinctly less than twice as long as high; posterior truncation very oblique, and gaping nearly to the full breadth of the united valves, but without reflexed edges, dorsal margins comparatively short, concave in outline, and rather distinctly inflected; basal margin forming a gentle curve from the sub-angular posterior basal extremity forward, and obliquely ascending anteriorly to the very short abruptly rounded anterior end; beaks very oblique, depressed, incurved, and located nearly over the anterior extremity; posterior umbonal slopes prominently rounded or sub-angular, and extending obliquely backward and downward to the posterior basal angle. Surface of cast showing obscure marks of growth parallel to the basal and truncated posterior margin, being rather abruptly flexed in crossing the umbonal ridge.

Length, 1.58 inches; height at posterior end of hinge, 0.92 inch; convexity near the middle, 0.78 inch; breadth of posterior gap, 0.56 inch.

I am not entirely sure that this is Professor Swallow's species, though it agrees pretty well with his description, and almost exactly with a tracing in my possession of one of Professor Swallow's drawings, understood by me to be from his type specimen. It will be distinguished from the last by its shorter and more rhombic form, more obliquely truncated posterior margin, and shorter, narrower, and more oblique anterior. Perfect specimens would probably show the usual granulations of the surface.

Locality and position.—The typical specimens described by Professor Swallow, were from the Middle Coal-Measure, at Lexington, Missouri. That here described and figured on Pl. II, was found in the Coal-Measures, at Plattsmouth, Nebraska, in bed No. 2, of the section at that place; but it was inadvertently omitted, among the fossils mentioned in the Plattsmouth section.

Genus ALLORISMA, King.

ALLORISMA (SEDGWICKIA) REFLEXA, Meek.

Pl. X, Fig. 15.

Shell rather small, longitudinally subovate, rather convex in the central region; nearly twice as long as high; anterior margin rounded in outline; base forming a subelliptic curve, excepting that it is slightly sinuous in advance of the middle; posterior margin compressed, and ascending obliquely from the base nearly half-way up, and a little straightened, or even slightly sinuous, in outline; above the middle, truncated obliquely forward and distinctly gaping or reflexed; dorsal margin concave in outline behind the beaks, and rounding forward into the anterior border in front of them; beaks depressed and placed near halfway between the middle and the front, with slight backward inclination, somewhat as in *Nuculana;* posterior dorsal part of the valves compressed, and provided with two obscure ridges nearly parallel with the cardinal edge, excepting that the lower, which is usually wider than the other, is more oblique. Surface ornamented with obscure lines of growth, and small, very regular concentric wrinkles, which become obsolete on the posterior dorsal portion of the valves, and quite distinct on all the more convex parts, where they curve parallel to the basal outline, being a little straightened, or even very slightly arched, as they cross an obscure, undefined concavity extending from near the beaks to the sinuous part of the base; crossing the concentric marking, exceedingly small, or almost microscopic, rather distant radiating lines may also be seen under a strong magnifier; on the anterior part of the valves, and near the beaks, these lines seem to be nearly or quite continuous, but on the posterior ventral portions they are seen to be composed of the usual minute granules.

Length, 1.50 inches; height about 0.55 inch; convexity near 0.38 inch.

This must be a very neat, pretty species, when found entire. It is remarkable for its concave dorsal margin behind the beaks and its abruptly gaping truncated posterior extremity. This gap, however, is entirely above the middle, and seems to have a somewhat upward direction. In its gaping posterior, and some of its other characters, it resembles the genus *Chænomya*, though its general physiognomy is different, being much more like that of *Lyonsia.* Specifically, it is perhaps most nearly related to the shells described by Dr. Shumard from the Upper Coal-Measures of Kansas, under the names *Leptodomus Topekaensis* and *L. granosus* (Trans. St. Louis Acad. Sci., 1, pp. 207 and 208), though, as near as I can determine from descriptions alone, it must be clearly distinct from both, on account of its concave dorsal outline, want of a posterior umbonal ridge, and in its gaping posterior extremity, &c.

It is quite probable that it was to this species that Professor Geinitz alludes, in connection with that I have described under the name *A. Geinitzii* (which he referred to *A. elegans*), as being an intermediate connecting link between the *A. Geinitzii* and *A. elegans.* He was probably led into this view by not having a sufficiently extensive series of the *A. Geinitzii* for study; otherwise he would certainly have seen that that little shell never attained more than one-fourth the size of this, and not only differs in general outline, and the absence of any traces of the distinct regular concentric wrinkles of this species, but in the constant possession of a sharply defined posterior umbonal carina, of which there is no trace in this shell. I probably must have seen altogether, at the locality and in the material collected, not less than one hundred specimens of *A. Geinitzii,* not one of which was even half as large as this, while they all possess the sharp carina, and show no tendency whatever to vary into the form or *A. elegans.* In short, it is remarkably constant in all of its characters, and perhaps the most strongly defined species in all of our rocks.

Locality and position.—Division C of the Nebraska City section. It also occurs in both the upper and lower divisions of the Coal-Measures of Illinois.

ALLORISMA (SEDGWICKIA) GEINITZII, Meek.

Pl. X, Fig. 16 *a*, *b*.

Allorisma elegans, Geinitz, 1866, Carb. und Dyas in Neb., p. 13, Tab. 1, Fig. 21; (not of King, 1844).
——*Geinitzii*, Meek, 1867, Am. Jour, Sci. and Arts, Vol. XLIV, new series, p. 170.

Shell small, rather compressed, longitudinally subovate, abruptly narrowed from the beaks posteriorly; umbonal slopes distinctly carinate from the beaks to the posterior basal angle; anterior side subtruncate, with an abrupt slope from the beaks obliquely forward above, and rounding into the base below; basal margin somewhat prominently rounded anteriorly, and nearly straight or faintly sinuous behind; posterior end compressed, its margin abruptly truncated vertically, so as to make its upper and lower parts nearly rectangular; cardinal margin sloping with a slightly concave outline, from the beaks to the truncated posterior end; beaks elevated, incurved, and placed about half-way between the middle and the anterior extremity of the valves. Surface ornamented with numerous, minute, closely crowded granules, which, on the umbones and other parts of the valves in front of the angular umbonal slope, show a tendency to arrange themselves in radiating lines, which are crossed by more or less distinct lines of growth; on the compressed corselet above and behind the umbonal carina, there are usually two or more obscure radiating ridges and furrows, crossed by moderately distinct, granular lines of growth, parallel to the truncated posterior margin.

Length of the largest example seen in a fine series of specimens, 0.50 inch; height of umbones in ditto, 0.30 inch; height of truncated posterior end same, 0.16 inch; convexity, about 0.13 inch.

This beautiful little shell was referred by Professor Geinitz to *Allorisma elegans* of King, but I have no doubt whatever in regard to its being a clearly distinct species, our figure 16 *a* of Pl. X, being the largest specimen of a series of about sixty examples; while *A. elegans* attains the size of figure 22 *b* of the same plate, copied for comparison from Professor King's typical figure. It is also easy to see from this and Fig. 22 *a* of same plate, copied from another figure of *A. elegans*, given by Professor Geinitz in his Dyas (Tab. XII, Fig. 14*b*) of a German specimen, that the form of our Nebraska shell is entirely different, so much so, indeed, as to suggest doubts whether it really belong to the same genus. In the first place it will be observed that our shell has its umbones and anterior ventral region much more prominent, and its posterior half more narrowed and abruptly truncated. Its umbonal ridge is likewise always *distinctly carinate.*

Professor Geinitz noticed some of these differences, but thought them not of specific importance, because *A. elegans* is variable, particularly in the umbonal angle, and he had seen a much larger specimen from Nebraska, with the umbonal ridge blunt. I can only say, however, that a series of about sixty specimens before me shows the Nebraska shell under consideration to be exceedingly constant in all its characters; while none of these specimens I have seen exceed in size that represented by Fig. 16 *a* of Plate X. The last-described species, however, from the same locality and position, is a larger somewhat similar shell, belonging to an entirely different species, with a rounded umbonal slope. This I suspect, as already suggested, is the one alluded to by Professor Geinitz.

Locality and position.—Very common in division C at Nebraska City, to which horizon I at one time supposed it to be confined. More recently, however, I have seen specimens of this shell, presenting *exactly* the same characters, from the Upper, Middle, and Lower Coal-Measures of Illinois.

ALLORISMA (SEDGWICKIA?) SUBELEGANS, Meek.

Pl. X, Fig. 14.

Shell small, rather compressed, less than twice as long as high, greatest convexity near the middle, and in advance of it; both extremities closed; anterior margin rounded into the base from near the middle, nearly straight and sloping forward from the beaks above; posterior extremity obliquely subtruncated, but connecting with the base and cardinal margin, without any defined angle; cardinal margin inflected, nearly straight, and sloping slightly from the beaks; base rather straight and subparallel to the hinge along the middle, but rounding up at the extremities; beaks rather broad, moderately prominent, and placed a little in advance of the middle; umbonal region somewhat compressed along the middle; posterior umbonal slopes moderately prominent, in consequence of the compression of the posterior dorsal region, but not carinated, nor even forming a defined ridge. Surface ornamented with concentric striæ, and small undulations of growth, which are distinctly and regularly defined on the umbones and more convex portions of the valves, but become abruptly obsolete on the compressed postero-dorsal region; extremely minute granules also cover the entire surface in very closely-crowded radiating rows.

Length, 0.56 inch; height, 0.32 inch; convexity about 0.20 inch.

This little shell is evidently allied to *A. elegans* of King, indeed, much more so than the last, which has been identified by Professor Geinitz, with that species. After a critical comparison, however, with Professor King's figure and description of *A. elegans*, as well as with those of German examples of the same published by Professor Geinitz, I do not feel warranted in identifying the form under consideration with that species. At any rate, it differs in having its umbones wider, and placed farther back, as well as more compressed, while its concentric undulations are more strongly and regularly defined on its umbonal and central regions. In fact, it seems to me to be as much like young examples of a Carboniferous shell figured by Professor McCoy, in his British Palæozoic Fossils, Pl. 3 F, under the name *Leptodomus variabilis*, particularly his Figure 8, of the plate cited.

That this shell is clearly distinct from the last is too obvious to require a careful comparison, especially when it is remembered that we have not less than fifty or sixty specimens of that shell, all agreeing with remarkable exactness with the figures given; while of the form here under consideration but the single example figured was found among all the collections yet obtained. It may be well to remark, further, that in addition to the manifest differences of form, the species under consideration has its surface granules much more minute, being, in fact, almost invisible, except when looked for in a good light, with a *strong* magnifier; while in the last they are visible, on well-preserved specimens, to good eyes without the aid of a magnifier of any kind.

Locality and position.—Same as last. It also occurs at various horizons in the Coal-Measures of Illinois.

ALLORISMA (SEDGWICKIA) GRANOSA, Shumard, sp.

Pl. II, Fig. 8.

Leptodomus granosus, Shumard, 1858, Transactions St. Louis Acad. Sci., Vol. 1, p. 207.
Compare *L. Topekaensis*, Shumard, 1858, ib., p. 208.

Shell very thin, approaching an irregular, oblong form, the length being less than twice the height, very convex, the most gibbous part being near the middle of the valves; beaks prominent, incurved, somewhat flattened on the outer side, and placed about half way between the middle and the front. Dorsal margin straight behind the beaks, and nearly parallel to the general outline of the base, inflected so as to form a distinct, flattened, lanceolate, lunule-like area, bounded on each side by a well-defined, subangular ridge; posterior side nearly or quite closed, obliquely truncated, with sometimes a faint sinuosity near the middle; anterior side rather abruptly sloping forward, and straightened above, and rounding into the base below, near which it seems to be a little gaping; base somewhat straightened, or even a little sinuous in outline, just in front of the middle, at the termination of a broad, very shallow concavity, extending a little obliquely downward and backward from the umbonal region; behind this rather prominent, thence ascending obliquely with a slightly convex outline, to the truncated posterior margin. Posterior umbonal slopes very prominently rounded above, and continued as a low undefined ridge, obliquely backward and downward; posterior dorsal slope, above the umbonal ridge, with an oblique, shallow, rounded sulcus, extending from the back part of the beaks to the middle of the truncated margin behind. Surface marked with fine lines of growth and small irregular concentric wrinkles, which latter are not defined on the posterior dorsal region above the umbonal ridge; crossing these are the usual radiating rows of minute granules.

Length, 2 inches; height, 1.15 inches; convexity, 1 inch.

Dr. Shumard described two species, *L. granosus* and *L. Topekaensis*, and without figures to illustrate the differences, I am left in some doubt which of his species this is, though it agrees best with the first. Both of these species belong to the same group as *Allorisima concava*, M. & H; *A. Geinitzii*, Meek; *Sanguinolites variabilis*, and *S. costellatus*, McCoy, and probably *A. elegans*, King. They differ from the typical forms of the genus *Allorisma*, in presenting a peculiar Lyonsia-like physiogomy, and should probably form at least a distinct section of the genus, if not an entirely different genus. McCoy referred his species to *Leptodomus*, but I cannot think them congeneric with the type upon which he *first* proposed to found that group. To me they appear more nearly allied to the original typical form of his *Sedgwickia*, which latter name I am inclined to retain for them, at least in a subgeneric sense.

Locality and position.—The specimen figured on Plate II was obtained at Rock Bluff, on the Missouri, in the Upper Coal-Measures. Dr. Shumard's typical specimen was found in the same horizon, on Verdigris River, Kansas.

ALLORISMA SUBCUNEATA, M. & H.

Pl. II, Fig. 10 *a*, *b*.

Allorisma subcuneata, Meek & Hayden, Decr. 1858, Proceed. Acad. Nat. Sci. Philad., p. 263; Palæont. Upp. Mo., 1864, Part 1, p. 37, Pl. 1, Fig. 10 *a*, *b*; Geinitz, 1866, Carb. und Dyas, p. 76.

? ——— *ensiformis*, Swallow, 1860, Trans. St. Louis Acad. Sci., Vol. 11, p. 653.

Shell attaining a large size, longitudinally elongated, or twice to three times as long as high, the proportional length increasing with age, greatest convexity a little in advance of the middle and in the umbonal region; cuneate and a little gaping behind, where the margin is more or less narrowly rounded in outline. Basal and dorsal margins nearly parallel, the latter being more or less concave in outline, or nearly straight, and inflected so as to form a lanceolate kind of false area, bounded by an obtuse ridge on each side, just outside of which there is a shallow undefined sulcus; basal margin slightly convex, or somewhat straightened along the middle, and sometimes very faintly sinuous just under the beaks, rounding up more abruptly before than behind; anterior margin very short, a little gaping and rather prominently rounded below; beaks convex, incurved, and placed near the anterior end, rather depressed, but rising moderately above the dorsal margin. Surface ornamented with fine striæ of growth, and well-defined concentric undulations usually most distinct and regular on the beaks and umbonal region.

Length of largest specimen seen, 4.81 inches; height from ventral to dorsal margin, near middle, 1.76 inches; convexity, 1.57 inch.

This fine species has neither umbonal ridge nor lunule, properly speaking, though there is an undefined excavation in front of the beaks, and an obscure ridge extending from the back part of each beak to the posterior extremity of the hinge. Like perhaps all other species of this and several allied genera, this species, when well preserved, has its entire surface covered with granules. These granules are rather scattering and, as usual, arranged in radiating rows; it is very rarely, however, that specimens are found in a condition to show these delicate surface characters, since there are usually no traces of them on casts of the shell.

The typical specimens upon which this species was founded were rather imperfect, and the one figured in the Palæontology of the Upper Missouri (Pl. 1, Fig. 10) has the posterior extremity broken away. The blank restoration of this wanting portion has the outline represented somewhat inaccurately, so as to give an unnatural straightness to the hinge. I have also since ascertained that the sinus of the pallial line, which is exceedingly obscure and very rarely seen, is there represented rather too angular in outline.

In many respects this species very closely resembles *A. regularis*, King, as figured on Pl. XIX, Fig. 6, of the Geol. Russia, Vol. II. Indeed, the resemblance is so close that at one time I was inclined to think they might possibly be the same. On comparison of the dorsal view, however, as represented by our Figure 10*a*, Pl. II, it will be seen that the inflected portion of its cardinal margin is much broader than in the Russian species, so as to form a broad lanceolate excavation, instead of a mere linear depression. In addition to this, our shell has a ridge along the hinge, on each side of the depression alluded to, not seen in the Russian shell.

Although Professor Swallow's diagnosis of his *A. ensiformis* does not agree in all respects with the shell under consideration, I am inclined to think this and the obscurity of his description has arisen from the accidental omission by the printer of a part of the description. At any rate, I have not much doubt, from a tracing of his typical specimen before me, that his shell and that under consideration are identical. It is rather longer in proportion to its height, and has its anterior extremity a little shorter than the specimen here figured, but a good series of specimens shows the species to vary in these respects.

Locality and position.—Rock Bluff, where it is quite abundant in bed No. 9 (see section of that place); two and one-half miles southwest of Nebraska City; Wyoming, &c., in beds referred by Professor Geinitz and Professor Marcou to the Dyas; also at Leavenworth and Atchison, Kansas; Plattsmouth, Nebraska, and in short at numerous places in the Coal-Measures of the Western States. It is not uncommon in both the Upper and Lower Coal-Measures of Illinois.

Genus PROTHYRIS, Meek.

PROTHYRIS ELEGANS, Meek.

Pl. X, Fig. 9 *a*, *b*.

Prothyris, Meek, 1869, Proceed. Acad. Nat. Sci. Phila., July, p. 172.
——— *elegans*, Meek, 1871, American Jour., Conchology, Vol. VII, p. 5, Pl. 1, Fig. 3.

Shell compressed, elongate-oblong, the length being about three and a half times the height; ventral and dorsal margins straight or a little arched; the latter with a faintly defined marginal furrow, below which there is usually an obscure ridge also parallel to the dorsal margin; posterior extremity obliquely subtruncate, the most prominent part being below the middle; beaks compressed, depressed, not distinct from the dorsal margin, and placed about one-eighth or one-ninth the length of the valves behind the anterior extremity; notch of the anterior margin well defined, and extending about half way up from the base, and near half the distance back from the front to the beaks; ridge from the inner angle of the notch narrow, flat, and widening slightly from above; anterior margin above the notch rounded, and having the appearance of a little flattened ear; surface striæ nearly obsolete on the upper half of the valves, and more distinct on the ventral and antero-ventral regions.

Length of largest specimen seen, 1.07 inches; height, 0.33 inch; convexity, about 0.08 inch.

The only other species of this genus known to me is from the lowest member of the Carboniferous system in Michigan and Ohio, known in the former State as the Marshall group, and in the latter as the Waverly group. This older species (*P. Meeki* of Winchell) is more convex, with the anterior ventral angle less sharply defined, and the posterior margin more sloping above.

Locality and position.—Nebraska City, division C. It also occurs at different horizons in the Coal-Measures of Illinois.

Genus SOLENOPSIS, McCoy.

SOLENOPSIS SOLENOIDES, Geinitz, sp.

Pl. X, Fig. 3.

Clidophorus solenoides, Geinitz, 1866, Carb. und Dyas in Neb., p. 25, Tab. II, Fig. 7.

Shell small, rather compressed, elongated, the length being about four times the height, narrowing posteriorly; cardinal margin nearly straight, erect, less than the entire length of the valves, with a faint external compression or shallow furrow just below it; basal margin broadly convex in outline, the most prominent part being in advance of the middle; posterior extremity very narrow, and faintly subtruncate; beaks much depressed and compressed, or scarcely distinct from the cardinal margin, placed within about one-eighth the entire length of the shell from the anterior extremity, and defined in front by a short vertical indentation; anterior side narrowly rounded, or with the upper side sometimes faintly truncated, with slight slope from the little indentation forward; surface with very fine, regular striæ of growth, which are nearly or quite obsolete, excepting on the lower half of the valves.

Length of one of the largest specimens, 0.60 inch; height, 0.15 inch; convexity, about 0.06 inch.

I have not seen the hinge of this shell, but, from its external characters, I can scarcely doubt that it really belongs to the genus *Solenopsis*. All the specimens I have examined, show an indentation just in advance of the beaks, somewhat like that left on internal casts of *Pleurophorus* by the internal ridge; but so far as I have been able to see, there would seem to be no impressions of long posterior lateral teeth like those of that genus, left in the matrix. Casts of the *exterior*, showing the finest lines of growth quite distinctly, indicate that the little indentation alluded to in front of the beaks must be impressed on the outside of the shell, though it doubtless also corresponds to an internal ridge.

Compared with *S. minor*, Professor McCoy, the type of the genus *Solenopsis*, the form here under consideration, seems only to differ specifically in being more attenuated and less distinctly truncated posteriorly. Among American species, it is most nearly related to *Solen scalpriformis* of Winchell, from the oldest division of the Carboniferous system in this country. Professor Winchell's species is larger, however, and agrees more nearly in form with Professor McCoy's type. I think these shells are all congeneric.

Locality and position.—Division C of the Nebraska City section. I have not seen it from any other locality or position in Nebraska, but it occurs in a lower position in the Coal-Measures of Illinois.

GASTEROPODA, Cuvier.

Genus DENTALIUM, Linnæus.

DENTALIUM MEEKIANUM, Geinitz.

Pl. XI, Fig. 16 *a*, *b*.

Dentalium Meekianum, Geinitz, 1866, Carb. und Dyas in Neb., p. 13, Tab. 1, Fig. 20.
Compare *D. ingens*, de Kon., 1844, An. Foss., p. 317, Pl. XXII, Fig. 2.

I have preferred to copy Professor Geinitz's figure of this species to figuring the only Nebraska specimen I have seen of it, because the latter is distorted and does not show the surface markings. From Professor Geinitz's figure, and the specimen alluded to, it is evident that the shell varied from 0.60 to 0.75 inch in length; is moderately curved, and increases rather rapidly in size from the smaller to the larger end, the aperture being nearly or quite circular. His figure shows that it has no longitudinal striæ, and its lines of growth to be fine, moderately distinct, and passing very obliquely around, with at regular intervals, a deeper sulcus. These lines, and occasional deeper grooves, are represented in his figure, apparently as if they passed spirally around; this, however, is certainly incorrect.

This species is evidently closely allied to *D. ingens*, de Kon., from the Carboniferous rocks of Belgium, from which it seems to differ only in its smaller size.

Locality and position.—Nebraska City bed C, of the section at that place. It also occurs near the middle of the Coal-Measures of Illinois.

Genus BELLEROPHON, Monfort.

BELLEROPHON CARBONARIUS, Cox.

Pl. IV, Fig. 16; Pl. XI, 11, *a*, *b*, *c*.

Bellerophon Urii, Norwood and Pratten, 1865, Jour. Acad. Nat. Sci. Philad., Vol. III, second series, p. 75, Pl. IX, Fig. 6 *a*, *b*, *c*; (not Fleming, 1828).
——— *carbonarius*, Cox, 1857, Kentucky Geol. Report, Vol. III, p. 562; Dana's Man. Geol., p. 349, Fig. 598.
——— *Blaneyanus*, McChesney, 1860, New Palæozoic Fossils, p. 60; and 1865, Pl. 2, Fig. 5, illustrations of same.
——— *carbonarius*, Geinitz, 1866, Carb. und Dyas in Nebraska, p. 6, Tab. 1, Fig. 8.

Shell globose, broadly rounded over the dorsum; umbilical impressions very small or closed. Aperture transversely sublunate, much wider than high, and strongly arched, but not expanding more rapidly than the uniform increase in the size of the volutions; inner lip nearly wanting or little developed; outer lip moderately thick near the umbilicus on each side, but thinner between; sinus shallow and rounded, its band obscure or scarcely visible on the costated part of the outer whorl, sometimes a little concave, or with, on each side, traces of a faintly marked ridge on the smooth part of the outer volution; surface, excepting on the terminal half or third of the body whorl, ornamented with about eighteen to twenty-five simple, rather distinct, revolving raised lines or costæ, some two or three of which are usually a little smaller and closer together on the mesial band than the others, while a few of these near the umbilicus on each side, just at the commencement of the smooth part of the outer whorl, are broken up into little elongated, oblique, node-like prominences; lines of growth obsolete.

Transverse and longitudinal diameters (of a medium-sized specimen), each 0.56 inch.

This shell is very nearly related to *B. Urii* of Fleming, and if it is possible that some slight details in the ornamentation of that species have been overlooked by those who have figured and described it, our shell may possibly not be distinct. As remarked by Professor Cox and Professor McChesney, the *B. carbonarius* differs from the published figures and descriptions of *B. Urii* in having about thirteen to eighteen costæ less, and in having a few of those on each side of the outer whorl, as stated above, near the umbilicus, broken up into little oblique, elongated ridges or interupted lines.

Locality and position.—This species is widely distributed and has a considerable vertical range in the Coal-Measures of the West. The specimen here figured on Pl. XI, was collected from division C of the section at Nebraska City, included by Professors Marcon and Geinitz in their Upper Dyas. We also have it from Mr. Morton's shaft, near Nebraska City, and from division B at that place, as well as from two miles and three-fourths west of Nebraska City, and numerous other places in Nebraska, at different horizons in the Coal-Measures. In Iowa, Kansas, Missouri, Illinois, Kentucky, and portions of Southwestern Indiana, it is a common Coal-Measure species. I have also identified it from the Lower Coal-Measures of West Virginia, among collections sent by Professor J. J. Stevenson, of Morgantown.

BELLEROPHON MONTFORTIANUS, N. & P.

Pl. XI, Fig. 15; and 12?

Bellerophon Montfortianus, Norwood & Pratten, 1855, Jour. Acad. Nat. Sci., Philad., Vol. III, sec. ser., p. 74, Pl. IX, Fig. 5 *a, b, c.*; Geinitz, 1866, Carb. und Dyas in Neb., p. 8, Tab. 1, Fig. 13.

Shell with inner volutions comparatively small; outer one greatly expanded at the aperture, both in length and breadth; umbilicus very small or closed; aperture large and transversely reniform; sinus of lip narrow and moderately deep, its band narrow, well defined, and a little raised in the middle of a rather deep, wide sulcus; outer lip thin and rounded in outline on each side of the sinus, but more or less thickened near the umbilicus; inner lip callous, particularly in the middle, where it often swells out in the form of a large prominent node in old specimens; surface beautifully ornamented with two sets of raised longitudinal lines, the larger being usually with several of the smaller between; crossing all of these, there are numerous, much more regular, crowded, very fine striæ, which curve gracefully as they cross the dorsal band;

on the expanded portion of the outer volution, the larger longitudinal lines increase in size, and become more spread out; that portion of the body volution not expanded, and the inner whorls are also ornamented by large, somewhat interrupted transverse ridges, extending from near the dorsal band to the umbilical region on each side; the finer markings already mentioned being defined on these ridges as on other parts of the surface.

Our specimens are too much broken and distorted to afford accurate measurements. One, however, shows that the aperture was not less than 1.13 inches in breadth, while the body whorl, just at the commencement of the expansion, is not more than 0.45 inch across in the same direction.

This beautiful species is remarkable for the great expansion of its aperture in proportion to the size of the inner whorls composing the body of the shell. It is very rarely the case that specimens are found with the expanded part of the outer volution entire, and it was doubtless from this reason that my friends Norwood and Pratten supposed the lip to be probably without the usual dorsal sinus, not having seen any specimens with the lip entire.

The figure I have given was drawn from an artificial cast made in a mould left in the fine clay matrix, so sharply defined as to show the most minute surface-markings. I also have before me several natural casts of the body part of the shell, more or less distorted, such as that figured by Professor Geinitz. From all of these specimens I am inclined to suspect that the enlarged figure, 14, given by Professor Geinitz on plate, 1, of his Carb. und Dyas (Fig. 12 of our Pl. XI), of the expanded part of a shell he refers to *Bellerophon interlineatus*, Portlock, may belong to the species under consideration. It certainly agrees well with the expanded portion of *B. Montfortianus* in surface-markings, that part of this species being entirely destitute of the large transverse ridges seen on other parts of the shell; while the longitudinal lines or costæ spread out there and always show from two to three or four finer lines between each two of the larger, which are here increased in size. His specimen evidently has the margin of the lip broken away so as not to show the sinus.

Locality and position.—The specimen figured, and some seven or eight other imperfect examples, were obtained from bed C of the Nebraska City section, at that place. It also occurs there in bed B, and at numerous other places in Nebraska, Kansas, Iowa, Missouri, Illinois, &c., at various positions in the Coal-Measures. Professor Stevenson also found it associated with the last in the Lower Coal-Measures of West Virginia.

BELLEROPHON MARCOUIANUS, Geinitz.

Pl. IV, Fig. 17; and pl. XI, Fig. 13 *a*, *b*.

Bellerophon Marcouianus, Geinitz, 1866, Carb. und Dyas in Neb., p. 7, Tab. 1, Fig. 12.

The specimens of this species yet known, either in the collections now under investigation or among those obtained by Professor Marcou, are too imperfect to give a correct idea of its characters. It is evidently, however, one of those rapidly expanding species, with a very small body and a greatly dilated mouth, like *B. Montfortianus* and *B. percarinatus*, Conrad. Yet it differs from these distinctly, in having no transverse ridges or nodes, and in being marked with exceedingly fine, regular, longitudinal striæ. It also shows very minute lines of growth, which, in crossing the band, bend backward so as to indicate a rather shallow sinus. The band forms a moderately prominent ridge, which, however, is not so abruptly raised, nor so roughened by the crossing of the marks of growth, in the specimens I have seen, as in that figured by Professor Geinitz. The longitudinal lines are extremely minute, and closely crowded on the band; and the umbilicus seems to be open.

Specimens with the lip entire may have attained a diameter of one inch.

Locality and position.—Nebraska City, division C, of the section seen at that place; likewise from Morton's shaft one mile and three-fourths west of there. Internal casts of a form probably identical with this also occur in the Upper Coal-Measures of Union County, Iowa, and at the same horizon in Kansas. It likewise occurs in the Coal-Measures of Illinois; and I think I have seen imperfect specimens of it among Professor Stevens's collections from the Lower Coal-Measures of West Virginia.

BELLEROPHON PERCARINATUS, Conrad.

Pl. XI, Fig. 14.

Bellerophon percarinatus, Conrad, 1842, Jour. Acad. Nat. Sci., Philad., Vol. VIII, Pl. 16, Fig. 5; Norwood & Pratten, 1854, Jour. Acad. Nat. Sci., Philad., Vol. III, new series, p. 74, Pl. IX, Fig. 4.

Of this species we have but a single distorted fragment, though it is quite sufficient to show its identity with the shell from the Illinois Coal-Measure, and the same horizon at other western localities, usually referred to Mr. Conrad's *B. percarinatus*. Mr. Conrad's species was originally described from Western Pennsylvania, where it was found associated with *Pleurotomaria sphærulata* and *P. tabulata*, Conrad, which are also common in Illinois, Missouri, Iowa, &c. Mr. Conrad's type specimen was probably distorted by lateral pressure, so as to make its mesial ridge more prominent than is natural in the western specimens. At any rate, I cannot think this a specific difference, since specimens from West Virginia agree well in this character with those found farther westward.

As known in the West, the shell under consideration is of medium, or rather above medium size, expands very rapidly at the aperture, which seems to have been much like that of *B. Montfortianus*, N. & P. Its umbilicus is closed by the rapid expansion of the body whorl; while its dorsal side, which is somewhat rounded, has three longitudinal nodose ridges, the middle one of which is the most prominent, the nodes being at the points where a series of transverse costæ, or ridges, cross over the dorsum from one umbilicus to the other. It seems to be otherwise only marked by small lines of growth.

Locality and position.—The specimen figured on Pl. XI, Fig. 14, is from division C of the Nebraska City section. We also have it from Brownville on the Missouri. The specimen figured by Norwood & Pratten is from the Upper Coal-Measures at Grayville, Illinois. It occurs at other localities in the same position in Illinois, Iowa, Missouri, &c.; also in the Lower Coal-Measures of West Virginia.

Genus PLATYCERAS, Conrad.

PLATYCERAS NEBRASCENSIS, Meek.

Pl. IV, Fig. 15 *a*, *b*.

Shell rather small, very obliquely elongate-conical, and strongly curved; apex free, oblique, somewhat obtusely pointed, curved and twisted to the right (looking at the shell with the apex behind); aperture subcircular or oval; lip thin and sharp, broadly sinuous nearly under the apex on the concave side, and sometimes provided with five or six smaller sinuses around the remaining sides; surface with more or less distinct marks of growth, usually strongly undulated parallel to the sinuses, and inequalities of the lip.

Length, 0.88 inch; breadth, 0.56 inch.

This species resembles more or less nearly several of those described from different horizons in the Western States, but after a critical comparison I have been unable to identify it with any of them.

Locality and position.—From a shaft one mile and three-fourths west of the Nebraska City landing. It also occurs near the middle of the Illinois Coal-Measures.

Genus MACROCHEILUS, Phillips.

MACROCHEILUS INTERCALARIS, *var.* PULCHELLUS, M. & W.

Pl. VI, Fig. 8.

Macrocheilus intercalaris, Meek & Worthen, 1860, Proceed. Acad. Nat. Sci. Philad., p. 467; 1866, Geol. Rep. Ill., Vol. II, p. 371, Pl. 31, Fig. 6 *a b*.
——— *pulchellus*, M. & H., 1860, ib.

This shell agrees closely with certain forms of *M. intercalaris*, M. & W., from the Upper Coal-Measures of Illinois, particularly with the form we at one time proposed to call *M. pulchellus*, which differs mainly from the typical *intercalaris* in having its spire a little more elevated, and its whorls slightly more convex. This, however, I do not now regard as a specific difference.

Locality and position.—Two miles and three-fourths southwest of Nebraska City. An imperfect cast of apparently a larger individual of the same species was also found at near the same horizon one mile and three-fourths west of Nebraska City, at Hon. J. S. Morton's place. The original typical specimens were found in the Upper Coal-Measures at Springfield, Illinois. I have also seen the same shell in a small collection belonging to Dr. R. P. Stevens, from near Pittsburg, Pennsylvania. I believe I have also seen it among the Lower Coal-Measure fossils from West Virginia.

Genus ORTHONEMA, M. & W.

ORTHONEMA SUBTÆNIATA, Geinitz, sp.

Pl. XI, Fig. 10.

Murchisonia subtæniata, Geinitz, 1866, Carb. und Dyas in Nebraska, p. 12, Tab. 1, Fig. 18.
Orthonema subtæniata, Meek, 1867, Am. Jour. Sci. & Arts, Vol. XLIV, second series, p. 178.

Although I have had an opportunity, through the politeness of Dr. White, to examine a fine specimen of this neat little shell, from the same horizon in Western Iowa, I have yet only seen imperfect fragments of it from Nebraska City; consequently I have preferred to give a copy of Professor Geinitz's figure, which appears to be an accurate representation of a perfect specimen. From this it may be seen to be a very small, elongate-conical, rather symmetrical shell, composed of eight or nine compactly wound, moderately convex whorls, the body one of which is ornamented with three, and those of the spire with each two, revolving linear costæ or ridges. The immediate apex is a little obtuse, the suture well defined, and the lines of growth minute, or scarcely visible without the aid of a magnifier.

Professor Geinitz refers this little shell to the genus *Murchisonia*, but I have elsewhere shown that it possesses neither the labial sinus nor the revolving band of that genus, but that it belongs to the genus *Orthonema* of Meek and Worthen. It is not generally the case that fossil shells of this, or even larger sizes, are found with the lip sufficiently well preserved and exposed to show whether it was provided with a sinus or not; but where the lines of growth can be seen, even by the aid of a magnifier, they enable us at once to decide this question, for in the genera *Murchisonia*, *Pleurotomaria*, and other similar types, these lines always show a curve corresponding to the outline of the sinus. On examining the specimen of this species, loaned to me by Dr. White, under a lens in a cross-light, I could see that the minute lines of growth show no such curve whatever, and that there is no spiral band corresponding to that seen in *Murchisonia*. Several species of this genus from the Coal-Measures of Illinois are also clearly without any traces of either sinus or revolving band.

Specifically, this shell seems to be somewhat related to *O. conica* of Meek and Worthen (Proceed. Acad. Nat. Sci. Philad., July, 1866, p. 270), from the Coal-Measures of Illinois. It is readily distinguished, however, by its more convex whorls, deeper suture, distinct revolving costæ, and the absence of a prominent angle around the base of the body whorl.

Locality and position.—Nebraska City, from bed C of the section at that place; also found by Dr. White at near the same horizon in Western Iowa.

Genus ACLIS, Loven.

ACLIS SWALLOVIANA, Geinitz, sp.

Pl. XI, Fig. 7 *a*, *b*.

Turbonilla (*Loxonema*) *Swallowiana*, Geinitz, 1866, Carb. und Dyas in Nebraska, p. 5, Tab 1, Fig. 19.

Compare *Turritella* ? ? *Stevensana*, Meek and Worthen, 1866, Ill. Report, Vol. II, p. 382, Pl. 27, Fig. 8; also, *Aclis minuta*, Stevens, 1858, Am. Jour. Sci., Vol. XXV, p. 259; and *Murchisonia minima*, Swallow, Trans. St. Louis Acad., Vol. I, p. 203.

Shell very small, subterete; volutions about seven, increasing very gradually in size; last one not larger in proportion to the gradual increase in the size of the shell than the others, all with a slightly flattened, outward sloping space just below the suture; surface below the flattened upper sloping space ornamented with comparatively distinct revolving lines, of which some five or six may be counted on the body whorl, and three or four on each of those of the spire; lines of growth very fine, and distinctly sigmoid.

Length, 0.15 inch; breadth, about 0.07 inch; angle of spire, about 20°.

I have some doubts in regard to the little shell represented by Fig. 7 *a* being specifically identical with the form figured by Professor Geinitz, since his figure represents the volutions as increasing more rapidly in size, so as to make the angle of the spire near 10° greater. It also represents the body whorl as being more produced below, and shows one or two more of the revolving lines on each of the volutions of the spire. In its general appearance it is more like the *Turritella* ? ? *Stevensana*, M. & W., though its spire is less attenuate, while its whorls are less numerous, and marked by a smaller number of revolving lines. Supposing Professor Geinitz's figure to be exact, I should think the form it represents, that represented by our figure 7 *a*; and the other described from the Coal-Measures of Illinois, may possibly belong to three distinct species. The great difficulty, however, of presenting enlarged figures of such small univalves with absolute exactness, is such as to leave doubts on this point, where we have not the means of making direct comparisons of specimens.

I am also in great doubt in regard to the generic affinities of these little shells, and have elsewhere shown that they cannot be properly referred to the recent genus *Turbonilla*, on account of their distinct revolving lines.* The revolving markings are also equally an objection to placing them in the genus *Loxonema*. It is very probable that they will be found to belong to an undescribed genus, when the aperture, lip, columella, &c., can be seen. The group is evidently represented by two or more species in the Coal-Measures of Illinois, one of which was described as an *Aclis* by Dr. Stevens some twelve or thirteen years since, and I even think it very probable that at least the form represented by our figure 7 *a* may be the same species described by Dr. Stevens. At any rate, it agrees with an Illinois specimen that was identified with his species, on comparison by a competent observer with his typical specimen; yet Dr. Stevens distinctly states that his species has twelve of the revolving lines.

Locality and position.—Nebraska City, bed C of the section taken there.

Genus STRAPAROLLUS, Montfort.

STRAPAROLLUS (EUOMPHALUS) RUGOSUS, Hall.

Pl. VI, Fig. 5 *a*, *b*; and Pl. XI, Fig. 4 *a*, *b*.

Euomphalus rugosus, Hall, 1858, Iowa Geol. Report, Vol I, Part II, p. 722, Pl. XXIX, Fig. 14 *a*, *b*, *c*; (not *E. rugosus*, Sowerby, 1847).

Serpula (*Spirorbis*) *planorbites*, Geinitz, 1866, Carb. und Dyas in Nebraska, p. 3, Tab. 1, f. 6; (not of Münster).

Shell small, planorbicular, with upper side concave (viewing it as dextral), and the lower nearly flat, or slightly concave; sometimes with both sides nearly equally concave; volutions four or five, increasing rather gradually in size, not embracing, and all exposed both above and below; all obliquely flattened on the periphery with a narrow ridge or angle around the lower edge, and flattened below; while on the upper side they slope rather abruptly inward from another angle around the upper outer edge, thus giving an irregular subquadrangular outline to the transverse section; aperture subcircular or a little oval within; surface with rather strong marks of growth that impart a rough, or sometimes slightly nodular appearance, to the ridge or angle around the lower outer edge of the periphery.

Breadth of a medium-sized specimen, 0.54 inch; thickness or height of same, 0.16 inch. Specimens are sometimes found, however, of double these dimensions.

As in many other cases, I am compelled to differ with Professor Geinitz, not only in regard to this shell being identical with *Spirorbis planorbites* of the Europian Permian, but in regard to its being a *Spirorbis* at all. On the contrary, it appears to me to be a mollusk, very closely allied to *Euomphalus*, if not really belonging to that group. The reasons for this opinion are, that, although a very common and widely distributed shell in our Coal-Measures (being often so abundant that hundreds of specimens may be picked up sometimes in an hour or so at a single

* Am. Jour. Sci. and Arts, Vol. XLIV, p. 171, 1867.

locality), not a single one has ever been found, so far as known to the writer, attached by the growth of the shell to any foreign body. Nor have I ever seen, among thousands of specimens, a single one showing any scar or mark of adhesion on either side. In addition to this, we now know, from the same horizon in Illinois, several allied species that no one would think of placing in any other genus, some of which are nearly as large as *E. pentangulatus*, and really connect the form under consideration generically with that shell almost beyond doubt.

In order that this shell may be more readily compared with *Spirorbis planorbites*, as illustrated by Professor Geinitz in his large work on the German Permian fossils, I have copied his figures on our plate VI. By comparing these figures with those of the shell under consideration the differences of form will be at once apparent, especially on comparing our figures 5 *a*, *b*, of the American shell with Fig. 16, *a b c d*, copied from the European form. From these it will be seen that instead of having the dorsal or outer side of its whorls obliquely flattened, as in the American shell, these figures show it to be narrowly rounded or subangular in the middle, and the whole outline of the transverse section of the whorls entirely different, while our shell is *remarkably constant* in this and nearly all of its characters. If required to select from the known European species the form most nearly allied to our shell, I would at once compare it with *Euomphalus quadratus* of McCoy, from the Carboniferous rocks of Ireland. By a glance at our figure 17, representing the last-mentioned species, it will be seen to be really nearly allied to the form under consideration, but still distinct specifically.

Professor Geinitz cites *Spirorbis permianus* and *S. helix* of King (see our figures 15 *a*, *b*, same plate, after Professor King) as synonyms of *S. planorbites*, which would make these, according to his views, also the same as that here described. To me, however, the last-mentioned forms seem to be clearly distinct, not only from the American form, but from the German specimens figured by Professor Geinitz as *S. planorbites*, though the foreign species both doubtless belong to the genus *Spirorbis*.

Locality and position.—The specimens figured on Pl. XI came from division C of the Nebraska City section, at that place. That represented on Pl. VI was found in division B of the same locality. We also found it at Rock Bluff and Aspinwall, on the Missouri, as well as at Cedar Bluff. It is common in the Coal-Measures of Kansas, Missouri, Iowa, and Illinois; being found both in the Upper and Lower Coal-Measures of the latter State, and it likewise occurs in the Lower Coal-Measures of West Virginia. It may also have been this species that Conrad described long back, from the Coal-Measures of Western Pennsylvania, under the name *Inachus catilloides*, as it resembles his figure nearly, though it does not agree so well with his description.

Genus PLEUROTOMARIA, Defrance.

Pleurotomaria Haydeniana, Geinitz.

Pl. XI, Fig. 5.

Pleurotomaria Haydeniana, Geinitz, 1866, Carb. und Dyas in Neb., p. 11, Tab. Fig. 15.

Of this little shell I have seen but a single specimen, and in endeavoring to clear it of the adhering matrix, it was broken to fragments. I have, therefore, no means of characterizing it, and merely reproduce Professor Geinitz's figure.

Locality and position.—Division C of the Nebraska City section.

PLEUROTOMARIA PERHUMEROSA, Meek.

Pl. IV. Fig. 13 *a*, *b*.

Shell subovate, not umbilicate; spire and aperture of nearly equal length; volutions about five and a half, increasing rather rapidly in size from the apex, last one comparatively large, and somewhat produced below, all convex, distinctly angular and shouldered, with a shallow, revolving impression on the outer vertical side just below the shoulder; upper side flattened and sloping a little outward from the suture to the nearly rectangular shoulder; suture well defined by the convexity of the whorls, but not channeled; aperture rather large, oval or a little higher than wide, rather narrowly rounded below, and somewhat angular on the upper outer side at the termination of the shoulder of the body whorl; sinus of the lip shallow and comparatively wide, placed just above the termination of the shoulder; its revolving band obscure, being neither impressed, raised, nor bounded by lines, extending from the sinus of the lip around the outer margin of the upper sloping flattened side of the whorls, near the angle; surface usually appearing nearly smooth, but showing small lines of growth under a magnifier, and a few obscure coarser revolving lines below the shoulder on the body whorl.

Length of largest specimen seen, 0.68 inch; breadth of same, 0.50 inch; angle of spire, 57° to 65°.

I am not acquainted with any species very closely allied to this, those most nearly resembling it in form being marked by more distinct revolving lines. Its most obvious characters are its nearly smooth surface, obscurely defined labial sinus and revolving band, with very distinctly shouldered whorls.

Locality and position.—One mile and three-fourths west of Nebraska City, from a shaft sunk on Hon. J. Sterling Morton's place; also quite abudant at Rulo, Kansas, in Upper Coal-Measures. It also occurs in the Coal-Measures of Illinois.

PLEUROTOMARIA INORNATA, Meek.

Pl. IV, Fig. 14.

Shell elongate, subovate, not umbilicate; spire elevated, conical, nearly twice as high as the aperture; volutions seven, increasing gradually in size, very convex and rounded, with a faint angle a little above the middle, which becomes more distinct on the upper turns; last one not much enlarged, nor extended much below; suture well defined by the convexity of the whorls; aperture oval, being a little longer than wide; sinus of the lip broad, shallow, and placed just above the angle, from which point its flat spiral band revolves immedidiately above the angle on all the whorls, without being elevated or depressed above the surface, or otherwise strongly defined; surface marked with fine, moderately distinct lines of growth, and very obscure traces of a few revolving lines below the angle of the whorls.

Length, 0.73 inch; breadth, 0.41 inch; angle of spire, about 45°

I am not acquainted with any species very closely allied to this. In some respects it is related to the last; particularly in having but very faint traces of revolving markings, and in the position of its sinus and band just above the angle of its whorls, as well as in the shallowness of its sinus, and the obscurity of its band. It will be at once distinguished, however, by its more slender form, more rounded whorls, and their much more obscure angle, which becomes nearly obsolete on the body volution. It is a rather slender elongated form for a species of this genus, thus approaching the genus *Murchisonia*.

Locality and position.—From Mr. Morton's shaft, one mile and three-fourths west of Nebraska City, where it was found associated with the last, in a bed of clay.

PLEUROTOMARIA GRAYVILLENSIS, N. & P.

Pl. XI, Fig. 9.

Pleurotomaria Grayvillensis, Norwood & Pratten, 1855, Jour. Acad. Nat. Sci., Philad., Vol. III, second series, p. 75, Pl. IX, Fig. 7; Geinitz, 1866, Carbonform. und Dyas in Nebraska, p. 9, Tab. I, Fig. 9.

Not having succeeded in finding specimens of the shell figured by Professor Geinitz under this name, at any of the Nebraska localities, I have merely reproduced his figure. In size, form, and general appearance, as well as in the nature and position of its spiral band, it agrees well with authentic examples of *P. Grayvillensis*, now before me from the Coal-Measures of Illinois. In one respect, however, if carefully drawn, it would seem to differ slightly; that is, in having its revolving striæ of uniform size, instead of consisting of a series of rather distant larger ones, with from about two to four very small ones between each two of the former. This, however, is a character in which shells of this kind often vary, and one that might be overlooked in specimens not showing the surface-marking very clearly.

Locality and position.—Nebraska City, from the upper part of division C of the section exposed there. The *P. Grayvillensis* is widely distributed in the Upper Coal-Measures of Illinois, Kentucky, Iowa, Missouri, Kansas, &c. It occurs in both the Upper and Lower Coal-Measures of Illinois, and in the Lower Coal-Measures of West Virginia.

PLEUROTOMARIA MARCOUIANA, Geinitz.

Pl. XI, Fig. 8.

Pleurotomaria Marcouiana, Geinitz, 1866, Carbonform. und Dyas in Nebraska, p. 10, Tab. 1, Fig. 10.

Having obtained no specimens of this shell in a condition to be drawn, I have carefully copied the figure given by Professor Geinitz. So far as this figure and the specimens obtained afford the means of comparison, I am inclined to regard it as a good species. In some respects it resembles *P. Grayvillensis*, though it is evidently distinct.

Locality and position.—Nebraska City landing, from the upper part of division C of the section.

PLEUROTOMARIA SUBDECUSSATA, Geinitz.

Pl. XI, Fig. 19.

Pleurotomaria subdecussata, Geinitz, 1866, Carbonform. und Dyas in Nebraska, p. 10, Tab. 1, Fig. 11.

This is another species of which I have only seen very imperfect specimens, and hence I have to depend almost entirely upon Professor Geinitz's figure (copied on Pl. XI) for a knowledge of its characters. It belongs to the trochiform section of the genus, including *P. turbiniformis*, M and W., *P. Riddelli*, Shumard, and *P. Missouriensis*, sp. of Swallow, all from the Coal-Measures. It seems, however, to be distinct from all of these, being very much smaller, and differing in the details of its surface-markings.

Locality and position.—Upper part division C, Nebraska City.

Genus MURCHISONIA, de Verneuil.

MURCHISONIA NEBRASCENSIS, Geinitz.

Pl. XI, Fig. 6.

Murchisonia Nebrascensis, Geinitz, 1866, Carbonform. und Dyas in Nebraska, p. 12, Tab. 1, Fig. 17.

Of this little shell I have seen no examples, and consequently only know it from Professor Geinitz's description and figure, which latter I have copied, in order to give American students who may not have access to his work the means of identifying it. It belongs to a strongly carinated section of the genus represented by several species (mainly undescribed) in our Western Coal-Measures.

The figure does not show the characteristic sinus of the lip, and the revolving band, of the genus *Murchisonia;* but from the general appearance of the shell, it probably belongs to that genus.

Locality and position.—Same as last.

CEPHALOPODA.

Genus ORTHOCERAS, Auct.

ORTHOCERAS CRIBROSUM, Geinitz.

Pl. XI, Fig. 18 *a*, *b*.

Orthoceras cribrosum, Geinitz, 1866, Carb. und Dyas in Neb., p. 4, Tab. 1, Fig. 5.

This is another form I have not seen from the Nebraska rocks, but it is common in the Upper Coal-Measures of Illinois, and also found in the same position in Iowa, Missouri, &c., and in the Lower Coal-Measures of West Virginia. It is probably identical with the Upper Coal-Measure species, *O. Knoxense*, of McChesney, or some of the forms described, but not yet figured, by Professor Swallow, from the same horizon, unless the peculiar surface-marking illustrated by Professor Geinitz is really the natural surface ornamentation of the shell. It is far more probable, however, that the pitting seen on Professor Geinitz's specimen is due to some accidental cause, such a style of marking being very unusual in this genus.

In the Illinois specimens I have seen, these markings are nearly always on one side only, or more numerous and more strongly defined on one side than the other, while in other specimens differing in no other respect, I could see no traces of them. From these facts I am inclined to think they were produced by some minute parasites, or boring animals, possibly on dead shells, as they were lying with one side exposed on the bottom of the sea.

Locality and position.—Division C of the Nebraska City section; also Upper and Lower Coal-Measures of Illinois, and Lower Coal-Measures of West Virginia.

Genus NAUTILUS, Linn.

NAUTILUS OCCIDENTALIS, Swallow.

Pl. XI, Fig. 17.

Nautilus occidentalis, Swallow, 1858, Trans. St. Louis Acad. Sci., 1, p. 175.
——— *quadrangularis*, McChesney, 1860, Descriptions New Palæozoic Fossils, pp. 65 and 66; illustrations same, 1865, Pl. III, Fig. 5 *a*, *b*.
——— *biserialis*, Hall, 1860, Supp. to Vol. I, part II of Iowa Geological Report, p. 2.

Although we have but the mere fragment of this shell from Nebraska City, figured on Plate XI, I can scarcely entertain the slightest doubt in regard to its identity with the form described by Professor Swallow, under the name *N. occidentalis*, and by Professor McChesney as *N. quadrangulus*, as well as that described by Professor Hall as *N. biserialis*. In this conclusion I am not led by comparisons of descriptions and figures alone, for I have carefully examined Professor Swallow's typical specimen, and have before me good tracings taken by his permission from drawings in his possession of the same. I have also carefully examined good specimens in Mr. Worthen's collection of the same shell, from the same locality and bed from which Professor McChesney's types were obtained.

Like other nodose *Nautili*, *Goniatites*, &c., this shell varies considerably in the prominence and form of its nodes, in different individuals, as well as often in the same individual at different ages, as may be seen by examining the different volutions of the same shell when broken apart. This is also illustrated by the two specimens figured by Professor McChesney. Professor Swallow's type is an internal cast, with less prominent nodes than either of those figured by Professor McChesney.

Well-developed specimens of this species are discoid in form, with a moderately wide, shallow umbilicus, showing nearly all of each inner volution. The whorls present a nearly quadrangular section, the sides and periphery being more or less flattened, with generally a slight concavity around the middle of the latter. Usually there are six rows of nodes; that is, one row of small, depressed nodes on each side around the umbilicus, another larger and more prominent series around each lateral angle of the periphery, and two others on the periphery or outer side. In old specimens those around the margins of the umbilicus are often obsolete, or nearly so. Generally the two rows around the middle of the periphery are smaller and less prominent than those on the lateral angles, and have between them a more or less marked mesial concavity, or furrow. The individual nodes of these two rows are also nearly always alternately arranged, so that those of one row stand opposite the intermediate spaces of the other. In some cases, however, these two mesial rows of the periphery are much more developed than those on the lateral angles, thus giving more convexity to the periphery than where they are smaller. The septa are moderately closely arranged, with a backward curve on the sides and periphery; while the siphuncle is very nearly central.

On specimens retaining the shell intact, very fine, regular lines of growth are seen, which cross the periphery with a rather deep, graceful, backward curve, indicating a deep, rounded sinus in the lip on the outer side.

I am not aware of any perfect specimens of this species having been found. Those figured by Professor McChesney are not entire at the aperture, and measure about 2.50 inches in their greatest diameter, and about 1.30 inches in thickness or convexity. The fragment we have figured from Nebraska City would indicate a size near one-third larger than these.

Locality and position.—Division C of the Nebraska City section. We also saw a fragment of it at a lower position, in the Coal-Measures at Rock Bluff. Professor Swallow's type was found in the valley of Cottonwood Creek, Kansas, in beds referred by him to the Lower Permian, but regarded by Dr. Hayden and myself as Permo-carboniferous. McChesney's specimens were found in the Upper Coal-Measures at Grayville, Illinois, and Mr. Worthen has it from the same horizon at La Salle, Illinois. I have likewise seen fragments of this shell in a small collection belonging to Dr. Stevens, from the Coal-Measures near Pittsburg,

Pennsylvania, and I suspect that some of the fragments found by Professor Stevenson in the Lower Coal-Measures of West Virginia may belong to the same species.

NAUTILUS PONDEROSUS, White, M. S.

Pl. III, Fig. 7 *a*, *b*.

Compare *Nautilus Illinoisensis*, McChesney, 1860, Descriptions New Palæozoic Fossils, p. 64; also *N. tuberosus*, McCoy, 1855, Brit. Pal. Foss., p. 562, Pl. 3 H, Fig. 15.

Shell attaining a large size, subdiscoidal; umbilicus large, or nearly equaling the dorso-ventral diameter of the outer volution near the aperture; volutions three, enlarging their diameter more than three-fold each turn; all broader transversely than dorso-ventrally; inner ones slightly embracing, while the last one is apparently merely in contact with the others near the aperture; each broadly flattened or a little concave on the periphery, and (particularly the last one) somewhat flattened between the periphery and the middle of each side, from which point the sides are broadly rounded into the umbilicus, the greatest transverse diameter being near the middle; ventro-lateral, or outer angles of the last whorl (in somewhat worn casts) each provided with obscure traces of about twenty wide, undefined nodes, scarcely perceptible to the eye; septa numerous, rather closely arranged, making a slight backward curve on each side, particularly between the middle and the outer angles, and crossing the broadly flattened dorsum with a strong backward curve; surface with distinct lines of growth, which curve strongly backward like the septa, in crossing the outer side. (Siphuncle unknown).

Greatest diameter of a specimen retaining apparently about one-third of the outer chamber, 12 inches; greatest thickness or convexity, 5.40 inches; breadth of umbilicus, 3.75 inches; dorso-ventral diameter of the body whorl at the widest part, 3.90 inches. Dr. White's type specimen, now in the Iowa State collection, is much larger than that from which the above measurements were taken.

This shell seems to be nearly related to *N. Tuberosus* of McCoy, as figured in his British Palæozoic Fossils, but differs in the greater dorso-ventral diameter of its whorls, which also increase more rapidly in size. The nodes of its outer angles are also much smaller and less prominent, being scarcely perceptible in casts.

I am in some doubt in regard to the generic relations of this species. The propriety of placing any of the old discoid *Nautili*, with a large open umbilicus, and distinctly quadrangular volutions, in the same section with the recent typhical *Nautilus*, is questionable.

Professor McChesney has not figured the type of his species, *N. Illinoisensis*, and I am, consequently, not positively sure that this may not be the same, as it agrees pretty well with his description, while I have seen from La Salle fragments of the form here described.

Locality and position.—Upper Coal-Measures, bed No. 2 of Plattsmouth section. The specimen of *N. Illinoisensis* described by Professor McChesney was found at apparently near the same horizon of the Coal-Measures at La Salle, Illinois. The fine specimen figured on our plate was obtained by Professor Thomas Egleston from the Plattsmouth exposure, and loaned to me for study and illustration. Dr. White's specimen came from the Upper Coal-Measures of Iowa.

ARTICULATA.

CRUSTACEA.

Genus CYTHERE, Müller.

CYTHERE NEBRASCENSIS, Geinitz.

Pl. XI, Fig. 2; and 3 *a*, *b*. ?

Cythere Nebrascensis, Geinitz, 1866, Carb. und Dyas in Nebr., p. 2, Tab. 1, Fig. 2.
——— *Cyclas?*, Geinitz, 1866, *Ib.* Fig. 3 and 4.

Although a number of the carapace-valves of these little crustacea are before me from the same bed in which those figured by Professor Geinitz under these names were obtained, none of them seem to agree *exactly* with his figures, though we undoubtedly have the same form from near the middle of the Upper Coal-Measures in Union County, Iowa, as well as of that he has referred with doubt to *C. cyclas*, Keyserling. The latter seems to me to be only the opposite valve of the *C. Nebrascensis*.

Locality and position.—Division C of the Nebraska City section.

CYTHERE, sp.

Pl. XI, Fig. 1 *a*, *b*, *c*, *d*.

Compare *Cypris subrectus*, Portlock, 1843, Report Londonderry, &c., p. 316, Plate XXIV, Fig. 136.

Carapace-valves oblong-subelliptic in the right valve, and broader elliptic in the left; right valve with dorsal margin straight, or faintly sinuous in outline; interior with a transverse ridge near the posterior end, and an oval depression occupying the whole end just behind it; exterior surface smooth, and usually with a broad, undefined, slightly more convex region at the posterior end, corresponding to the depression within; left valve smooth, most convex behind the middle; interior with a transverse ridge near the posterior end as in the other valve, but having a depression or concavity immediately in front of it as well as behind; dorsal margin a little thickened.

Length, 0.03 inch; breadth of right valve about half the length; breadth of the left about three-fifths the length.

The right valve of this little species agrees very nearly in form with *Cypris subrectus* of Portlock, and might with more propriety be referred to that species than the form, that has been by Professor Geinitz considered doubtfully identical with *C. cyclas* of Keysering, can be referred to that Russian Permian species. The valves of these little crustacea, however, present generally so few characters that not much reliance can be placed on identifications of species, unless by one who has made an especial study of the fossil forms of the group.

Locality and position.—Division C of the Nebraska City section; also associated with the last in the Upper Coal-Measures of Union County, Iowa.

Genus PHILLIPSIA, Portlock.

PHILLIPSIA, sp.

Pl. III, Fig. 1 *a*, *b*.

Phillipsia, sp., Geinitz, 1866, Carb. und Dyas in Nebr., p. 1, Tab. 1, Fig. 1.

I have no specimens of this species for study and description, and therefore merely reproduce Professor Geinitz's figures, in order to present as full an illustration as possible of the characteristic fossils of these rocks. It is probably identical with some of the forms already described, but not yet figured, from the Upper Coal-Measures of the West.

Locality and position.—Upper Coal-Measures, Plattsmouth, Nebraska.

PHILLIPSIA SCITULA, M. & W.

PLATE VI, Fig. 9.

Phillipsia scitula, Meek & Worthen, 1865, Proceed. Acad. Nat. Sci. Philad., p. 270. Compare *P. Ciftonensis*, Shumard, 1858, Trans. Acad. Sci. St. Louis, Vol. I, p. 226.

The only specimen of the little trilobite in our collections from Nebraska was not observed until the whole collection had been looked over several times, owing to the fact that it is crushed, and, being incrusted with shaly matter, was placed in a tray with flattened specimens of *Retzia punctulifera*, from the same bed, under the supposition that it was a crushed example of that little shell. On carefully cleaning it, however, it was found to belong to the above species of *Phillipsia*. It was evidently folded with the head and pygidium together, and crushed laterally by accidental pressure, so as to become nearly flat. After careful cleaning, however, it is not difficult to make out enough of its specific characters, by the aid of a magnifier, to leave no doubt of its identity.

It is possible that this is the same form described by Dr. Shumard from the Upper Coal-Measures of Kansas, under the name of *P. Cliftonensis*, but as he only had the pygidium, and had no means of characterizing its other parts, I do not feel warranted in referring it to his species. If they are identical, however, and their identity can be in any way established, his name will have to take precedence, as it was first published.

Locality and position.—Nebraska City, division B. The specimen originally described was from the Upper Coal-Measure at Springfield, Illinois; and it also occurs in the Lower Coal-Measures of that State.

PHILLIPSIA MAJOR, Shumard.

Pl. III, Fig. 2 *a*, *b*, *c*.

Phillipsia major, Shumard, 1858, Trans. St. Louis Acad. Sci., Vol. I, p. 225.

Pygidium semi-elliptical, very convex, a little longer than wide, narrowing posteriorly, the lateral margins being straightened near the middle; posterior extremity narrowly rounded; mesial lobe strongly elevated above the lateral ones, distinctly compressed, and longitudinally furrowed on each side; strongly arched longitudinally, rounded on top, and gradually tapering posteriorly to its rather prominent, somewhat obtuse extremity, which terminates at about one-fifth the entire length of the pygidium from the posterior margin; segments twenty-two or twenty-three, not arching forward or backward, a few of the anterior ones only faintly defined by slender linear furrows, those farther back more distinct, all becoming obsolete on the flattened or furrowed sides of the lobe; lateral lobes wider than the axial one, rounding down abruptly on each side, and sloping more gradually behind, into a smooth border, which continues all around the free margins, but is broader and more flattened behind; segments twelve or thirteen, moderately oblique, very short behind, and gradually increasing in length anteriorly; all ending abruptly on reaching the rather broad, smooth, sloping, marginal zone; surface smooth, or only showing obscure traces of fine granules, with minute scattering pits.

Length of pygidium, 0.77 inch; breadth, 0.72 inch; height, about 0.35 inch.

It is possible that this may be distinct from Dr. Shumard's *P. major*, but it agrees so nearly with his description, that there seems scarcely room to doubt its identity with that species. His type was larger, and proportionally somewhat wider, but agrees well in all its other details, while the slight difference of proportional length and breadth of the pygidium may be due to a difference of age or sex.

It is also closely related to *P. Sangamonensis*, M. & W., described from the Upper Coal-Measures of Illinois, but not only differs in being proportionally a little longer, but also in having its axial lobe proportionally larger. Still it may be the same.

Locality and position.—Dr. Shumard's specimens were from the Upper Coal-Measures of Clinton County, Missouri, and on Vermillion River, 12 miles south of Lecompton, Kansas. Our specimen represented by Fig. 2 *a*, *b*, of Pl. III, was found on the Missouri, in the same horizon at Bellevue, Nebraska. That represented by Fig. 2 *c* was found by Professor Egleston at Plattsmouth.

DESCRIPTIONS OF FOSSIL FISHES FROM THE UPPER COAL-MEASURES OF NEBRASKA.

By Orestes H. St. John.

Genus CLADODUS, Agassiz.

Cladodus mortifer, N. and W.

Pl. III, Fig. 6 *a*, *b*, and Pl. VI, Fig. 13 *a*, *b*, *c*, *d*.

Cladodus mortifer, Newberry and Worthen, Geol. Illinois, Vol. II, p. 22, Pl. I, Fig. 5; St. John, 1870, Proceed. Am. Philos. Soc., Vol. XI, p. 431.

In the collection there are fragments of three individuals of the above species—two showing the base with portions of the crown, and one preserving about a third of the lower portion of the median cusp of a very large specimen. There can be no doubt that the teeth before me are referable to the above species; but as they exhibit characters not shown in the imperfect specimen figured and described by Messrs. Newberry and Worthen, a short description of the Nebraska teeth is here appended.

Description.—The base of the tooth is semi-elliptical in outline, obtusely angular behind, with low protuberances rising at the angles upon the superior inner margin, the outer margin interrupted by a broad, shallow sinus, at either angle of which, immediately beneath the smaller lateral denticles, an obtuse node projects downward, similar to those upon the upper opposite side of the root, the presence of which would seem to have been designed to lend additional strength to the muscular attachment of the tooth upon its cartilaginous support; median cone cervical, regularly tapering, recurved, inequally compressed, with acute lateral edges; striæ sharp, interrupted, separated by wide plane spaces, less numerous upon the strongly compressed anterior face, and confined to the lower half of the cusp; lateral denticles two upon either side, strong, with sharp cutting edges, and strong sharp striæ or ridges.

Breadth of base twice its length, and equal to the entire height of the tooth.

This species, so far as we at present know, is restricted to the Upper Coal-Measures. The single type specimen from which the species was originally described was found in the Upper Coal Strata near Springfield, Illinois; and in the prosecution of the geological survey of Iowa, Dr. White has brought to light the same species from the Upper Coal-Measures of the southwestern portion of the State. I have also found this species in the same formation at Manhattan, Kansas.

Compared with other species, the present one is probably more closely related to *C. mirabilis*, Agassiz, from the Mountain Limestone, Ireland, than with any other to which I am acquainted. It differs, however, in being less robust, and more symmetrical in its general proportions.

Formation and locality.—Upper Coal-Measures, bed B, Nebraska City section, Nebraska.

Genus DIPLODUS, Agassiz.

DIPLODUS COMPRESSUS, Newb.

Pl. IV, Fig. 19 *a*, *b*.

Diplodus compressus, Newberry, Geol. Illinois, Vol. II, p. 60, Pl. IV, Fig. 2; St. John, 1870, Proceed. Am. Philos. Soc., Vol. XI, p. 432.

The single specimen of *Diplodus* in the collection is probably referable to the form described by Dr. Newberry under the name *D. compressus*.

Description.—The tooth is of medium size; base slightly narrower than long, broadly rounded in front, and terminating in an obtuse point behind; under surface slightly raised in the middle; anterior extremity produced into a large obtuse tubercle projecting slightly outward and downward, with a flattened, sharply defined, obovate, pad-like projection upon the upper surface of the posterior extremity, marked upon either side by a shallow groove terminating above in a little pit, which is entirely separated from the bases of the crown cusps; in this latter respect, as Dr. Newberry has remarked, offering marked contrast to *D. gibbosus*, Agassiz, from the Mountain Limestone of Europe; cusps three, median one rudimentary, slender, compressed, with finely crenulated lateral edges, base well defined from the general surface and terminating in a slight protuberance in the osseous root in front; the apices of the lateral cusps are broken away in the specimen before me; they are strongly compressed, smooth, with sharp, beautifully crenulated cutting edges, unequally divergent; left one—viewed from before—most inclined from a vertical line and broadest at base; transverse section of both lenticular.

In the collection of the Iowa State geological survey, there is a tooth from the Upper Coal-Measures of Southwestern Iowa, which is doubtless specifically identical with the Nebraska specimen, though possessing some slight differences. In the Iowa specimen the base has, as in the above-described tooth, a lozenge-shaped outline, its posterior extremity is more abruptly truncated, and the pad-like elevation surmounting its surface is elliptical with its longer axis transverse to the root—in other respects the same as the Nebraska tooth; viewed in front, the right lateral cone is the strongest and most inclined laterally, and the bases on the anterior face are swelled out, producing an angular ridge or buttress, which, however, is lost both in the crown above and in the root below. These two individuals are the only ones I have had opportunity to examine, and comparing them with the excellent description and figures of *D. latus*, Newb., I cannot doubt but that they are distinct from that species. The present species is described from the Coal-Measures of Ohio and Southwestern Indiana, the latter locality holding a stratigraphical position probably below the Nebraska horizon.

Formation and locality.—Upper Coal-Measures, Rulo, Nebraska.

Genus PETALODUS, Agassiz.

PETALODUS DESTRUCTOR, N. and W.

Pl. III, Fig. 5.

Petalodus destructor, Newberry and Worthen, Geol. Illinois, Vol. II, p. 35, Pl. II, Figs. 1, 2, 3; St. John, 1870, Proceed. Am. Philos. Soc., Vol. XI, p. 433.

The collection contains a large, almost perfect specimen of the above species, which presents the following characters:

Description.—The crown is sharp, compressed, gradually thickening toward the base; crest more or less gently arched from the lateral extremities, obtusely acuminate at the apex, and distinctly striated for the space of a line or less, below which the striæ are lost in the dense enamel-like coating which covers both faces of the crown; posterior face of crown rhombic, outline of base similar to that of crest, and bordered by five strongly marked imbricating folds, which are conspicuously arched downward in the middle, and more or less deflected at the lateral extremities; anterior face broadly rhomboidal, basal fold consisting of four or five obscurely marked imbrications, gently curved downward in the middle, and again at the lateral extremities; the upper edges of the imbricating folds are minutely crenulated; root broad, compressed at the edges, rapidly tapering from the lateral shoulders, and terminating in a blunt rounded point. Upon much-worn surfaces the crown is finely punctate.

	Inches.
Length, nearly	2.00
Greatest breadth of crown, about	1.60
Height of anterior face of crown	0.95
Height of posterior face	1.28
Breadth of root across the lateral shoulders, about	1.10

This species bears a striking resemblance to *Petalodus acuminatus*, Agassiz, from the Mountain Limestone of Europe; but, at the same time, it possesses characters which readily distinguish it from that species. The present species differs mainly in the more tapering root, the coronal band upon the inner face is more strongly curved downward in the middle, and the crown is relatively higher. This species was originally described from the Upper Coal-Measures of Central Illinois. I have seen a fine specimen of the same species in the collections of the museum at Cambridge, from similar horizons in Southwestern Indiana, and also from the Upper Coal-Measures of Central Iowa.

Formation and locality.—Upper Coal-Measures, Rock Bluff, Nebraska.

Genus PERIPRISTIS, Agassiz, M. S.

Peripristis (Agassiz), St. John, Proceed. Am. Philos. Soc., Vol. XI, p. 434.

Generic characteristics.—Teeth small or of medium size, possessing the general characteristics of the petalodonts. Crown compressed, acuminate, serrate, more or less curved laterally; extremities on the inner face connected by a raised transverse shoulder, in which the crown terminates below, and which gives rise to a more or less profound coronal cavity. Root well developed, entire, as in *Petalodus*. The surfaces of the crown and coronal cavity are covered by a dense and highly polished layer of ganoine, which forms an imbricated band at the base.

The above generic designation was suggested by Professor Agassiz for the reception of a group of peculiar teeth, of which we have at least two representative species—that of *P. semicircularis* being regarded as the type. These forms certainly possess features which are widely at variance with the typical species of the genus *Ctenoptychius*, as represented by *C. apicalis*, Agassiz; and in their description of the following species, Messrs. Newberry and Worthen have also referred to the remarkable characters which distinguish it from the typical species of *Ctenoptychius*. The central coronal cavity and the prominent transverse ridge in which the root is terminated above on the posterior aspect are peculiarities which do not appear in any of the numerous other genera comprised in the groups of petalodonts.

The genus is Carboniferous, ranging from the Lower Carboniferous to the Upper Coal-Measures inclusive.

PERIPRISTIS SEMICIRCULARIS, N. & W., sp.

Pl. III, Figs. 3, 4; and Pl. IV, Fig. 20 *a*, *b*.

Ctenoptychius semicircularis, Newberry and Worthen, Geol. Illinois, Vol. II, p. 72, Pl. IV, Figs. 18, *a*, *b*.

Description.—Tooth small, broadly obovate in outline, crown much compressed and strongly curved laterally, giving the crest a semicircular outline viewed from above; cutting edge divided into seven to nine denticulations, the median lobe strongest, lateral ones gradually decreasing in size toward the lateral extremities, where they are scarcely relieved from the edge; the calcigerous tubes slightly diverge on nearing the edge, producing a minute radiated striation of the denticulations like that observed in the even crest of *Petalodus*, and when the crown is much worn the surface is finely punctate; outer face of crown very low in proportion to its breadth, base sharply beveled, coronal band narrow, imbrications very obscure or obsolete, gently descending in the middle, and slightly curved downward at the lateral extremities; upon the posterior face the base of the crown is defined by a conspicuous transverse ridge, which unites the lateral extremities, and gives origin to a deep central coronal cavity; the enamel-like coating lines the walls of the cavity, and, spreading over the gently and regularly downward arched transverse shoulder, it forms a thin coronal band with one or two faint imbrications upon its external inflexed border. The root is nearly as wide and much thicker than the crown, tapering rapidly, and rounded at its extremity; anterior side convex or ridged, posterior face slightly concave transversely; both surfaces more or less roughened.

	Inches.
Greatest length	0.77
Greatest breadth at the lateral angles of the crown	0.72
Height of crown upon its anterior face	0.32
Depth of the coronal cavity from the apex of the median denticulation, about	0.45
And from the transverse shoulder, about	0.20

The collection contains a perfect specimen of the above-described species, from Bellevue, Nebraska, imbedded in a matrix of limestone, but exhibiting the entire posterior aspect of the tooth without a blemish; and I owe to the kindness of Hon. J. Sterling Morton, of Nebraska City, another equally perfect specimen, obtained from a shaft excavation near the city, which shows the anterior face of the tooth. I think there can be no question as to their specific identity with the form described by Messrs. Newberry and Worthen from the Upper Coal-Measures of Illinois.

I am acquainted with but a single other form to which this species seems to be closely related, and that is from the Mountain Limestone of Yorkshire, England. Specimens of the latter species are in the extensive collections of the Museum of Comparative Zoölogy at Cambridge. The English specimens are, however, markedly distinct specifically from the American; they are less curved laterally, and possess some sharp, thick serrations on either side of the median cusp; the crown is relatively higher, and the coronal band on the outer face is more deeply arched downward in the middle, is wider and more distinctly imbricated; the coronal cavity of the inner face is shallower, and the transverse shoulder less prominent. I am not aware that the English species is described.

Formation and locality.—Upper Coal-Measures; Rockbluff and Bellevue; and from Morton's Shaft near Nebraska City, Nebraska.

GENUS CHOMATODUS, Agassiz.

CHOMATODUS ARCUATUS, St. John.

Pl. VI, Fig. 14 *a*, *b*.

Chomatodus arcuatus, St. John, 1870, Proceed. Am. Philos. Soc., Vol. XI, p. 435.

A fragment of limestone from Bennett's Mill, near Nebraska City, preserves the impression of a tooth of the genus *Chomatodus*, which seems to be distinct from all the species of this genus heretofore described from the Coal-Measures and Lower Carboniferous. The impression presents almost the entire figure of the anterior face, from which the following description is given:

Description.—Tooth large, laterally elongated, moderately thick (?), extremities rounded; crown slightly arching from the lateral angles and curved laterally; anterior face slightly convex vertically and rounded at the crest, which was probably more or less obtuse; the anterior face of the crown was apparently undulated along its crest; the obscure sulci may have reached half the distance from the crest toward the base, and at the median line a very shallow depression, about as high as it is wide at the base, reaches upward about two-thirds the height of the crown, and seems to interrupt the continuity of the basal folds, which, however, may not be persistent or of specific importance; basal band narrow, linear, with two or three imbricated folds, and parallel with the base of the root; surface coarsely punctate. Root nearly as wide as the crown, its anterior face deeply channeled by an angular transverse furrow, with a low ridge traversing the lower portion from one extremity to the other, below which it is beveled to the outer basal edge.

	Inches.
Greatest breadth, about	1.60
Height	0.50
Greatest height of anterior crown face	0.22

In outline the above species bears a somewhat marked resemblance to *C. loriformis*, N. and W., from the Keokuk limestone; but it differs from that form in having the anterior face of the crown relatively higher, its crest undulated and less parallel, and its bow-shaped outline viewed from above as well as in the more vertical concavity of the outer aspect of the root. It is not improbable that the basal angle of the posterior crown face was quite prominent, and the vertical concavity of that face of the crown must have been considerable, judging from the arched character of the opposite face, and in this respect somewhat resembling *C. cinctus*,

Agassiz, though the present species is not acuminate, the coronal band not nearly as wide as in that species, and the tooth is not as thick and massive.

Formation and locality.—Upper Coal-Measures, Bennett's Mill, near Nebraska City; where it was found in place by the writer.

GENUS XYSTRODUS, Agassiz, M.S.

XYSTRODUS ? OCCIDENTALIS, St. John.

Pl. IV, Fig. 18 *a*, *b*, *c*, *d*.

Xystrodus ? occidentalis, St. John, 1870, Proceed. Am. Philos. Soc., XI, p. 436.

The collection affords an interesting little deltoid tooth, which I believe has not been heretofore described. Unfortunately, the specimen is quite imperfect, and, although its specific characters permit of description, its generic affinity remains somewhat in doubt.

Description.—Terminal tooth small, subtrigonal in outline, little narrower than long, but slightly enrolled, flattened or gently depressed above; the straight side is abruptly beveled, and from its edge the crown gently inclines to the opposite oblique margin, which is very slightly raised; the border extremity is thickened, forming a well-defined continuous marginal border, which rapidly descends upon the inner side and gently slopes into the shallow, depressed space in front; toward the terminal extremity the tooth becomes exceedingly thin, and in the specimen before me the pointed end and outer margin are broken away. The superior surface is coarsely punctate, as is also the straight articular margin. Distance between the angles of the broader extremities, 0.38 inch.

The tooth above described possesses some characters which seem to connect it more closely with *Xystrodus*, Agassiz, M.S., than with any other genus with which I am acquainted. Its general depressed triturating surface, and but slightly convoluted terminal extremity, are strongly suggestive of this relation. The genus *Xystrodus* was established by Prof. Agassiz for the reception of *Cochliodus striatus* and two or more other European species from the Mountain Limestone.

Formation and locality.—Upper Coal-Measures, Aspinwall, Nebraska.

DELTODUS? ANGULARIS, N. and W.

Pl. VI, Fig. 18 *a*, *b*.

Deltodus angularis, Newberry and Worthen, Geol. Illinois, Vol. II, p. 97; Pl. IX, Fig. 1; St. John, 1870, Proceed. Am. Phil. Soc., Vol. XI, p. 437.

Description.—Terminal tooth small, obliquely triangular in outline, thick, but slightly enrolled; the broader extremity has a sigmoidal curvature, terminating in an acute point at the oblique posterior extremity; straight side forming an angle of about 55° with the oblique margin, abruptly truncated, with a narrow sulcus about the middle of the beveled articular face extending from the inner angle to the pointed end, below which the tooth apparently expands into a thin, narrow border similar to that upon the opposite side; the articular margin is bordered by a prominent, flattened ridge, which occupies about one-third the surface of the crown and gradually narrowing as it approaches the terminal point; a sharp, narrow keel rises from the oblique margin, rapidly converging and decreasing in prominence toward the apical end, and separated from the broad, flattened prominence of the straight margin by an equally broad, deep, angular furrow; along the oblique side the tooth was slightly expanded into a thin marginal border. The crown surface is beautifully granulo-punctate, the broader extremity very faintly marked by longitudinal sigmoid lines of growth, and the broad mesial depression is traversed by very obscure undulations parallel with the oblique keel. Under surface longitudinally undulated, smooth.

Length of tooth along the straight margin, about 0.52 inch; greatest distance between the acute and obtuse angles of the broader extremity, 0.48 inch.

The collection affords but a single example of this handsome form. The specimen before me has a remarkable resemblance to the posterior teeth of *Deltoptychius*, Agassiz,* M.S., founded upon *Cochliodus acutus* of the Irish Mountain Limestone; but we do not at present possess the materials fully to demonstrate this identity. The Nebraska tooth, however, is evidently identical with the form described by Messrs. Newberry and Worthen, from stratigraphically corresponding horizons in Illinois; and Dr. White has discovered the same, or a very closely allied species, in the Upper Coal-Measures of Southwestern Iowa.

Formation and locality.—Upper Coal-Measures, bed B, Nebraska City: where it was discovered by Dr. White.

PART III.

ENTOMOLOGY,

BY

S. H. SCUDDER.

NOTES ON THE ORTHOPTERA COLLECTED BY DR. F. V. HAYDEN IN NEBRASKA.

BY SAMUEL H. SCUDDER.

I.—LIST OF THE SPECIES OBSERVED IN NEBRASKA.

1. GRYLLUS NEGLECTUS, Scudd.—Abundant along the Platte, up Loup Fork, Pawnee Reserve.
2. GRYLLUS PERSONATUS, Uhl.—A few specimens were taken on the banks of the Platte.
3. NEMOBIUS VITTATUS, Harr.—The specimen obtained seems to be referable to Harris's species; it was found on the banks of the Platte.
4. ŒCANTHUS NIVEUS, Sew.—Judging from poor specimens, this seems to be identical with our eastern species—1♂ 1♀. Banks of the Platte.
5. CEUTHOPHILUS GRACILIPES, Scudd. (?)—3♂ 1♀, mostly immature, were taken at Nebraska City.
6. UDEOPSYLLA ROBUSTA, Scudd.—3♂ 3♀, from Nebraska City and the banks of the Platte.
7. XIPHIDIUM SALTANS, nov. sp.—Similar to *X. brevipennis*, Scudd., but with much shorter wings. In the male the stridulating field occupies one-half of the tegmina. The projection of the vertex is unusually great; the ovipositor is very long and slender, slightly curved.

   Length from tip of vertex to extremity of abdominal segments, 0.5 inches; length of tegmina, ♂ 0.14 inch, ♀ 0.06 inch; length of hind femora, 0.41 inch; length of ovipositor, 0.52 inch.

   3♂ 1♀ from the banks of the Platte.
8. CONOCEPHALUS ATTENUATUS, nov. sp.—Tubercle of vertex bordered beneath with black, long, slender, conical, bluntly pointed, with a slight basal tooth on the under side; tegmina very slender, pointed at the tip, more than twice the length of the abdomen; hind legs slender, the femora furnished beneath, externally, with a row of exceedingly short and fine, distant, spinous hairs; internally with a row of four or five very short and distant spines, directed backward; ovipositor very slender, barely reaching the tip of the wing-covers at rest.

   Length of tubercle, measured beneath, 0.08 inch; length of hind femora, 0.85 inch; length of tegmina, 2 inches; length of ovipositor, 1.3 inches.

   1♂, 2♀, and several immature specimens were taken on the banks of the Platte.
9. ANABRUS SIMILIS, nov. sp.—This species is very closely allied to *A. purpurascens*, Uhl.; it differs from the latter principally in its slightly shorter and more curved ovipositor.

   Length of pronotum, 0.46 inch; length of hind tibiæ, 0.74 inch; length of ovipositor, 0.72 inch.

   1♂ 3♀. Banks of the Platte.

10. OPSOMALA BIVITTATA, Serv.—2♀ from Nebraska City.
11. BRACHYPEPLUS MAGNUS, Girard.—1♂ 8♀. One specimen from Nebraska City, the remainder from the banks of the Platte.
12. STENOBOTHRUS GRACILIS, nov. sp.—Vertex of the head broad, swollen, with elevated anterior border; foveolæ extremely shallow, long, triangular, with the base toward the eye; lateral carinæ of pronotum rather prominent, regularly curved, approximate in the middle; median carina slight; hind border of pronotum a little angulated; middle of the lower border produced into a rounded projection.

Slightly mottled, pale reddish brown, a band of the deeper tint just behind the eyes, bordered above by the yellowish lateral carinæ of the pronotum; tegmina nearly uniform in color, but with two or three small spots in the central field.

Length from vertex to tip of tegmina, 0.65 inch. 1♂ taken on the banks of the Platte.

13. ACRIDIUM EMARGINATUM, Uhl, MSS.—Pronotum light greenish red, lighter on sides; a distinct yellow band from the vertex of the head to the tip of the pronotum; antennæ yellowish, changed into brown at tips; tegmina brownish-red, inner border margined with dull yellowish; hind femora yellowish, with a few black points on the upper outer carina; hind tibiæ dark brownish-red, armed with black-tipped, yellow spines; abdomen yellow, with transverse rows of black dots at the tips of the segments.

Length from vertex to tip of tegmina, ♂ 1.4 inches, ♀ 1.75 inches. 3♂ 3♀ along the banks of the Platte River.

14. CALOPTENUS FEMUR-RUBRUM, Burm.—1♂ 2♀ only were collected, from Rulo, Southern Nebraska, July 9th; from Nebraska City and from the banks of the Platte River. For further remarks on this species see below.
15. CALOPTENUS SPRETUS, Uhl.—Rulo, Southern Nebraska, on the banks of the Missouri River, in great quantities, July 9th; other specimens were brought from the banks of the Platte River and from Nebraska City. For further remarks on this species see the second division of my paper.
16. CALOPTENUS BIVITTATUS, Uhl.—1♂ 3♀, taken along the banks of the Platte River and at Nebraska City.
17. PEZOTETTIX SPECIOSA, nov. sp.—Pronotum rugose, the anterior half with irregular transverse, the posterior half with irregular longitudinal lines; the raised portions pale yellowish; the depressed parts deep reddish; tegmina immaculate, yellowish, with the longitudinal vein of inner half pinkish.

This species is placed here provisionally, as it does not strictly appertain to *Pezotettix*.

Length of pronotum, ♂ 0.23 inch, ♀ 0.30 inch; length of tegmina, ♂ 0.4 inch, ♀ 0.6 inch; length of hind femora, ♂ 0.56 inch, ♀ 0.64 inch.

1♂ 2♀ from Nebraska City and the banks of the Platte River.

18. ŒDIPODA TRIFASCIATA.—*Gryllus trifasciatus*, Say. Amer. Entom., Pl. XXXIV. 1♂ from Nebraska Territory.
19. ŒDIPODA COLLARIS, nov. sp.—Dark reddish-brown; lower half of head, base of hind femora, and a broad band along the posterior edge of the pronotum, pale clay-yellow; hind tibiæ reddish; tegmina mottled somewhat uniformly with fuscous blotches, which form three distinct bands, the outer of which is sometimes lost in the nearly equal mottling of the tip; wings pale yellow at base, some-

what obfuscated at tip with a broad median blackish band, occupying the middle third of the wing at the costal border, crossing the wing at right angles with uniform breadth, (excepting a spur thrown out toward the base, as in *Œ. xanthoptera*,) and then, with decreasing width, following the curve of the wing to the inner angle; pronotum with a prominent median ridge throughout its extent.

Length from vertex to tip of tegmina, ♂ 1.2 inches, ♀ 1.6 inches; 2 ♂ 2♀ from Nebraska City and from the banks of the Platte.

20. ŒDIPODA TENEBROSA, nov. sp.—Nearly uniform dark fuscous, front of head with somewhat pale markings; pronotum above dull, dirty yellow, with dots and streaks of fuliginous; hind femora banded with yellow just before the tip; hind tibiæ with a dull reddish band next the base, followed by a darker band; tegmina almost uniformly mottled with fuscous, the outer half interspersed with paler spots; wings, pale reddish at base, the whole outer border obscured by a broad, dark, fuliginous band, slightly paler at the extreme tip, projecting sharply inward almost to the base of the wing, near the costal margin; antennæ fuscous, basal third paler; pronotum with a slightly raised median carina.

Length from vertex to tip of tegmina, ♂ 1.2 inches, ♀ 1.25 inches; 2 ♂ 2 ♀, Nebraska City and along the banks of the Platte.

21. ŒDIPODA HALDEMANII, nov. sp.—This species is closely allied to *Œ. corallipes*, Hald., but differs from it in the greater rugosity of the pronotum and in the greater separation and distinctness of the markings on the tegmina. The wings are pale yellow at base, hyaline with black veins at tip and across the middle, have a rather narrow fuliginous band, curving regularly around to the inner angle, broken at the division of the central and inner areas of the wing, and above this projected strongly and broadly inward, almost to the very base of the wing; at the costal border the band becomes almost piceous.

Length from vertex to tip of tegmina, ♂ 1.25 inches, ♀ 1.8 inches; 1 ♂ 4 ♀, Nebraska City and the banks of the Platte.

22. ŒDIPODA CAROLINA, Burm.—2 ♂ 5 ♀ from Nebraska City and the banks of the Platte.

23. DIAPHEROMERA VELII, Walsh.—1 ♂ 1 ♀ from the banks of the Platte River.

24. STAGMATOPTERA MINOR, nov. sp.—This little *Mantis* is a miniature of *S. carolina*, but, on a front view, the vertex of the head is less swollen and, indeed, nearly flat; the eyes are not so prominent, and instead of being regularly rounded, are slightly flattened above, in continuation of the contour of the vertex; head and pronotum mottled with black and dark, dirty yellowish, the pronotum with raised black points along the outer border; tegmina dark brown, obfuscated on the apical half; stigma of the same color as its surroundings and therefore inconspicuous.

Length of pronotum, 0.25 inch; length of tegmina, 0.28 inch; breadth of head, including eyes, 0.17 inch.

Three specimens from Nebraska City.

25. TEMNOPTERYX MARGINATA, nov. sp.—Pronotal shield piceous, bordered laterally with a wide yellowish band; furnished, also, with a few widely and irregularly scattered punctulations; tegmina piceo-ferruginous, the outer margin with a yellowish band in continuation of that on pronotal shield, reaching half-way to the apex; femora beneath luteous; tibiæ and tarsi castaneus.

Length of pronotal shield, 0.17 inch; breadth of the same at the posterior border, 0.22 inch; length of tegmina, 0.26 inch; breadth of the same in the middle, 0.16 inch.

From the banks of the Platte.

II.—REMARKS ON THE DEVASTATING GRASSHOPPERS OF NORTH AMERICA.

At least three kinds of migratory grasshoppers, two of which belong to the genus *Caloptenus*, are found in the United States.

One, *Caloptenus femur-rubrum*, Burm., is found in all the States east of the Mississippi, unless we except the southernmost, as well as in those bordering it on the west. It was long ago described by De Geer. This seems to be the least destructive kind, and only on extraordinary occasions has been known to migrate in clouds, or even extend its ravages over a wide stretch of country.

The second, *Caloptenus spretus*, Uhl., a much more destructive species, occurs over the whole of the vast region west of the States bordering the Mississippi, to the very base of the Rocky Mountains, from Texas on the south, past the boundaries of the United States, to the Saskatchewan River on the north. The species, which received its name in manuscript from Mr. Uhler, has been briefly described by Mr. Walsh. This entomologist discusses at length, in the Practical Entomologist for October, 1866, the probability of its permanent foothold in Kansas and Nebraska. While we agree with him in his conclusions, we would by no means assent to a line of argumentation founded upon his hypothesis of its origin. "It is evidently a strictly alpine insect," says Mr. Walsh. He bases this assertion solely upon the statement of Drs. Velie and Parry, who assured him that, in Colorado, "it breeds in the mountains, and comes down into the settlements in vast swarms through the cañons, or deep perpendicular cuts, leading from the mountains to the more level country." Mountain valleys, with a southern exposure, will, however, breed animals which perhaps could not endure the rigors of the open plain, so that this evidence lacks sufficiency. Moreover, we have two facts of a different nature—first, that when regions outside the natural limits of the range of this pest are invaded, the inclemency alone of a succeeding spring can prevent the eggs or newly hatched young from coming to maturity; secondly, that the insect has often bred, year after year, in a region south of, and much milder, than Kansas and Nebraska, viz, Central Texas. The actual limitations of its natural abode I do not know, nor have we as yet sufficient data for a reasonable hypothesis. Indeed, we ought to be supplied with an extensive series of simultaneous observations made over the whole of the ravaged districts, stating the exact time and direction of each flight, and the direction of the wind at the time. By collating these data we might be able to arrive at some clear conclusions concerning the original habitation of the species.*

---

* Since this article left my hands, Mr. Walsh has reiterated his arguments, with a great deal of force and originality, in the American Entomologist, Vol. I, pp. 73–76, and in his First Annual Report on the Noxious Insects of the State of Illinois, pp. 82–103. The former article, however, has brought out the adverse statement of two observers in Colorado, the seat of the grasshoppers, (see American Entomologist, pp. 94–95,) one of whom, Mr. Byers, states explicitly that it is "common to *all* this western or *rainless* region, one-third of the United States, but its breeding-place is upon the hot, parched plains and table-lands, from four to six thousand feet above the sea, instead of in the cañons of the mountains. The greater the heat the more they flourish."

A third species of grasshopper, unnamed as yet, belonging to the genus *Œdipoda*, appears to be the insect which has ravaged the cultivated districts of California and Oregon, and the neighboring States and Territories; it probably ranges over the whole extent of country west of the Rocky Mountains and included within the limits of the United States. Mr. A. S. Taylor, in one of his articles in the California Farmer, subsequently communicated to the Smithsonian Institution and published in their Report for 1858, describes the grasshopper as found near Monterey, and it is doubtless the migratory species which ravaged the State. It is a species of *Œdipoda*, which, from the devastating nature of its ravages, may be called *Œdipoda atrox*, or the terrible grasshopper.* To the best of my knowledge, it is the only species of the genus which has anywhere proved seriously and persistently injurious to crops. Several species of the closely allied genus *Pachytylus* have ravaged the fields of Eastern Europe and Asia; and it is interesting, in a zoölogical point of view, to find that California, whose insect fauna bears a much more general resemblance to the peculiar types of the Old World than to those characteristic of the opposite border of the New World, should in this case also harbor a devastating grasshopper so much more nearly allied to the destructive species of the Mediterranean than to those found upon the same continent with itself.

For the elucidation of the history of these three species, and to show the similarity of their devastations, I have brought together all the original matter I could obtain, as well as a few already published statements, and will present them under the heading of each species.

A.—CALOPTENUS FEMUR-RUBRUM, Burmeister.

Its ravages in New England.—Although the ravages of locusts in America are not followed by such serious consequences as in the Eastern Continent, yet they are sufficiently formidable to have attracted attention, and not unfrequently have these insects laid waste considerable tracts, and occasioned no little loss to the cultivator of the soil. Our salt-marshes, which are accounted among the most productive and valuable of our natural meadows, are frequented by great numbers of the small red-legged species, (*Acrydium femur-rubrum*,) intermingled occasionally with some larger kinds. These, in certain seasons, almost en-

---

*We subjoin a description of this species:

ŒDIPODA ATROX, nov. sp.—Head uniform, pale brownish yellow; the raised edge of the vertex dotted with fuscous; a dark fuscous spot behind the eye, broadening posteriorly, but not extending upon the pronotum. Antennæ as long as the head and pronotum together, dull honey-yellow, growing dusky toward the tip. Pronotum dark-brownish yellow, the sides darker anteriorly; median carina extending the whole length of the pronotum, moderately raised, cut once by a transverse line a little in advance of the middle; lateral carinæ prominent, extending across the anterior two-thirds of the pronotum; anterior border of the pronotum smooth, very slightly angulated; posterior border delicately marginate, bent at a very little more than a right angle, the apex rounded; tegmina dull-yellowish on the basal half, with distinct fuscous spots; toward the apex obscurely fuscous, with indistinct fuscous markings; humeral ridge yellowish, and, when the tegmina are in repose, inclosing a brownish fuscous triangular stripe; the spots are scattered mostly in the median field, consisting in the basal two-fifths of the tegmina of small roundish spots and one larger longitudinal spot in the middle of the basal half; there is a large irregular spot in the middle of the tegmina, and beyond a smaller transverse spot, followed by indistinct markings; wings hyaline, slightly fuliginous at the extreme tip; the veins, especially in the apical half, fuscous; legs uniform brownish fuscous; apical half of spines of hind tibiæ black.

Length of body, 0.9 inch; of tegmina, 0.9 inch; of body and tegmina, 1.125 inches; of pronotum, 0.2 inch; of hind femora, 0.5 inch.

It bears a strong resemblance to *Œdipoda pellucida*, Scudd., common in Northern New England.

tirely consume the grass of these marshes, from whence they then take their course to the uplands, devouring, in their way, grass, corn, and vegetables, till checked by the early frosts, or by the close of the natural term of their existence. When a scanty crop of hay has been gathered from the grounds which these puny pests have ravaged, it becomes so tainted with the putrescent bodies of the dead locusts contained in it that it is rejected by horses and cattle.

At various times they have appeared in great abundance in different parts of New England. It is stated that, in Maine, during dry seasons they often appear in great multitudes and are the greedy destroyers of the half-parched herbage. In 1749 and 1754 they were very numerous and voracious; no vegetables escaped these greedy troops; they even devoured the potato tops; and in 1743 and 1756 they covered the whole country and threatened to devour everything green. Indeed, so great was the alarm they occasioned among the people, that days of fasting and prayer were appointed on account of the threatened calamity. The southern and western parts of New Hampshire, the northern and eastern parts of Massachusetts, and the southern part of Vermont, have been overrun by swarms of these grasshoppers and have suffered more or less from their depredations.

Among the various accounts which I have seen, the following, extracted from the Travels of the late President Dwight, seems to be the most full and circumstantial:

Bennington, Vermont, and its neighborhood, have for some time past been infested by grasshoppers of a kind with which I had before been unacquainted. At least, their history, as given by respectable persons, is in a great measure novel. They appear at different periods, in different years; but the time of their continuance seems to be the same. This year (1798) they came four weeks earlier than in 1797, and disappeared four weeks sooner. As I had no opportunity of examining them, I cannot describe their form or their size. Their favorite food is clover and maize. Of the latter they devour the part which is called the silk, the immediate means of fecundating the ear, and thus prevent the kernel from coming to perfection. But their voracity extends to almost every vegetable, even to the tobacco plant and the burdock. Nor are they confined to vegetables alone. The garments of laborers, hung up in the field while they are at work, these insects destroy in a few hours; and with the same voracity they devour the loose particles which the saw leaves upon the surface of pine boards, and which, when separated, are termed saw-dust. The appearance of a board fence, from which the particles had been eaten in this manner, and which I saw, was novel and singular; and seemed the result, not of the operations of the plane, but of attrition. At times, particularly a little before their disappearance, they collect in clouds, rise high in the atmosphere, and take extensive flights, of which neither the cause nor the direction has hitherto been discovered. I was authentically informed that some persons employed in raising the steeple of the church in Williamstown, were, while standing near the vane, covered by them, and saw, at the same time, vast swarms of them flying far above their heads. It is to be observed, however, that they customarily return and perish on the very ground which they have ravaged.

Through the kindness of the Rev. L. W. Leonard, of Dublin, New Hampshire, I have been favored with specimens of the destructive locusts which occasionally appear in that part of New England, and which, most probably, are of the same species as the insects mentioned by President Dwight. They prove to be the little red-legged locusts, whose ravages on our salt-marshes I have already recorded.

In the summer of 1838, the vicinity of Baltimore, Maryland, was infested by insects of this kind; and I was informed by a young gentleman from that place, then a student in Harvard College, that they were so thick and destructive in the garden and grounds of his father, that the negroes were employed to drive them from the garden with rods; and in this way they were repeatedly whipped out of the grounds, leaping and flying before the extended line of castigators like a flock of fowls. Some of these insects were brought to me by the same gentleman, on his return to the university, at the end of the summer vacation, and they turned out to be specimens of the red-legged locusts already mentioned.—(*Dr. T. W. Harris, in his Report upon the Insects of New England Injurious to Vegetation.*)

Its ravages on Cape Cod.—In some dry seasons within my recollection they have been so numerous here as to destroy almost every green thing. I am not able to give the precise year when their depredations were committed; I only know that in times of great drought their ravages have been very great.—(*From an account by Mr. F. Scudder, Barnstable, June* 22, 1868.)

Its ravages in Maine.—The year 1821 or 1822 was an unusually dry season during the summer months. In the early part of the season fires had raged extensively on newly cleared lands. In June there appeared an immense number of red-legged locusts (*Acridium femur-rubrum*) on the farm where I then resided in the town of Pownal, Cumberland County, Maine.

The land is a light sandy soil, in places merging into a sandy loam, which, but a few years before, had supported a dense growth of spruce and hemlock; this had been burned over, leaving the ground covered with a heavy scurf of vegetable matter. It was apparently in this scurf that the locusts had laid their eggs the previous year.

During the haying season the weather was dry and hot, and these hungry locusts stripped the leaves from the clover and herdsgrass, leaving nothing but the naked stems. In consequence, the hay crop was seriously diminished in value. So ravenous had they become that they would attack clover, eating it into shreds. Rake and pitchfork handles, made of white ash, and worn to a glossy smoothness by use, would be found nibbled over by them if left within their reach.

As soon as the hay was cut and they had eaten every living thing from the ground, they removed to the adjacent crops of grain, completely stripping the leaves; climbing the naked stalks, they would eat off the stems of wheat and rye just below the head, and leave them to drop to the ground. I well remember assisting in sweeping a long cord over the heads of wheat after dark, causing the insects to drop to the ground, where most of them would remain during the night. During harvest-time it was my painful duty, with a younger brother, to pick up the fallen wheat heads for threshing; they amounted to several bushels.

Their next attack was upon the Indian corn and potatoes. They stripped the leaves and ate out the silk from the corn, so that it was rare to harvest a full ear. Among forty or fifty bushels of corn spread out in the corn-room, not an ear could be found not mottled with detached kernels.

While these insects were more than usually abundant in the town generally, it was in the field I have described that they appeared in the greatest intensity. After they had stripped everything from the field they began to emigrate in countless numbers. They crossed the highway and attacked the vegetable garden. I remember the curious appearance of a large flourishing bed of red onions, whose tops they first literally ate up, and, not content with that, devoured the interior of the bulbs, leaving the dry external covering in place. The provident care of my mother, who covered the bed with chaff from the stable-floor, did not save them, while she was complimented the next year for so successfully sowing the garden down to grass. The leaves were stripped from the apple-trees. They entered the house in swarms, reminding one of the locusts of Egypt, and as we walked they would rise in countless numbers and fly away in clouds.

As the nights grew cooler, they collected on the spruce and hemlock stumps and log fences, completely covering them, eating the moss and decomposed surface of the wood, and leaving the surface clean and new. They would perch on the west side of a stump where they could feel the warmth of the sun, and work around to the east side in the morning as the sun reappeared. The foot-paths in the fields were literally covered with their excrements.

During the latter part of August and the first of September, when the air was still dry, and for several days in succession a high wind prevailed from the northwest, the locusts frequently rose in the air to an immense height. By looking up at the sky in the middle of a clear day, as nearly as possible in the direction of the sun, one may descry a locust at a great height. These insects could thus be seen in swarms, appearing like so many thistle-blows as they expanded their wings and were borne along toward the sea before the wind; myriads of them were drowned in Casco Bay, and I remember hearing that they frequently dropped on the decks of coasting vessels. Cart-loads of dead bodies remained in the fields, forming in spots a tolerable coating of manure.

It was an object of curiosity to me, then a boy, to catch some of the largest locusts, and turn up their wings to find the little red parasite which covered their bodies; this might have done something toward hastening their destruction, although it did not prevent the ravages on the crops.

During the years necessary to clear up the forests on the sandy lands in the vicinity, it was no uncommon thing to have the crops seriously injured by these locusts, but never, to my knowledge, to the extent described above.

In response to my special inquiries concerning the flight of these insects, my correspondent replied as follows:

I do not remember ever to have witnessed the flight of these grasshoppers to any extent, except during the year mentioned, and the preceding one. Nor do I ever recollect a time when the wind blew so steadily for days in succession, from the northwest, generally rising soon after mid-day, and going down with the sun. I have no meteorological record, but speak from memory.

The town of Pownal was principally settled after the opening of the present century. As the lands were cleared, the Canada thistle and other species sprang up in great quantities; when they ripened, the winds spoken of as occurring at that time carried off immense numbers of the thistle-blows to the ocean. I was wont to spend hours in my boyhood lying on the ground and directing my eyes as near as I could to the sun, to watch the thistle-blows as they passed across or near its disk. I think I could have seen them in this situation several hundred feet high. I injured my eyes permanently by indulging in this amusement. Whether the grasshoppers ever rose to so great a height I do not know, but I think that they generally flew at a lower level. * * * Although they would rise in clouds as one approached them, it was only an occasional one that would rise higher, and fly off before the wind, and then only when the wind was blowing freshly. They did not fly with their heads directly before the wind, but seemed to rise in the air, set their wings in motion, and suffer themselves to be borne along by the current. They generally, perhaps always, rose in the afternoon, when the sun was hot and the wind blowing freshly.—(*From accounts furnished by Dr. N. T. True, Bethel, Maine, February* 28 *and March* 10, 1868.)

Its ravages in Ohio and Pennsylvania.—Mr. Schenck, of Franklin, Warren County, Ohio, writes to the Ohio Farmer that the grasshoppers are making their appearance in vast numbers. He says, "Last year we had millions of them; this year we have hundreds of millions." For five years, he says, they have been increasing on his farm, and he fears that unless some means are discovered for their destruction they will totally ruin his own and his neighbors' clover-fields.

The speed of the Central Railroad locomotives is considerably decreased by the immense swarms of grasshoppers between Lancaster and Philadelphia. One engineer stated that his train was forty minutes behind, owing to the number of grasshoppers on the track, and that he used twenty buckets of sand, which was thrown on the rail in front of the driving-wheels, to enable him to get along at all. Improbable as this story may appear, its truth is vouched for by the engineer above alluded to.—(*Quoted by Mr. A. S. Taylor in the California Farmer*, *April* 22, 1859.)

B.—CALOPTENUS SPRETUS, Uhler.

Notes upon CALOPTENUS SPRETUS, *Uhler*, *as observed during its invasion of Western Iowa in August and September*, 1867.

By September 3, these grasshoppers had reached the vicinity of Redfield, in the central part of Dallas County, and were sufficiently numerous to attract general attention. I first detected them September 1st, while collecting insects in the afternoon. The next day they were very common, and on the day succeeding rose before me in my walks in swarms. At this time I did not notice their arrival in immense flights, but in Guthrie County, in the valley of the Beaver, they were said to have appeared in *myriads* at about 11 a. m., September 5th, coming from the northwest and alighting so thickly as literally to cover the ground. They gradually dispersed, and after a day or two were much less common. At noon of the 13th I observed the lighting of an immense flight, in the same vicinity, likewise coming from the northwest. They filled the air quite thickly, and, with their wings glistening in the sunlight, resembled immense snow flakes. Their progress was readily traced from the north and west, and their origin was undoubtedly Dakota and Nebraska, where my friend Mr. O. H. St. John observed them in abundance in the larval state, in May. He also noticed them at or near Sioux City, and at other points on the Iowa side of the Missouri River. Their progress eastward was gradual; they appeared at Exira, in Audubon County, only forty or fifty miles west of Redfield, about August 20th, or nearly two weeks earlier than at the latter place.

After a few days their havoc with the vegetation began to be painfully apparent; in ten days they had stripped the leaves from the corn, the potatoes, and the white-willow hedges; several species of *Ambrosia* and *Chenopoda* presented but bare stems, and even the resinous foliage of the *Helianthi* did not wholly escape. I was obliged at once to suspend for the season my herbarial collections, from the ragged condition of the foliage of almost all species. Yet they displayed discrimination in selecting their food, leaving sorghum wholly untouched while denuding maize growing by its side; melons themselves were eagerly devoured while the vines that bore them were scarcely mutilated; every cabbage patch was quickly disposed of; they even attacked the *ears* of corn, eating off the outer husks, and devouring an inch or more of the end of the ear.

They commenced pairing soon after their arrival, but oviposition was delayed for some days. I think the exact time of this may depend upon the weather, or rather upon the condition of the ground. For a time after their arrival the weather was intensely dry—the culmination of a very protracted and severe drought. Although they commenced pairing a week or ten days previously, none were seen depositing their eggs until the occurrence of a heavy rain, which moistened and of course somewhat softened the hard, parched ground. This happened on the 17th, when oviposition became at once general, and was being carried on in every favorable place when I left the State, September 23d.

For breeding sites they generally select spots devoid of vegetation, preferring paths, road-sides, and other hard, bare ground, the compactness of the earth in such places doubtless tending to preserve the eggs from being crushed, or from disturbances to which in loose earth they would be liable. Into this hard ground the female, by diligent working, inserts her somewhat extensible abdomen its whole length, making a cylindrical cavity about an inch and one-half in depth. At the bottom of this the eggs are laid; they form a mass three-fourths of an inch to an inch in length, and about one-fourth of an inch in diameter. The eggs are elongated, fifteen to twenty-five in number, and when first laid are of a light yellow color, which soon deepens almost to an orange. Around the eggs a quantity of white or light-pinkish frothy substance is deposited, which soon hardens to form a compact egg-case. Usually the hole is filled with it, the summit of the egg-case being about half an inch or more below the surface. Ordinarily the holes are nearly vertical, but sometimes more or less inclined. During oviposition, as at other times, these grasshoppers are eminently gregarious; every bare space of ground of considerable size is completely covered with them, and each square inch made the receptacle of several clusters of eggs; the surface of the ground appears thickly perforated with small holes.

It is difficult to give a just conception of their numbers. Any one who has once seen them would feel no surprise in hearing of locomotives stopped in rising grades by their abundance on the rails, as actually happened in Iowa. The tires of ordinary vehicles were covered with their crushed remains, while it was impossible to walk without treading upon them; swine feasted upon them, especially when they were slightly stiffened by cold in the early morning; hazel and other slender bushes I have seen bending under their weight; and closely-grazed hill-sides appeared as white as though sown with plaster from the glistening of their wings. They were most partial to cultivated districts; the vicinity of valleys along the streams was next in favor. On the high, broad, wild prairie they were rarely met with, and I never saw them in such places in numbers sufficient to attract attention.

I left the invaded district too soon to state their duration. At New Jefferson, in Greene County, September 22, I saw immense numbers in the air, as though new accessions were coming; at the same time the ground was thickly strewn with the dead. Apparently they live but a very short time after depositing their eggs. So far as I could learn, they did not proceed eastward beyond the middle of the State.

Two other invasions of Western Iowa by immense numbers of a migratory grasshopper—probably this same species—have occurred since its settlement: one three years ago, when they only devastated the western tier of counties; the other ten years previous to that, when they swept over the entire State, although, from all accounts, in less numbers than recently. In both cases they deposited their eggs, and greatly terrified the inhabitants by their prospective abundance. In both cases the eggs were hatched during the following spring, but while the larvæ were very young cold rains came on and destroyed them all. Hence this spring the farmers of Iowa may in like manner be relieved of them; if not, their ravages during the coming summer must be fearful. It was the universal testimony that this species of grasshopper is not ordinarily seen in Iowa. Indeed, nearly every one with whom I conversed claimed never to have seen it before.—(*From an account furnished by Mr. J. A. Allen, Cambridge, Massachusetts, March* 10, 1868.)

Its ravages in British America.—Three kinds of grasshoppers made their appearance here in 1864 and 1865: 1st, a grasshopper yellow on the belly, and about three inches in length, stout in proportion. From one of these I pressed out as many as one hundred eggs; there were very few of this species, [probably *Œdipoda corallipes*, Hald., which occurs at Red River.] 2d, a grasshopper one and one-half inches in length with two yellow stripes, commencing at the antennæ and running down on the upper wings, meeting at the tips when the wings are closed. I have dug up some of their nests and found sixty-two eggs in one, and as many as seventy-two in another; this species was not very numerous. [*Caloptenus bivittatus*, Uhler.] 3d, the kind which I now forward, [*Caloptenus spretus*, Uhl.] The male is one inch in length; the female from one and one-fourth to one and one-half inches in length; they deposit from twenty to thirty-one or thirty-two eggs in a nest. They visited here in 1857, laying their eggs in great numbers, in the Assinniboin region, destroying the crops of 1858. In 1864 and 1865 they came again in swarms and destroyed many fields. The eggs deposited in 1865 produced their young from the 20th of May to the 10th of June, and moulted for the last time from the 20th of July to the 10th of August; a few days after they fled away. On the 12th of last August (1867) they fell in immense numbers, swarm following swarm, until late in September; as soon as they fell they commenced the work of reproduction, which they continued during August and September, and the first week of October; this labor performed, they took their flight to the south. A correspondent at Carleton House, on the Saskatchewan River, in speaking of these grasshoppers, says, "The day in which you say they made their appearance at Red River, (August 12, 1867,) I left Carleton House for Fort Pit, (on the north branch of the Saskatchewan River,) and all along the route they were very numerous; about the latter end of August they were flying south in great numbers." They also occurred last autumn at Fort Pelly, in the Swan River district, (near the head-waters of the Assiniboin,) and they have doubtless deposited their eggs in all the barren ground (prairie) between Fort Pit and Red River. They did not extend to the Winnipeg River, but myriads of them were drowned in Lakes Mantowaba and Winnipeg.—(*From a letter of Mr. Donald Gunn, Red River, April* 17, 1868.)

Its ravages in Nebraska.—The last day of August, 1866, near the middle of the afternoon, quite a number of grasshoppers were seen alighting, and that number rapidly increased till a little before sunset. The next morning they appeared much thicker, but were only so from having crawled more into the open air to sun themselves. About 9 o'clock they began to come thicker and faster from a northerly direction, swarming in the air by myriads, and making a roar like suppressed, distant thunder. By looking well up to the sun they could be seen to good advantage, and could be seen as high as the eye could discover an object so small, in appearance like a heavy snow-storm, each hopper very much like a very large flake, save that it passed by instead of falling. The number was beyond imagination—the leaves of the timber in this section of the Territory would be but little in comparison. The air was literally full of them, and continued so till along in the afternoon, when the air was free of them; countless millions having passed on, leaving other countless millions covering the earth to devour vegetation. Sunday and Monday being cloudy and damp, they contented themselves by devouring every eatable thing that came in their way; but Tuesday brought a repetition of the scene of Saturday. Since then they have not flown so much, and at this writing there are millions of them in this neighborhood, fortunately working their way a little east of south. * *

Their present visitation may be for some good, but I am too blind to see it. Their ravages here have drawn down many a hearty yet uncouth expression of disgust and hate from honest and hard-working farmers. Go into the gardens and see them stripped of nearly every vestige of vegetation, both stock and fruit; go into the field and see the vines of all sorts stripped of all their leaves, and eaten to the ground; go and see the corn, as completely naked as if some violent storm had torn every blade from the stalks, leaving it looking like a lot of degenerate hoop-poles; go into the orchards and timber, and see many of the smaller trees, especially, almost bare, the leaves having been devoured by these ravaging creatures. Many a sad sight and many a downcast countenance now fill the roll.—(*Mr. S. C. Maxima, quoted by Mr. B. D. Walsh in the Practical Entomologist, October*, 1866.)

Its ravages in Kansas.—The grasshoppers sent herewith are popularly known as the "Mormon," "Western," or "Colorado" grasshopper. Last month they made their appearance in the frontier settlements of Kansas and Nebraska. To-day I was expecting specimens to send you, and they came—not a pill-box full, but in clouds. As high as the eye could reach, the air was filled with them; and they came down glittering in the sunlight like huge flakes of snow, and at once commenced their vocation of destroying every green thing. Indian corn, however, seems their favorite food, and they promise to be as destructive to it as their neighbors, the spearmen, (*Doryphora decemlineata*,) have been to the potato. On the Nemaha the late corn has been entirely destroyed by them. Even where some men hastily cut up and shocked their corn, the grasshoppers continued their depredations until only the bare stalk remained. Wheat, when sown, was eaten up if left uncovered.

In many places the ground is fairly honey-combed by their egg-cells, which are from three-tenths to five-tenths of an inch in depth. The common length of the egg-cells is one and three-tenths of an inch; but by calling on a number of boys for a large one and a small one, I found the extremes to be one and six-tenths and nine-tenths of an inch.—(*Professor W. I. Robertson, quoted by Mr. B. D. Walsh in the Practical Entomologist, October*, 1866.)

Its ravages in Minnesota.—For two years in succession—in 1856 and 1857—the grasshoppers destroyed our crops, and many resolved then to keep two years' supply of produce on hand afterward. One fact I noticed: although they ate the bark from saplings and consumed our corn, tobacco, &c., ate holes in clothes hanging out to dry, and destroyed boots and shoes when they lit on them in the house, yet pease they avoided, and it was an odd sight to see the fields stripped, even of the weeds, and the pea patch left undisturbed.—(*Mr. O. H. Kelley in the Country Gentleman, quoted by Mr. B. D. Walsh in the Practical Entomologist, October*, 1866.)

C.—ŒDIPODA ATROX, Scudder.

Its ravages on the Pacific coast.—The Sacramento Union of the same date (July, 1855) states that the "most remarkable circumstance we have ever been called on to notice in this locality was the flight of the grasshoppers on Saturday and yesterday. For about three hours, in the middle of the day, the air, at an elevation of about two hundred feet, was literally thick with them, flying in the direction of Yolo. They could be the more readily perceived by looking in the direction of the sun. Great numbers fell upon the streets on Saturday, absolutely taking the city by storm, and yesterday they commenced the wholesale destruction of everything green in the gardens of the neighborhood. Their flight *en masse* resembled a thick snow-storm, and their depredations the sweep of a scythe."

The Shasta Courier, printed in the Northern Sacramento Mountains, remarks that "on Wednesday last (September 19, 1855) an immense flight of grasshoppers passed over this place, flying westward. The greater portion of them flew very high, and could only be seen by shading the eyes from the sun. They were as thick in the heavens as flakes of snow in a winter's storm."

The Sacramento Valley papers mention that whole orchards, gardens, and vineyards have been consumed by them. Entire fields of young grain, of crops and vegetables, have been eaten up within the space of a single day, leaving the ground like a wilted, blackened desert. In some parts of the valley they annoyed the passengers and horses of the public stages to such an extent as to cause the greatest inconvenience, and appear in some cases to have positively endangered human life.

A gentleman who resided in Colusa County, in the Sacramento Valley, in the summer of 1855, informs me that these insects appeared to rise out of the eastern boundaries of the valley, where it is hot, dry, and sandy, and that on some days they filled the air so as to obscure the sun. They consumed all garden vegetables, the leaves and bark of the elder-tree, and the young leaves and bark of the small branches of the cotton-wood and willow, and even the soft, green parts of the tules or bullrushes. In Stony Creek, in the same county, their dead bodies were seen at one time completely covering the surface of the water for miles in extent. In some parts of this valley they ate through gauze and textile coverings of all kinds, which had been used to shield animals and plants from their attacks.—(*Mr. A. S. Taylor in the Smithsonian Report for* 1858, *and in part in the California Farmer, January* 15, 1858.)

INDEX.

O

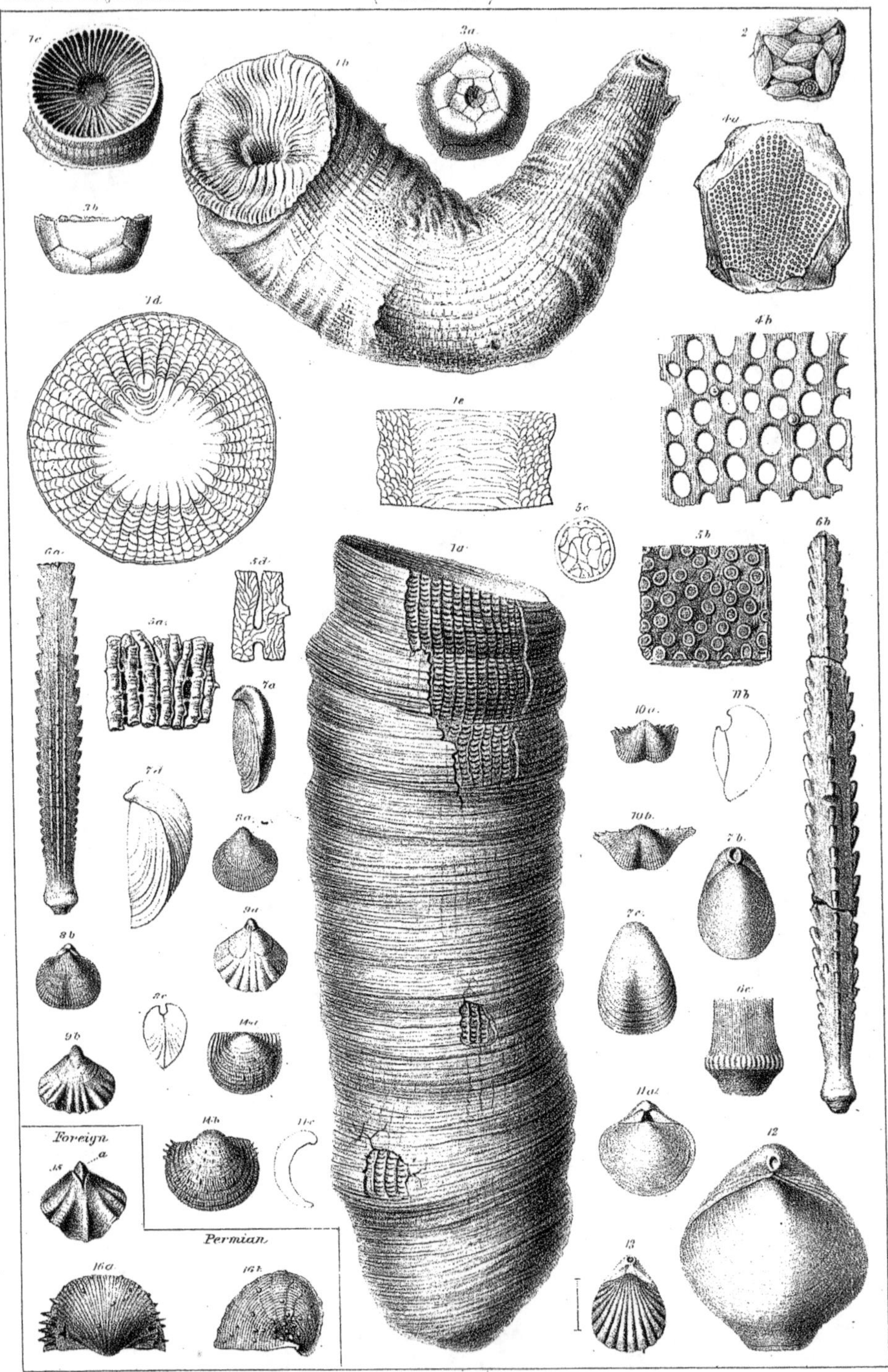

F.B. Meek & H.W. Elliott, del. On Stone by Sebotsan. T. Sinclair's lith. Phila.

EXPLANATIONS OF PLATE I.

Page.

Fig. 1.—Campophyllum torquium 145

1a. A nearly straight part of a large specimen, incomplete at both extremities, slightly reduced in size. (Rock Bluff.)

1b. A smaller flexuous specimen, side view, showing also the calice with its margins broken away; one-eighth diameter less than natural size. The epitheca is eroded in places so as to expose the septa.

1c. A small specimen showing the calice nearly complete, with a small septal fossula above, and a round central depression.

1d. Transverse section of the large specimen, showing the two sets of septa, the thin dissepiments between, and the central area occupied by the smooth tabulæ. The short septum above indicates the position of the fossula.

Fig. 2.—Fusulina cylindrica 140

A small portion of the weathered surface of limestone from Plattsmouth, almost entirely composed of these little foraminiferous shells; the figure being slightly reduced in size.

Fig. 3.—Erisocrinus typus 146

3a. View of under side of the cup, slightly reduced in size, and showing the small basal, larger subradial, and first primary radial pieces.

3b. A side view of the same.

Fig. 4.—Fenestella, sp 152

4a. A portion slightly reduced from Professor Geinitz's figure representing the natural size.

4b. Non-poriferous side a little reduced from Professor Geinitz's magnified figure of same.

Fig. 5.—Syringopora multattenuata 144

5a. A small portion, one-eighth diameter less than natural size, showing the wrinkled, rather flexuous corallites with their connecting tubes.

5b. A transverse section of a portion of the corallum where the corallites are rather widely separated; one-eighth diameter less than natural size.

5c. Transverse section of one of the corallites magnified so as to show the very short septa around the wall, and sections of the irregular tabulæ within.

5d. Longitudinal section of a portion of two of the corallites, a little magnified, showing the oblique irregular tabulæ.

Fig. 6.—Archæocidaris? triserrata 151

6a. Part of a primary spine, convex side, magnified a little less than two diameters, and showing the lateral serrated margins with three mesial rows of little crenated ridges.

6b. The opposite side of another specimen, magnified slightly less than two diameters, showing a mesial and two lateral serrated carinæ.

6c. A portion of the articulating end of the same, enlarged about four diameters, to show its milled ring and minute longitudinal striæ.

Fig. 7.—Terebratula bovidens 187

7a. A medium-sized specimen, profile view.

7b & c. Dorsal and ventral views of same.

7d. Side view of a larger specimen from Kansas.

Fig. 8.—Orthis carbonaria 173

8a. Ventral view of a rather large specimen slightly less than natural size.

8b. Dorsal view of same.

8c. Outline profile view of same.

Fig. 9.—Rhynchonella Osagensis 179

9a. Dorsal view of an internal cast, nearly natural size.

9b. Ventral view of same, showing it to be a true *Rhynchonella.*

Fig. 10.—Chonetes Verneuiliana 170

10a. The usual form, and a little less than the natural size, of the species.

10b. Another specimen unusually extended on the hinge line.

Page.

Fig. 11.—SPIRIFER (MARTINIA) PLANOCONVEXUS 184

11*a*. Dorsal view, magnified a little less than two diameters.

11*b*. Outline profile view of same.

Fig. 12.—ATHYRIS SUBTILITA 180

A rather large, well-developed specimen, dorsal view.

Fig. 13.—RETZIA (EUMETRIA) PUNCTULIFERA 181

Dorsal view, magnified rather less than two diameters.

Fig. 14.—PRODUCTUS PERTENUIS 164

14*a*. View of a cast of the outside of the concave or dorsal valve, one-eighth diameter less than natural size.

14*b*. View of outside of ventral valve of another specimen about natural size.

14*c*. Section in outline of the two valves united, showing the distinct concavity of the dorsal valve.

FOREIGN SPECIES FOR COMPARISON.

Fig. 15.—CAMAROPHORIA SCHLOTHEIMI 179

An internal cast (slightly reduced in size from Professor King's figure), showing (at *a*) the cast of the rostral chamber, separated by the deep slit on each side left by its walls. For comparison with Fig. 9*b*, in which there are no traces of the walls of such a chamber.

Fig. 16.—PRODUCTUS CANCRINI 165

16*a* & *b*. Posterior and lateral views, slightly reduced from de Verneuil's figure; for comparison with Fig. 14.

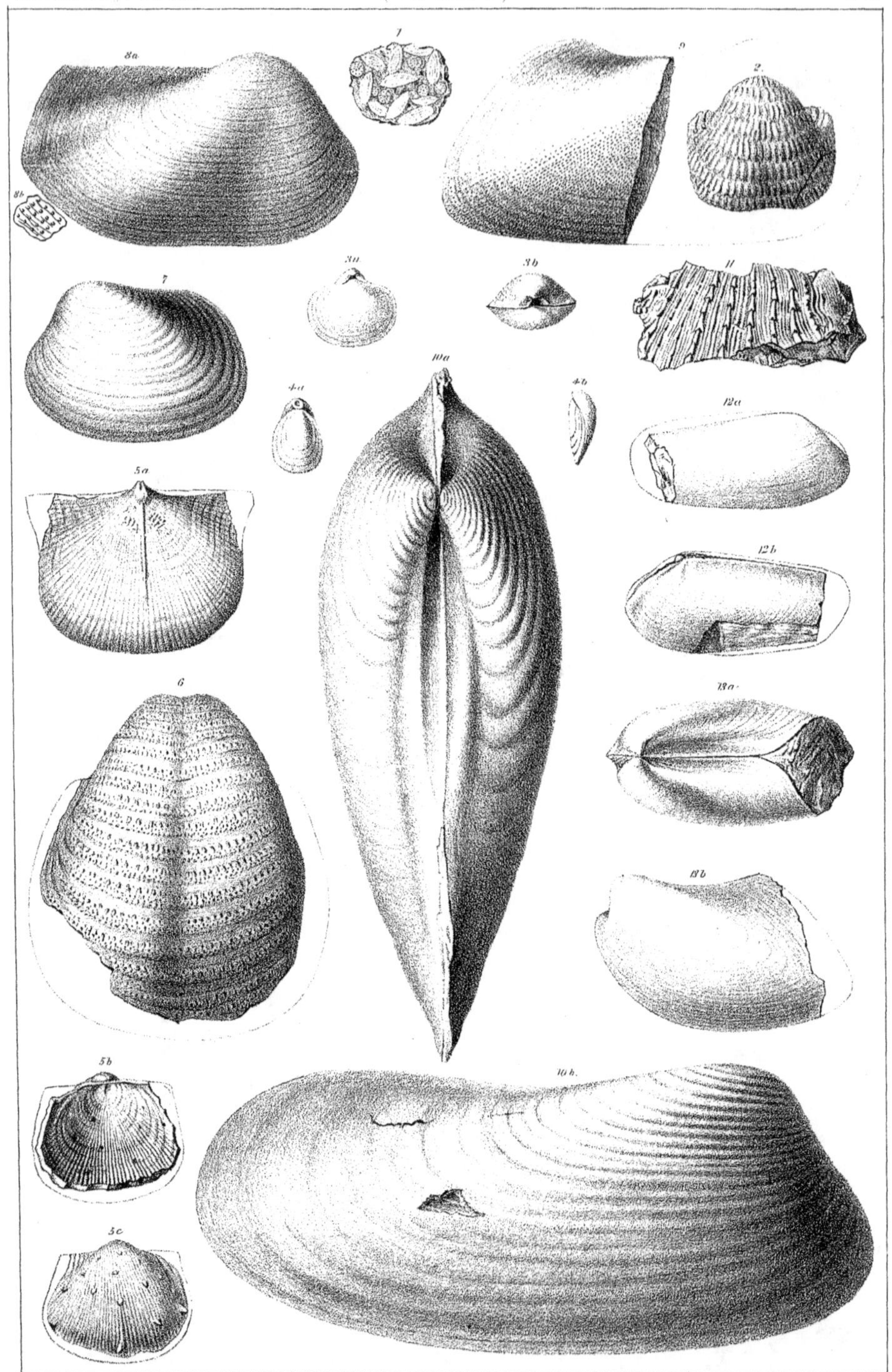

F.B. Meek & H.W. Elliott, del. On Stone by Thootson. T. Sinclair's lith. Phil.

EXPLANATIONS OF PLATE II.

Page.

Fig. 1.—FUSULINA CYLINDRICA.. 140
A fragment of limestone, showing sections of *Fusulina*, one-eighth diameter less than natural size. (Rock Bluff.)

Fig. 2.—PRODUCTUS NEBRASCENSIS.. 165
A cast of the interior of a ventral valve (reduced one-eighth diameter in size) as seen when broken from a hard limestone matrix, so as to leave the shell and spines in the latter. (Bellevue.)

Fig. 3.—SPIRIFER LINEATUS?*
3*a*. Dorsal view of a small specimen of the shell divested of its minute hair-like spines.
3*b*. Same, cardinal view. (Platte River.)

Fig. 4.—TEREBRATULA BOVIDENS... 18,
4*a*. Dorsal view of a small specimen, of nearly natural size.
4*b*. Profile view of same. (Rock Bluff.)

Fig. 5.—PRODUCTUS PRATTENIANUS 163
5*a*. Interior of an imperfect dorsal valve from Plattsmouth, showing the mesial ridge, and very short cardinal process (nearly natural size).
5*b* & *c*. Dorsal and ventral views of a smaller specimen, from Rock Bluff, with the ears, margins, and spines broken away (a little less than natural size).

Fig. 6.—PRODUCTUS PUNCTATUS.. 169
An imperfect ventral valve, partly exfoliated, so as to remove the small spines, ears and margins. A little less than natural size. (Plattsmouth.)

Fig. 7.—EDMONDIA SUBTRUNCATA... 215
View of cast of left valve, one-eighth diameter less than natural size. (Rock Bluff.)

Fig. 8.—ALLORISMA (SEDGWICKIA) GRANOSA.. 220
View of right valve; mainly a cast, but with some remaining portions of shell. One eighth diameter less than natural size. (Rock Bluff.)

Fig. 9.—CHÆNOMYA LEAVENWORTHENSIS.. 216
Posterior portion of the two valves united. Slightly less than natural size. (Rock Bluff.)

Fig, 10.—ALLORISMA SUBCUNEATA.. 221
10*a*. Dorsal view of an internal cast of a large specimen, one-eighth diameter less than natural size.
10*b*. Side view of same. (Rock Bluff.)

Fig. 11.—PSEUDOMONOTIS, sp... 200
A fragment of a left valve (cast) showing the larger costæ, with vaulted scale-like laminæ of growth, and smaller ribs between; slightly reduced in size. (Cedar Bluff.)

Fig. 12.—SOLENOMYA, sp.†
12*a*. An imperfect cast, slightly reduced in size.
12*b*. Opposite view of same. (Rock Bluff.)

Fig. 13.—CHÆNOMYA MINEHAHA?... 217
13*a*. Dorsal view of a cast, with some portions of the posterior margins wanting.
13*b*. Side view of same. (Plattsmouth.)

---

* This is the shell described by Professor McChesney (Palæozoic Fossils, page 43, 1860) under the name *Spirifer perplexus*; it is always small, and may be distinct from *S. lineatus*. The desiription of it was inadvertently omitted in our text.

† This cast is too imperfect for description or identification. It does not differ, however, so far as can be seen, from a form common in the Coal-Measures of Illinois, but seems to be different from the Nebraska City specimen, referred (erroneously, as I think) by Professor Geinitz to *S. biarmica*, de Verneuil

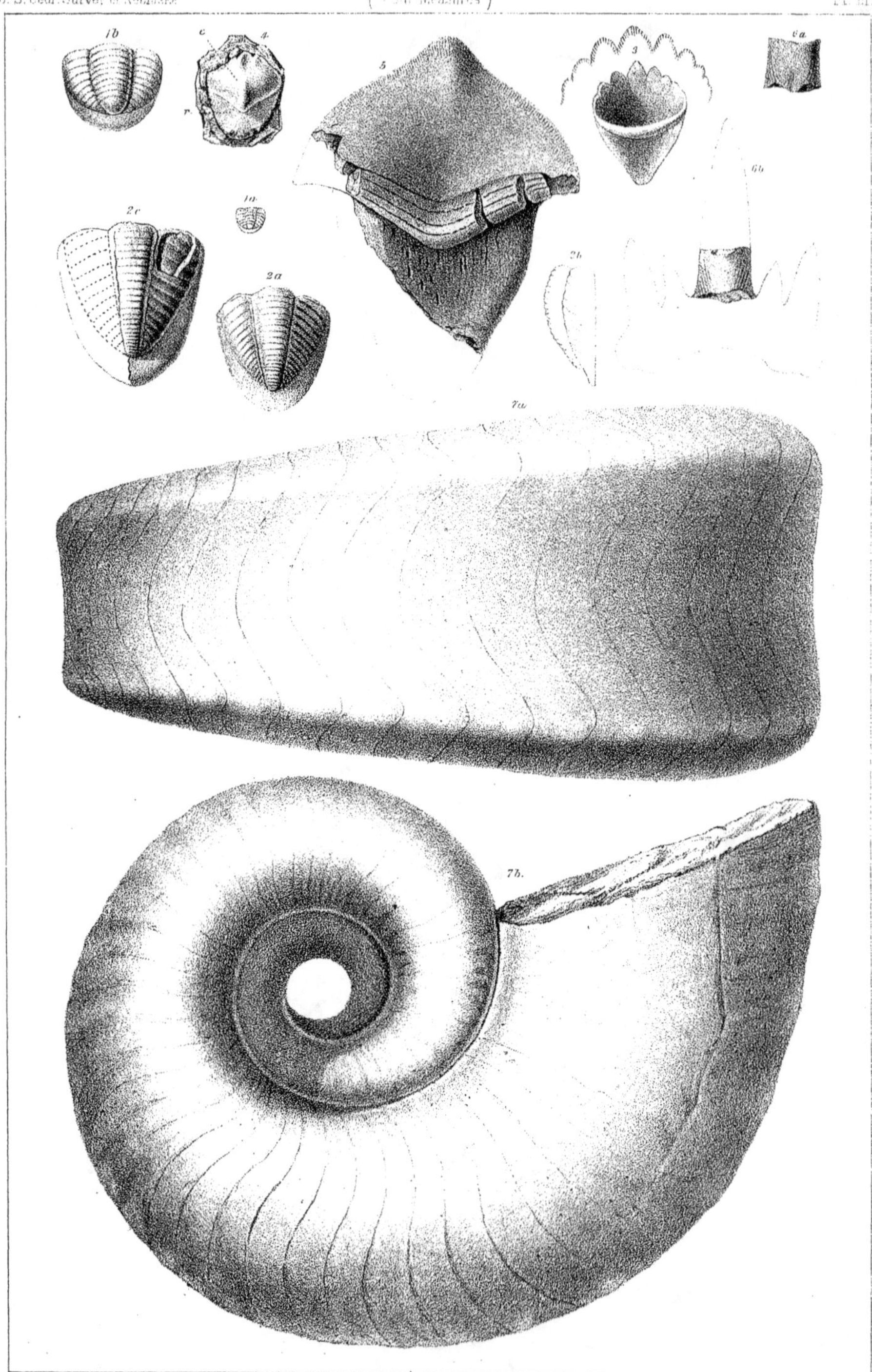

F. B. Meek & H. W. Elliott, del. On Stone by Ibbotson. T. Sinclair's lith. Phil.

EXPLANATIONS OF PLATE III.

Page.

Fig. 1.—PHILLIPSIA, sp. 237
1a. Pygidium one-eighth diameter less than natural size.
1b. Same, enlarged. (Both from Geinitz, Plattsmouth.)

Fig. 2. PHILLIPSIA MAJOR 238
2a. Pygidium (cast) one-eighth less than natural size, from Bellevue.
2b. An outline profile view of same.
2c. Part of the pygidium of a larger individual, from Plattsmouth, slightly reduced in size. [Segments of lateral lobes not oblique enough in the figure.]

Fig. 3.—PERIPRISTIS SEMICIRCULARIS 242
3a. Posterior view of a tooth and root, from Bellevue, showing the profound concavity of the crown, and the serrated edge, slightly less than natural size, with an enlarged outline of the latter above.

Fig. 4.—PERIPRISTIS SEMICIRCULARIS 242
An internal cast of the deep concavity of the crown (*c*), and a portion of the root (*r*), slightly reduced in size. (Rock Bluff.)

Fig. 5. PETALODUS DESTUCTOR 241
A specimen consisting of nearly an entire crown, and a portion of the root, posterior view, represented slightly less than natural size. (Rock Bluff.)

Fig. 6.—CLADODUS MORTIFER 239
6a. A fragment of the main cusp, near base, slightly reduced in size.
6b. Outline restoration of the whole tooth.

Fig. 7.—NAUTILUS PONDEROSUS 236
7a. View of the flattened or slightly concave ventral or outer side of a large specimen, reduced to slightly less than two-fifths diam. of natural size.
7b. Side view of some. (Plattsmouth.)

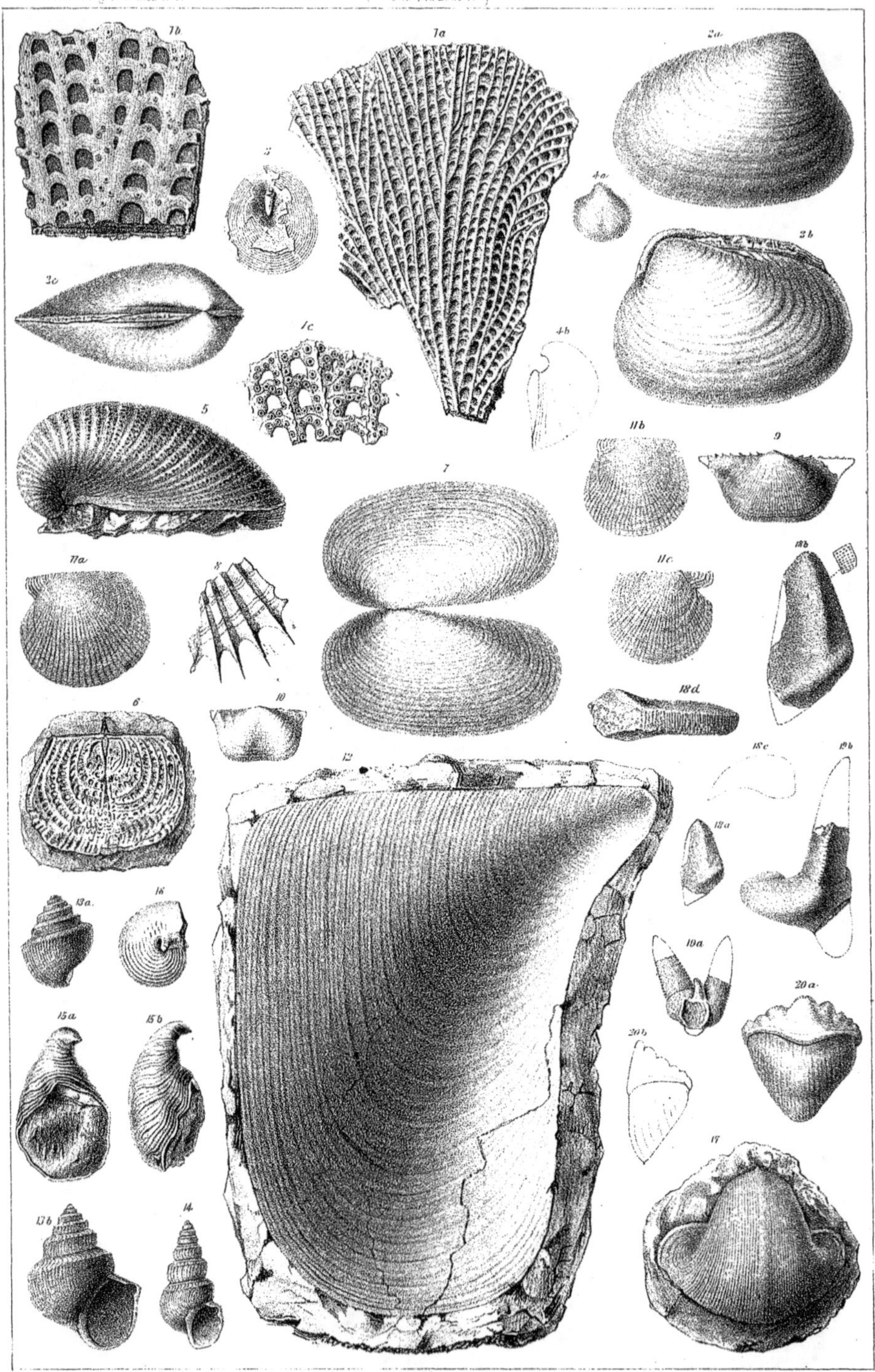

F. B. Meek & H. W. Elliott del. On Stone by Ibbotson. T. Sinclair's lith. Phil^a

EXPLANATIONS OF PLATE IV.*

Page.

Fig. 1.—SYNOCLADIA BISERIALIS 156

1*a*. A portion of the polyzoum, one-eighth diameter less than natural size, non-poriferous side.

1*b*. A portion of same, magnified so as to show the fine striæ, and irregularly scattered dimorphous pores, same side.

1*c*. Poriferous side of a fragment, magnified.

Fig. 2.—EDMONDIA ASPINWALLENSIS 216

2*a*. Side view of internal cast, one-eighth diameter less than natural size.

2*b*. View of the left valve, with the beak and dorsal margin of the right valve seen above, in consequence of a slight displacement.

2*c*. Dorsal view of the cast of both valves united. (Aspinwall.)

Fig. 3.—ORBICULOIDEA, sp 158

An impression of the outside of the under valve in the matrix (slightly less than natural size), with portions of the thin shell adhering, so as to show the smooth inside, and a prominent internal ridge (*d*) corresponding to a deep external furrow, with a small, round, or oval foramen at its outer end.

Fig. 4.—SPIRIFER (MARTINIA) PLANOCONVEXUS 184

4*a*. Ventral view, slightly less than natural size.

4*b*. Outline profile view of same, magnified.

Fig. 5.—PRODUCTUS PUNCTATUS 169

Side view of an accidentally compressed ventral valve, a little reduced in size.

Fig. 6.—PRODUCTUS NEBRASCENSIS 165

An impression in the matrix of the inner side of a ventral valve, one-eighth diameter less than natural size.

Fig. 7.—EDMONDA REFLEXA? 213

A slightly reduced view of the two valves opened and spread out upon the matrix. [This figure does not show the marks of growth strong enough, and has the beaks a little too far forward.]

Fig. 8.—AVICULOPECTEN CARBONIFERUS 193

A fragment of a left valve, enlarged nearly two diameters, and showing the costæ and marginal digitations.

Fig. 9.—CHONETES GRANULIFERA 170

A ventral view, slightly reduced in size.

Fig. 10.—CHONETES GLABRA 171

A ventral view, slightly reduced in size.

Fig. 11.—AVICULOPECTEN? WHITEI 195

11*a*. Left valve, flattened by pressure; one-eighth diameter less than natural size.

11*b*. Another of the same, of smaller size.

11*c*. A flattened right valve, showing the deeper sinus under the anterior ear; slightly reduced in size.

Fig. 12.—MYALINA SUBQUADRATA 202

An impression of the left valve (reduced one-eighth diameter) in the clay matrix, with adhering portions of the thin outer fibrous layer of shell remaining.

Fig. 13.—PLEUROTOMARIA PERHUMEROSA 232

13*a*. A somewhat distorted specimen the distortion making the spire appear rather shorter than natural (a little reduced in size).

13*b*. Another specimen from Rulo, enlarged nearly two diameters, and somewhat restored.

Fig. 14.—PLEUROTOMARIA INORNATA 232

View of side and aperture, slightly reduced in size.

---

* All the specimens figured on this plate, not otherwise designated, are from a shaft sunk to a depth of about 100 feet, at an elevation of 73 feet above the Missouri, one mile and three-fourths west of Nebraska City landing.

	Page.
Fig. 15.—PLATYCERAS NEBRASCENSIS	227
15*a*. View of aperture, and laterally curved apex.	
15. A side view, showing the undulated lip and marks of growth.	
Fig. 16.—BELLEROPHON CARBONARIUS	224
A side view, a little reduced in size, of a specimen from Mr. Morton's shaft.	
Fig. 17.—BELLEROPHON MARCOUIANA	226
A cast somewhat flattened and distorted by pressure, with much of the expanded lip also broken away; enlarged nearly two diameters.	
Fig. 18.—XYSTRODUS ? OCCIDENTALIS	244
18*a*. Upper view, a little less than natural size.	
18*b*. The same, magnified nearly two diameters, and a portion of the pitted surface still further magnified.	
18*c*. Outline transverse section across the widest part.	
18*d*. Side view of same. (From Aspinwall.)	
Fig. 19.—DIPLODUS COMPRESSUS	240
19*a*. Anterior view, a little less than natural size, of a specimen, with the cusps, and anterior tubercle of the root in part broken away.	
19*b*. Lateral view of same, magnified.	
Fig. 20.—PERIPRISTIS SEMICIRCULARIS	242
20*a*. View of the convex side of the root and crown, slightly less than natural size.	
20*b*. An outline lateral view of the same.	

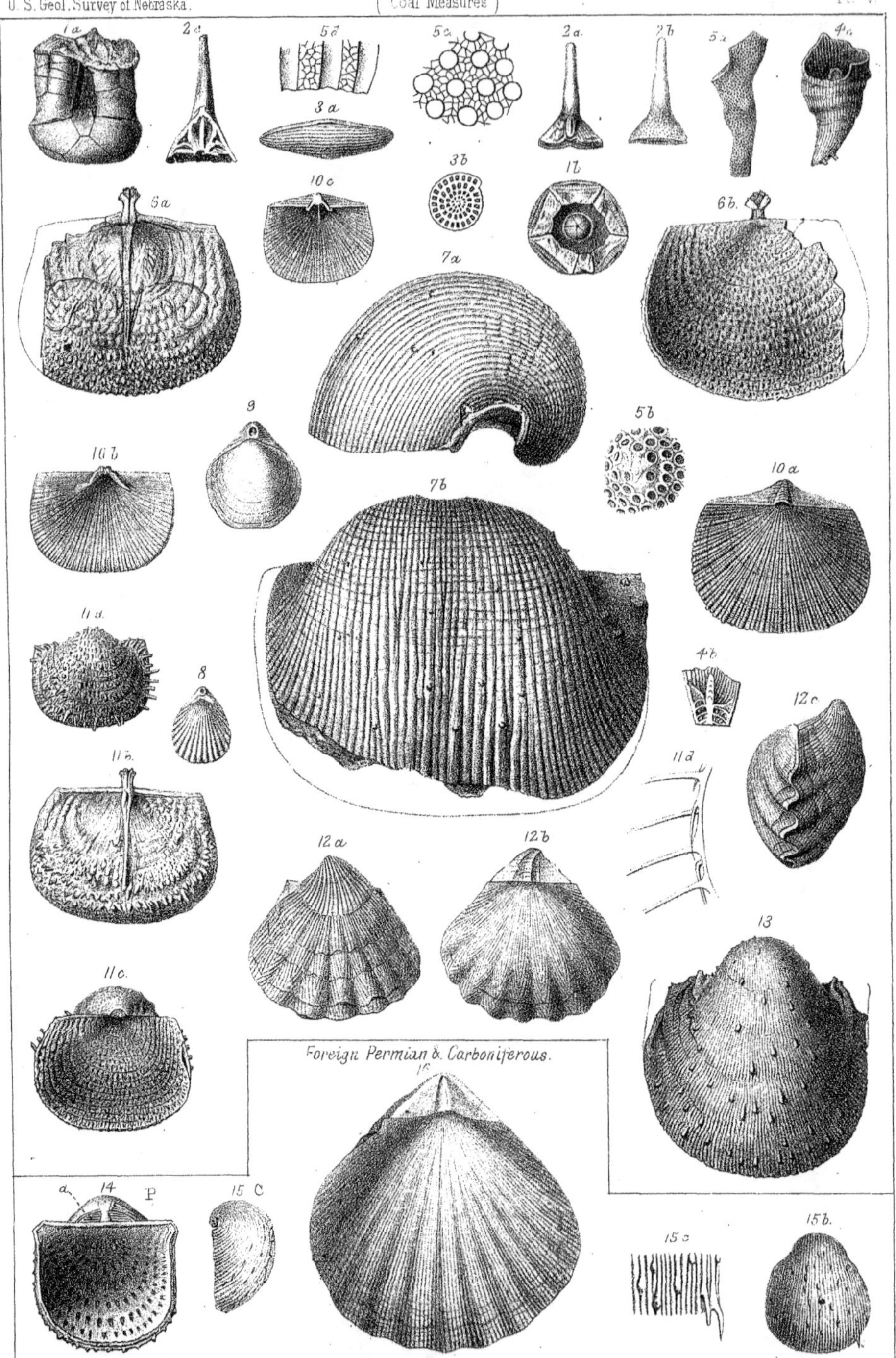

F. B. Meek & H. W. Elliott, del. On Stone by Tobbotson. T. Sinclair's lith. Phil.^a

EXPLANATIONS OF PLATE V.

Page.

Fig. 1.—SCAPHIOCRINUS? HEMISPHÆRICUS 147

1*a*. Posterior view of body and arm-bases, including a few pieces above the bifurcation on the second primary radial pieces; all reduced one-eighth diameter in size.

1*b*. Another specimen, with the second radials and arms removed, so as to show the upper and inner side of the cup in which the internally protuberant base is seen in the form of a little depressed cone within.

Fig. 2.—ZEACRINUS MUCROSPINUS 149

2*a* & *b*. Two views of the spine-like second radial pieces, reduced one-eighth diameter, from Professor Geinitz's figures.

2*b*. Upper side of the same, reduced from Professor McChesney's figures of an unworn specimen.

Fig. 3.—FUSULINA CYLINDRICA 140

3*a*. A specimen magnified nearly three diameters.

3*b*. A transverse section of the same, somewhat more enlarged, to show the involuted and septate character of the interior.

Fig. 4.—LOPHOPHYLLUM PROLIFERUM 141

4*a*. A side view, slightly less than natural size, of a specimen with the walls of the cup partly broken away, so as to show the end of the columella within.

4*b*. A longitudinal section of another specimen of the same species, slightly reduced in size, showing the transverse arching plates, columella, &c.

Fig. 5.—FISTULIPORA NODULIFERA 143

5*a*. A specimen nearly natural size, incrusting a piece of another little coral of ramose form.

5*b*. One of the nodes or prominences of the same, enlarged so as to show the pores with their prominent margins, all directed a little outward from the middle of the prominence, which is nearly without pores.

5*c*. A transverse section of a portion of the same species, more highly magnified, to show the finely cellular tissue between the cell-tubes.

5*d*. A magnified longitudinal section of a piece of the same species, cutting through the cell-tubes and exposing the intercellular tissue.

Fig. 6.—PRODUCTUS SYMMETRICUS 167

6*a* & *b*. Slightly reduced interior and exterior views of a dorsal valve, with the trifid cardinal process, which does not show its entire length, owing to its strong curve. In Fig. 6*a*, the reniform scars are put in from another specimen of the same species, not being defined in that figured.

Fig. 7.—PRODUCTUS SEMIRETICULATUS 160

7*a*. Side view (slightly reduced in size) of a specimen with the anterior margin not quite complete.

7*b*. Ventral view of another somewhat larger specimen, also reduced a little in size.

Fig. 8.—RETZIA PUNCTULIFERA 181

Dorsal view of a small specimen, enlarged to nearly two diameters.

Fig. 9.—ATHYRIS SUBTILITA 180

Dorsal view of a medium-sized specimen, slightly reduced in size.

Fig. 10.—HEMIPRONITES CRASSUS 174

10*a*. Dorsal view of a medium-sized specimen (slightly reduced in size), showing the area of the ventral valve with its false deltidium.

10*b*. Interior of a smaller ventral valve, showing the cardinal process (also slightly reduced).

Fig. 11.—PRODUCTUS NEBRASCENCIS 165

11*a*. Ventral view of a small specimen (reduced one-eighth diameter in size), showing spine bases.

Page.

FIG. 11.—11*b*. Interior of a larger dorsal valve, also reduced in size, showing the bifid cardinal process, and the divided mesial ridge, internal asperities, &c.

11*c*. Dorsal view of another specimen (slightly reduced in size), showing the cardinal margin without traces of an area, and the slightly truncated beak of the ventral valve.

11*d*. A diagram profile view, enlarged, of a piece of the shell of the ventral valve, to show the two sets of spines.

Fig. 12.—MEEKELLA STRIATO-COSTATA 175

12*a*. Ventral view of a rather small specimen, enlarged to nearly the medium adult size of the species.

12*b*. Dorsal view of same, showing area and false deltidium.

Fig. 13.—PRODUCTUS PRATTENIANUS 163

A view of the ventral side of a moderately large specimen, slightly reduced in size, from Professor Geinitz's figure, with the ears and spines broken away. (The restored outline of the ears has been added from our specimens.)

FOREIGN SPECIES FOR COMPARISON.

Fig. 14.—STROPHALOSIA HORRESCENS 167

Dorsal view, showing hinge line and cardinal area (*a*), for comparison with Fig. 11*c*. (Slightly reduced in size, from Professor Geinitz's Animalischen Ueberr. der Dyas.)

Fig. 15.—PRODUCTUS KONINCKIANUS 164

15*a*. Side view, slightly reduced in size.

15*b*. Ventral view of same, slightly reduced in size.

15*c*. An enlargement of a piece of the surface, showing the striæ and short spines magnified. All after Mr. Davidson. For comparison with Fig. 13.

Fig. 16.—MEEKELLA PECTINIFORMIS, (=*Streptorhynchus pectiniformis*) 175

Dorsal view, showing area and deltidium of ventral valve (a little reduced in size, from Mr. Davidson). For comparison with Fig. 12*b*.

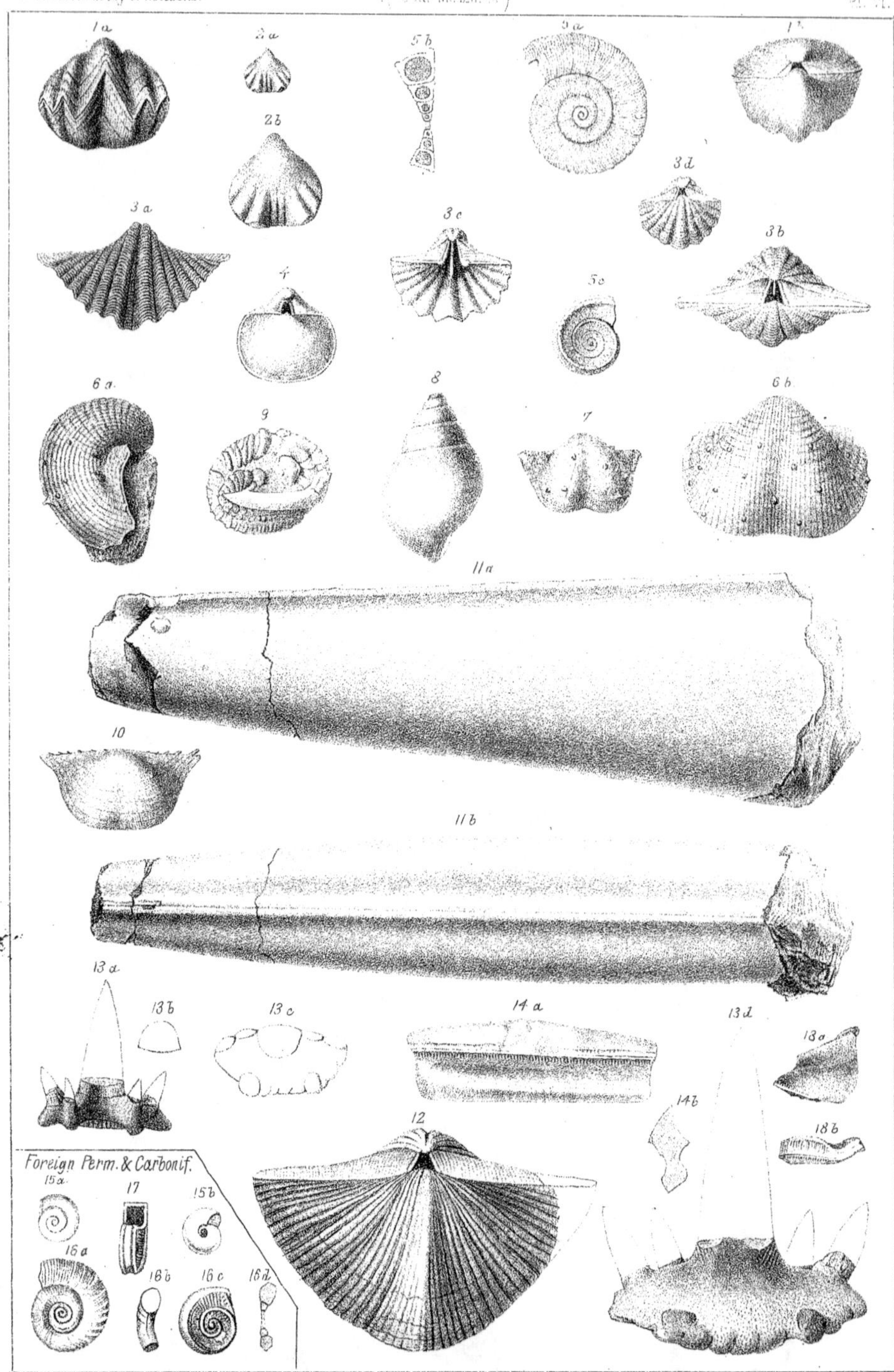

F. B. Meek & H. W. Elliott, del. On Stone by Roberson. T. Sinclair's lith. Phila.

EXPLANATIONS OF PLATE VI.

Page.

Fig. 1.—SYNTRILASMA HEMIPLICATA 177
 1*a*. Front view, a little reduced in size.
 1*b*. Cardinal or posterior view of same.

Fig. 2.—RHYNCHONELLA OSAGENSIS 179
 2*a*. Ventral view, reduced one-eighth diameter in size.
 2*b*. Same, enlarged nearly two diameters of natural size.

Fig. 3.—SPIRIFERINA KENTUCKENSIS 185
 3*a*. A specimen greatly extended on the hinge line; enlarged nearly to two diameters.
 3*b*. Cardinal or posterior view of same.
 3*c*. Cardinal and internal view of another specimen of ventral valve enlarged, showing internal lamina.
 3*d*. A very small individual, with hinge line short; magnified two diameters.

Fig. 4.—SPIRIFER (MARTINIA) PLANOCONVEXUS 184
 Dorsal view, magnified nearly two diameters.

Fig. 5.—EUOMPHALUS RUGOSUS, (Hall; not Sowerby) 230
 5*a*. Left view, magnified nearly two diameters.
 5*b*. Section of same, showing the quadrangular form of the volutions, and the concavity of the right side.
 5*c*. Right or concave side of another specimen, one-eighth diameter less than natural size.

Fig. 6.—PRODUCTUS COSTATUS ? 159
 6*a*. Profile side view, reduced one-eighth diameter in size.
 6*b*. Ventral view, same.

Fig. 7.—PRODUCTUS LONGISPINUS ? 161
 Ventral view, reduced one-eighth diameter in size.

Fig. 8.—MACROCHEILUS INTERCALARIS, *Var.* PULCHELLUS 228
 A side view, reduced one-eighth diameter in size.

Fig. 9.—PHILLIPSIA SCITULA 238
 A crushed specimen, enlarged about two diameters.

Fig. 10.—CHONETES GRANULIFERA 170
 A ventral view, slightly reduced in size.

Fig. 11.—PINNA PERACUTA 198
 11*a*. Side view (slightly reduced in size) of an internal cast, incomplete at both ends.
 11*b*. Dorsal view of same.

Fig. 12.—SPIRIFER CAMERATUS 183
 A dorsal view, nearly natural size, showing area and foramen.

Fig. 13.—CLADODUS MORTIFER 239
 13*a*. A specimen, consisting of the root of a tooth, with the bases of the cusps remaining; one-eighth diameter less than natural size.
 13*b*. An outline section of the main cusp, at base.
 13*c*. An outline view from above.
 13*d*. An opposite view of the root and bases of cusps, enlarged.

Fig. 14.—CHOMATODUS ARCUATUS 243
 14*a*. View of anterior side, one-eighth diameter less than natural size.
 14*b*. Profile or section of the same.

Fig. 18.—DELTODUS ? ANGULARIS 244
 18*a*. View of upper surface, somewhat reduced in size.
 18*b*. Profile view of same.

FOREIGN SPECIES FOR COMPARISON.

Fig. 15.—SPIRORBIS PERMIANUS 231
 15*a*. View of left or attached side, magnified nearly three times, for comparison with 5*a* and 16*a*.
 15*b*. Right or free side of same, for comparison with 5*c* and 16*c*: after Professor King.

Page.

Fig. 16.—SPIRORBIS PLANORBITES 230

16*a*. Left view, magnified nearly three diameters, for comparison with Fig. 5*a*.

16*b*. A fragment of the same, showing a section of whorls, for comparison with 5*b*.

16*c*. Another specimen of same, magnified two diameters, right side for comparison with 5*c*.

16*d*. Section of last, for comparison with 5*b*: all after Professor Geinitz.

Fig. 17.—EUOMPHALUS QUADRATUS 231

An oblique anterior view. (Slightly reduced from Professor McCoy's figure.

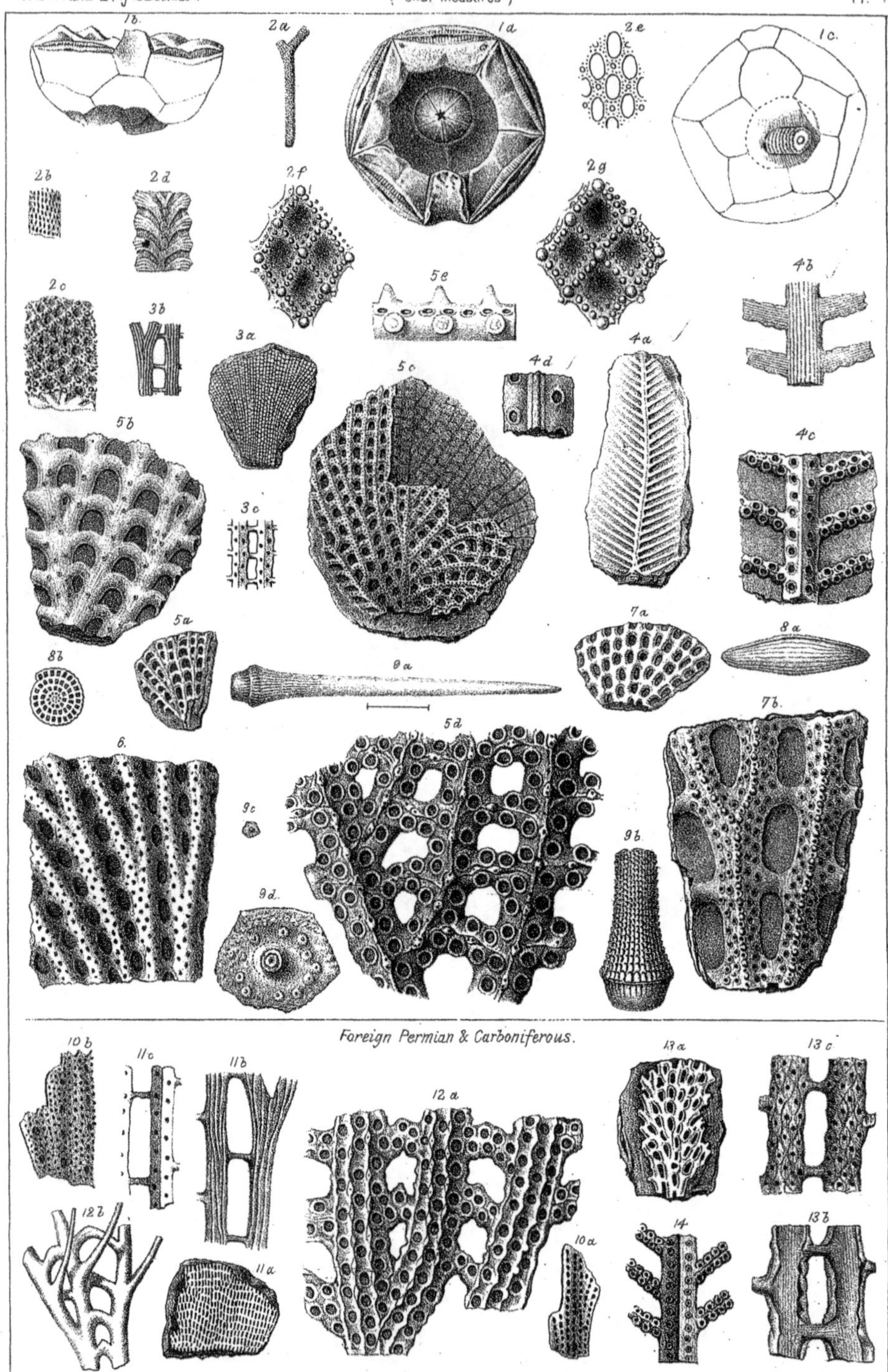

F. B. Meek & H. W. Elliott del. On Stone by Ibbotson. T. Sinclair's lith. Phila.

EXPLANATIONS OF PLATE VII.

Page.

Fig. 1.—SCAPHIOCRINUS? HEMISPHÆRICUS 147
- 1*a*. View of the body from above, slightly reduced.
- 1*b*. View of the same from below.
- 1*c*. Posterior side of same in outline; all slightly reduced from Professor Geinitz's figures.

Fig. 2.—RHOMBOPORA LEPIDODENDROIDES 141
- 2*a*. Branch, one-eighth diameter less than natural size.
- 2*b*. Part of same, magnified a little less than three diameters.
- 2*c*. Same, magnified about six diameters, to show the angular outlines of the calices, and the granules at their angles and along their margins.
- 2*d*. Longitudinal section, magnified nearly five diameters, showing the calices (c) *apparently* without tabulæ or sépta, ascending from an imaginary axis, and the interspaces (i), with longitudinal sections of very minute intermediate pores.
- 2*e*. Portion of surface, ground off a little obliquely, showing the calices within rather more oval in outline than natural, with the very minute pores of the interspaces, apparently corresponding to the marginal granules of the exterior.
- 2*f*. Four of the calices, greatly magnified, to show their rhombic outline and marginal granules.
- 2*g*. Same, from another part of the same species, where they differ a little in form.

Fig. 3.—FENESTELLA SHUMARDI? 153
- 3*a*. Fragment of the polyzoum slightly less than natural size, non-poriferous side.
- 3*b*. Fragment of the same, magnified to show the form and proportions of the fenestrules, branches, and dissepiments, comparatively course striæ, &c.
- 3*c*. Same, poriferous sides.

Fig. 4.—GLAUCONOME TRILINEATA, Meek 157
- 4*a*. Slightly less than natural size, non-poriferous side.
- 4*b*. Same, enlarged to show striæ.
- 4*c*. Poriferous side enlarged.
- 4*d*. Fragment of main stem, greatly magnified, to show the three mesial lines.

Fig. 5.—SYNOCLADIA BISERIALIS, (=*S. Cestriensi*?) 156
- 5*a*. Fragment, slightly less than natural size, non-poriferous side.
- 5*b*. Same, magnified about five diameters, to show striæ and dimorphous pores.
- 5*c*. Poriferous side of another specimen, a little enlarged.
- 5*d*. Part of same, greatly magnified, to show the arrangement of the pores, the mesial carina, and spine-like projections, with their minute perforations, &c.
- 5*e*. Side view of one of the branches, much enlarged, to show the elevation of the spine-like projections along the middle, on the poriferous side.

Fig. 6.—POLYPORA 155
- Enlarged poriferous side, slightly reduced from Professor Geinitz's figure.

Fig. 7.—POLYPORA SUBMARGINATA 154
- 7*a*. Fragment showing non-poriferous side, slightly less than natural size.
- 7*b*. Part of same, magnified, showing the poriferous side, with its mesial carina and row of node-like granules.

Fig. 8.—FUSULINA CYLINDRICA 140
- 8*a*. Specimen magnified a little more than three diameters.
- 8*b*. Transverse section of the same, mangnified between four and five diameters, so as to show the internal structure.

Page.

Fig. 9.—EOCIDARIS HALLIANA 152

9*a*. One of the primary spines, magnified a little less than six times.

9*b*. Part of same, still more greatly magnified, to show the surface sculpturing.

9*c*. One of the interambulacral plates of nearly natural size.

9*d*. Same, magnified; (all after Geinitz).

FOREIGN SPECIES FOR COMPARISON.

Fig. 10.—POLYPOR BIARMICA 155

10*a*. A piece of the polyzoum, slightly less than natural size.

10*b*. Magnified portion of same, to compare with Fig. 6. Both from Count Keyserling's illustrations of the species.

Fig. 11.—FENESTELLA PLEBEJA 153

11*a*. Fragment, nearly natural size, to compare with Fig. 3*a*.

11*b*. A piece of the same, non-poriferous side, greatly enlarged, to compare with Fig. 3*b*.

11*c*. Another fragment, greatly enlarged, showing poriferous side, for comparison with Fig. 3*c*. (All from McCoy's figures of the typical specimens.)

Fig. 12.—SYNOCLADIA VIRGULACEA 156

12*a*. Fragment, greatly enlarged, showing the poriferous side, for comparison with Fig. 5*d*.

12*b*. Non-poriferous side, magnified to show spine-like processes, for comparison with Fig. 5*b*. (All from Professor King's figures.)

Fig. 13.—POLYPORA MARGINATA 154

13*a*. Fragment, slightly less than natural size, nonporiferous side, for comparison with Fig. 7*a*.

13*b*. Piece of same, magnified, to show surface striæ of non-poriferous side, &c.

13*c*. Poriferous side of same, magnified, to compare with Fig. 7*b*. (All from Professor McCoy's original figures.)

Fig. 14.—GLAUCONOME GRANDIS 157

14. Fragment, greatly enlarged, showing poriferous side, for comparison with Fig. 4*c*. (From Professor McCoy's figure.)

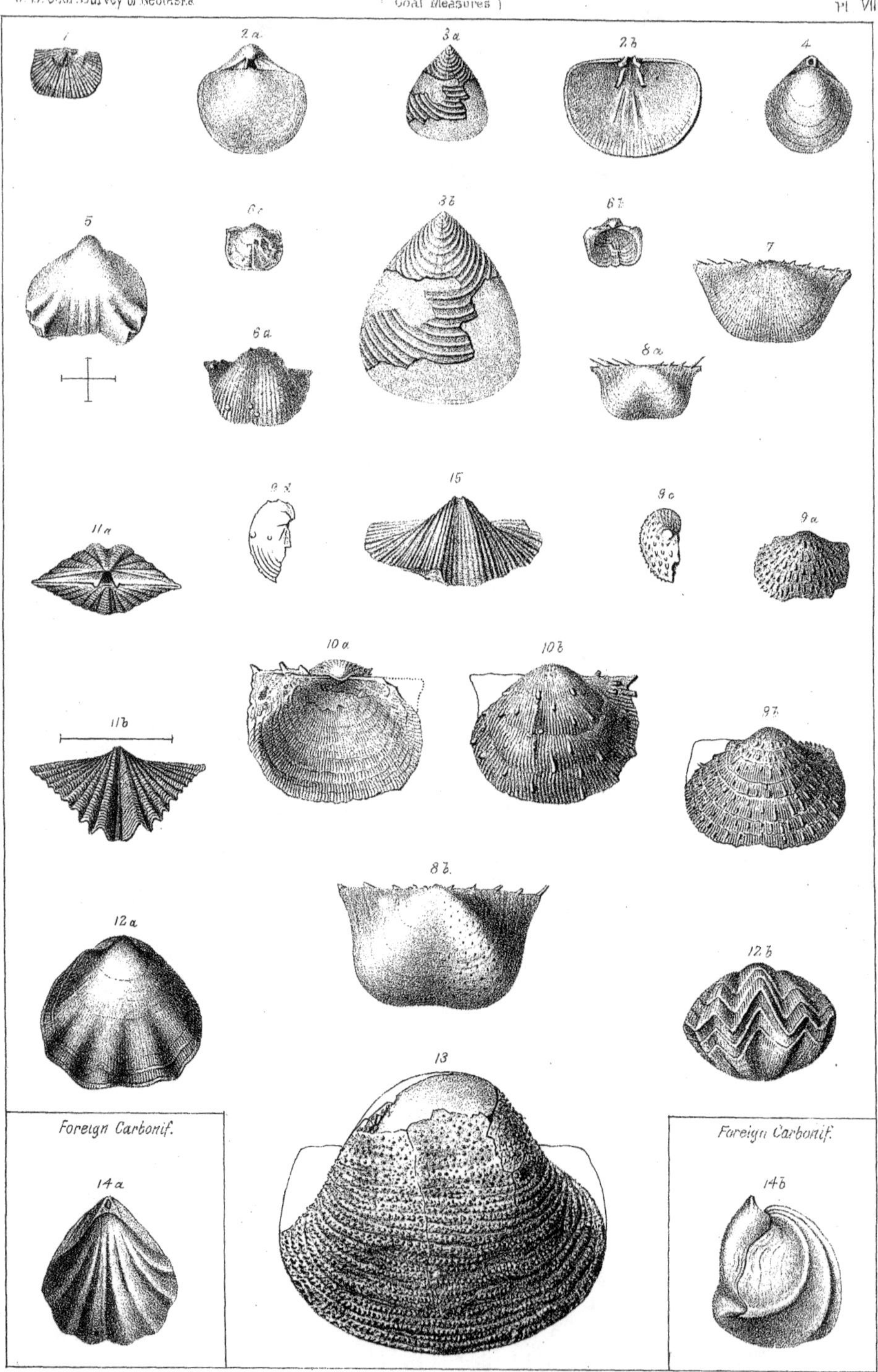

On Stone by Ibbotson.

T. Sinclair's lith. Phila

EXPLANATIONS OF PLATE VIII.

Page.

Fig. 1.—HEMIPRONITES CRASSUS 174
A very small young specimen.

Fig. 2.—SPIRIFER (MARTINIA) PLANOCONVEXUS 184
2a. Dorsal view, showing area and foramen of ventral valve magnified to nearly two diameters.
2b. Interior of dorsal valve, showing socket processes and crura for the attachment of the spires, still more enlarged.

Fig. 3.—LINGULA SCOTICA, *var.* NEBRASCENSIS 158
3a. View of dorsal valve, slightly less than natural size.
3b. Enlargement of same.

Fig. 4.—ATHYRIS SUBTILITA 180
Small specimen, dorsal view; slightly reduced in size from Professor Geinitz's figure of a specimen from bed C, at Nebraska City.

Fig. 5.—RHYNCHONELLA OSAGENSIS 179
Ventral view (enlarged nearly two and one-half diameters); after Professor Geinitz's figure of a rather broad, flattened specimen from bed C, at Nebraska City.

Fig. 6.—PRODUCTUS LONGISPINUS? 161
6a. Ventral view, slightly less than natural size.
6b & *c*. Dorsal and ventral views of a young example; after Geinitz.

Fig. 7.—CHONETES GRANULIFERA 170
Ventral view, nearly natural size, showing a few very scattering large perforations.

Fig. 8.—CHONETES GLABRA 171
8a. Ventral view, nearly natural size.
8b. Ditto of another specimen, enlarged nearly to two diameters, to show what appears to be the bases of minute, scattering spines, connected with large, distantly-separated perforations, arranged in quincux.

Fig. 9.—PRODUCTUS PERTENUIS 164
9a, c. Ventral and lateral views of an imperfect specimen, nearly natural size.
9b. The same enlarged. Both figured by Professor Geinitz, from bed C, Nebraska City.

Fig. 10.—PRODUCTUS PRATTENIANUS 163
10a & *b*. Dorsal and ventral views of a young shell; specimen nearly natural size; figured by Professor Geinitz, from bed C, at Nebraska City.

Fig. 11.—SPIRIFERINA KENTUCKENSIS 185
11a. Posterior or cardinal view, showing area and foramen, beaks, &c. (Enlarged about one and one-half diameters).
11b. Ventral view of a compressed specimen, enlarged nearly two diameters. Both from Professor Geinitz's figures of specimens from bed C, Nebraska City.

Fig. 12.—SYNTRILASMA HEMIPLICATA 177
12a & *b*. Ventral and front views; slightly reduced from Professor Geinitz's figures of a Nebraska City specimen.

Fig. 13.—PRODUCTUS SYMMETRICUS 167
Ventral view of an imperfect specimen from bed C, Nebraska City, slightly less than natural size.

FOREIGN SPECIES FOR COMPARISON.

Fig. 14.—RHYNCHONELLA ANGULATA 178
14a, b. Dorsal and side views for comparison with Fig. *12a, b*; both slightly reduced from Mr. Davidson's figures.

Fig. 15.—SPIRIFER CAMERATUS 183
An imperfect, rather small, specimen, from bed C, Nebraska City.

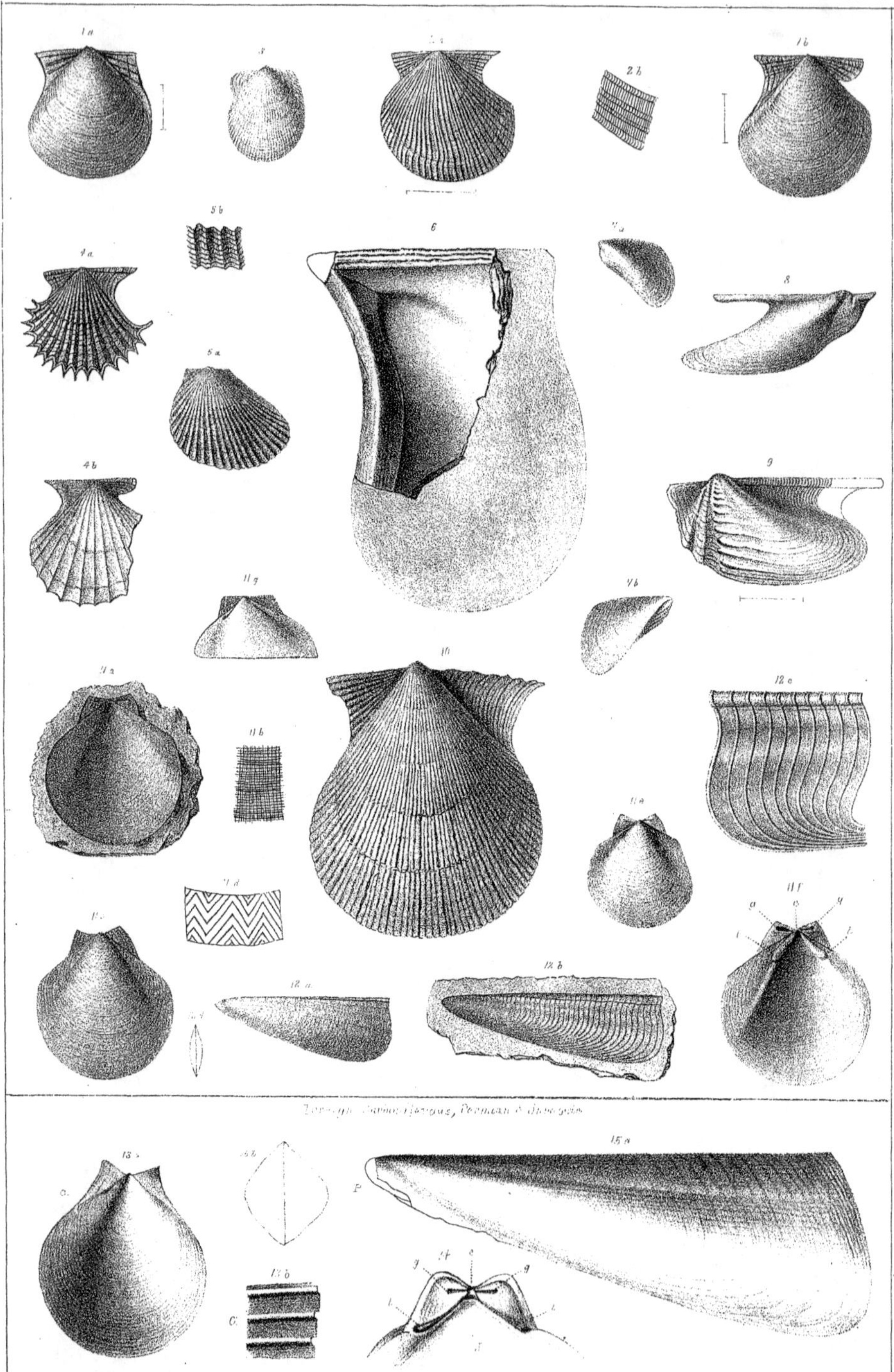
Foreign Carboniferous, Permian & Jurassic

EXPLANATIONS OF PLATE IX.

Page.

Fig. 1.—AVICULOPECTEN NEGLECTUS.. 193
1*a*. Internal cast of a left valve, enlarged nearly to three and one-half diameters. (From Geinitz.)
1*b*. Ditto of a right valve, from bed C, at Nebraska City, enlarged nearly three and one-half diameters.

Fig. 2.—AVICULOPECTEN COXANUS.. 196
2*a*. Lelt valve enlarged about two diameters.
2*b*. Enlargement of surface markings of same.

Fig. 3.—PSEUDOMONOTUS RADIALIS ??.. 201
View of cast of left valve, slightly less than natural size.

Fig. 4—AVICULOPECTEN CARBONIFERUS.. 193
4*a*. Left valve, one-eighth diameter less than natural size.
4*b*. Internal cast of a right valve a little less than two diameters.

Fig. 5.—LIMA RETIFERA.. 188
5*a*. Cast of right valve, one-eighth diameter less than natural size.
5*b*. Costæ of same, a little enlarged.

Fig. 6.—MYALINA SUBQUADRATA.. 202
A fragment of a right valve, showing cardinal area, and interior, nearly natural size, from Professor Geinitz's figure of a specimen from bed C, at Nebraska City. The dimly-shaded outline is here added to show the most usual form of the entire shell.

Fig. 7.—MYALINA SWALLOVI.. 201
7*a*. Left view of a small specimen, slightly less than natural size.
7*b*. Right view of a larger individual, slightly reduced in size.

Fig. 8.—AVICULA LONGA.. 199
Right valve, enlarged nearly two diameters.

Fig. 9.—AVICULA ? SULCATA.. 200
Left valve enlarged to nearly three and a half diameters; from Professor Geinitz's figure.

Fig. 10.—AVICULOPECTEN OCCIDENTALIS.. 191
Left valve slightly reduced in size; from Professor Geinitz's figure.

Fig. 11.—ENTOLIUM AVICULATUM.. 189
11*a*. Mould of the exterior (nearly natural size) of left valve.
11*b*. Enlarged external radiating and concentric striæ.
11*c*. Cast from the same.
11*d*. Enlarged zigzag marking sometimes very dimly seen on the surface of internal casts.
11*e*. Cast of a smaller specimen, left valve, with posterior margin a little truncated.
11*f*. Interior of another specimen, left valve, showing the diverging hinge, teeth (t, t), cartilage pit (c), and transverse furrow (g, g), in the ears, for the articulation of the margin of the other valve.
11*g*. Cast of the interior of opposite valve, showing its cardinal line to be straight, apparently for articulation in the furrow (g) of Fig. 11*f*.

Fig. 12.—AVICULOPINNA AMERICANA.. 197
12*a*. An internal cast, enlarged to nearly two diameters, showing the beaks not to be quite terminal.
12*b*. Mould of the outside of another specimen, somewhat enlarged.
12*c*. Surface markings, enlarged.
12*d*. Section to show convexity of valves.

FOREIGN SPECIES FOR COMPARISON.

Fig. 13.—ENTOLIUM SOWERBYI.. 191
13*a*. External view, for comparison with figure 11*c*.
13*b*. Enlarged surface striæ of same. Both a little less than natural size, from Professor McCoy's Carb. Fossils of Ireland.

Fig. 14.—ENTOLIUM DEMISSUM.. 190
Internal cast of hinge, for comparison with figure 11*f*; slightly reduced in size; from Quenstedt.

Fig. 15.—AVICULOPINNA PINNÆFORMIS.. 197
15*a*. View of left valve, showing traces of radiating striæ; a little reduced from Professor Geinitz's figure of a German specimen.
15*b*. Section of the two valves united, taken from a fragment sent from Germany by Professor Geinitz to Professor Worthen. For comparison with Fig. 12.

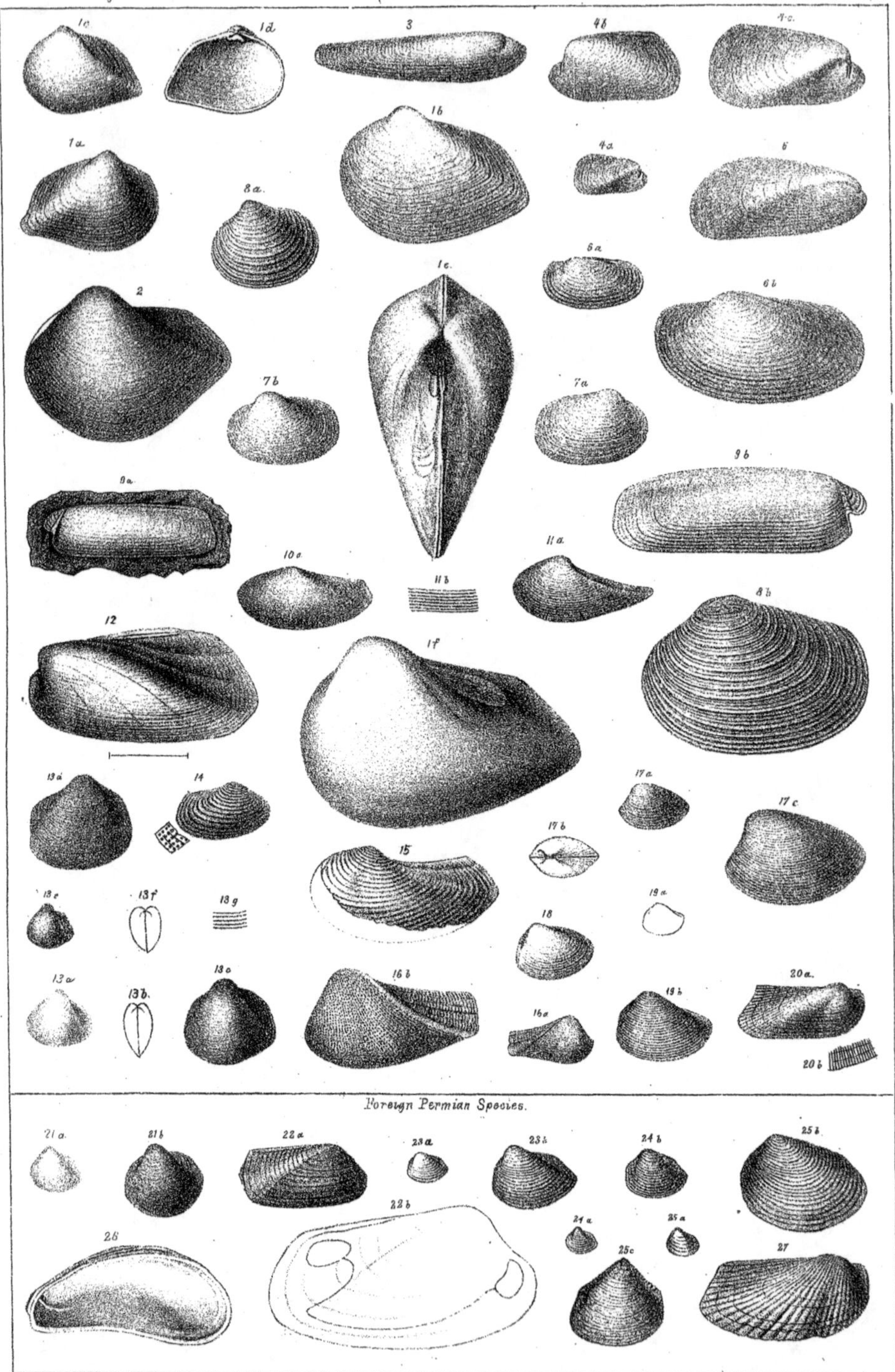

F.B. Meek & H.W. Elliott, del. On Stone by Ibbotson. T. Sinclair's lith. Phil[a]

EXPLANATIONS OF PLATE X.

Page.

Fig. 15.—ALLORISMA (SEDGWICKIA) REFLEXA 217
An imperfect left valve, nearly natural size.

Fig. 16.—ALLORISMA (SEDGWICKIA) GEINITZII 219
16*a*. View of right valve, slightly less than the natural size, of one of the largest specimens among an extensive collection.
16*b*. Left valve of another individual, enlarged a little less than two diameters, to show the fine surface granules.

Fig. 17.—NUCULA VENTRICOSA 204
17*a*. Left view, a little less than natural size.
17*b*. Dorsal outline view of same.
17*c*. Left view of same, enlarged to nearly two diameters.

Fig. 18.—NUCULA BEYRICHI ? ? 203
An internal cast, enlarged to nearly three diameters. From Nebraska City.

Fig. 19*a*.—An outline, nearly natural size, of a different form, from Professor Geinitz's figure, referred by him to *N. Beyrichi,* from the same bed as the last, at Nebraska City.
19*b*. The same, enlarged to show the striæ; (from Geinitz.)

Fig. 20.—MACRODON TENUISTRIATA 207
20*a*. Right valve, nearly natural size.
20*b*. Enlargement of the minute radiating striæ on the anterior part of the shell.

FOREIGN SPECIES FOR COMPARISON.

Fig. 21.—SCHIZODUS ROSSICUS 209
21*a*. Left view, slightly less than natural size, for comparison with Fig. 13*a*.
21*b*. A different form, supposed to be a variety of the same; both from de Verneuil.

Fig. 22.—ALLORISMA ELAGANS 219
22*a*. Right valve, nearly natural size (from Professor Geinitz), of a German specimen, for comparison with Fig. 16*a*, *b*.
22*b*. Internal cast of same, slightly reduced in size, from Professor King. Compare with Fig. 16*a*, *b*.

Fig. 23.—NUCULA BEYRICHI 204
23*a*. From a German specimen, slightly less than natural size. Compare with Figs. 18 and 19.
23*b*. Same, enlarged.

Fig. 24.—NUCULA BEYRICHI 204
24*a*, *b*. After Geinitz, from German specimen, nearly natural size, and enlarged. Compare with Figs. 18 and 19.

Fig. 25.—NUCULA BEYRICHI 204
25*a*. Nearly natural size, from von Schauroth's original figure.
25*b*. Same, from same, enlarged. Compare with Figs. 18 and 19.
25*c*. Internal cast of same, from same.

Fig. 26. CARDIOMORPHA ? (MYTILUS) PALLASI 212, 213
Hinge and interior of right valve, nearly natural size. For comparison with Fig. 12.

Fig. 27. MACRODON STRIATA 207
Right view, after Professor Geinitz, from German specimen, for comparison with Fig. 20*a*.

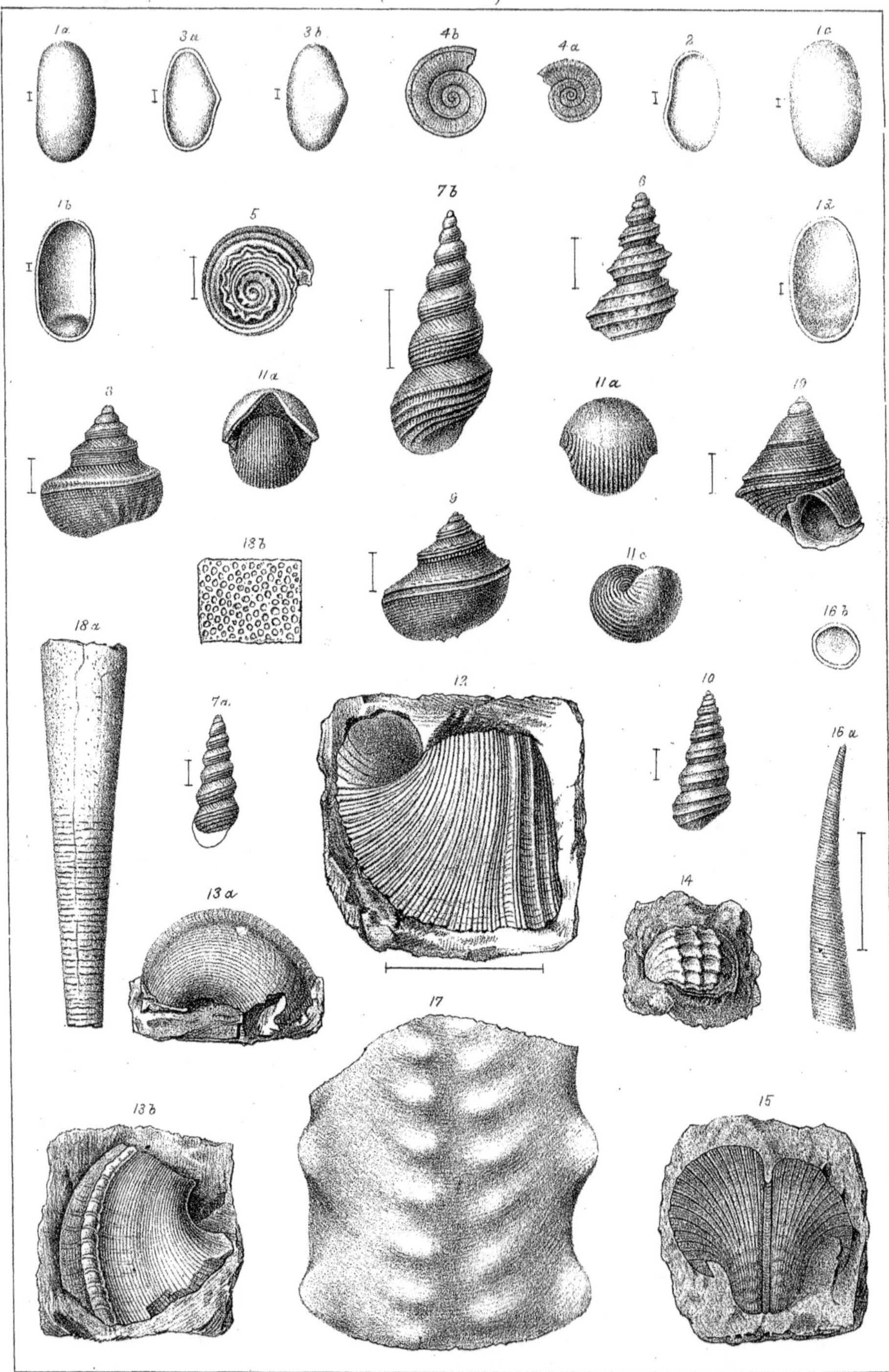

On Stone by Ibbotson. T. Sinclair's lith Phil.ª

EXPLANATIONS OF PLATE XI.

Page.

Fig. 1.—CYTHERE, sp. 237
1*a*, *b*. Outside and inside view of right valve, greatly enlarged.
1*c*, *d*. Outside and inside of left valve, greatly enlarged.

Fig. 2.—CYTHERE NEBRASCENSIS. 237
Inside, greatly enlarged, left valve; from Geinitz.

Fig. 3.—CYTHERE, sp. 237
3*a*, *b*. Outside and inside of apparently the right valve of the same species; from Geinitz.

Fig. 4.—EUOMPHALUS RUGOSUS, (Hall; not Sowerby). 230
4*a*. Left side of a rather small specimen, natural size.
4*b*. Right side of another specimen, somewhat enlarged.

Fig. 5.—PLEUROTOMARIA HAYDENIANA. 231
Enlarged nearly to three diameters; from Professor Geinitz's figure.

Fig. 6.—MURCHISONIA NEBRASCENSIS. 234
Enlarged less than three diameters; from same.

Fig. 7. ACLIS? SWALLOVIANA. 229
7*a*. A very small imperfect specimen, enlarged less than five diameters.
7*b*. A large perfect specimen, enlarged less than three and a half diameters (after Professor Geinitz).

Fig. 8.—PLEUROTOMARIA MARCOUIANA. 233
Enlarged less than four diameters; from same.

Fig. 9.—PLEUROTOMARIA GRAYVILLENSIS. 233
Enlarged between three and three and half diameters; from same.

Fig. 10.—ORTHONEMA SUBTÆNIATA. 228
Enlarged between five and five and a half diameters; from same.

Fig. 11.—BELLEROPHON CARBONARIA. 224
11*a*. Front view, nearly natural size.
11*b* & *c*. Posterior and lateral views of same; from division C, at Nebraska City.

Fig. 12.—BELLEROPHON, sp. 225
The expanded part of body whorl enlarged about two diameters.

Fig. 13.—BELLEROPHON MARCOUIANUS. 226
13*a*. Side view of an imperfect specimen, consisting of a portion of the expanded body volution, enlarged a little less than two diameters.
13*b*. An oblique view of a similar specimen, a little more enlarged; from Professor Geinitz.

Fig. 14.—BELLEROPHON PERCARINATUS. 227
A portion of the body volution, slightly less than the natural size.

Fig. 15.—BELLEROPHON MONTFORTIANUS. 225
Body volution and expanded lip nearly entire, slightly less than the natural size.

Fig. 16.—DENTALIUM MEEKIANUM. 224
16*a*. Side view, enlarged between two and three diameters.
16*b*. Enlarged section or aperture of same; both from Professor Geinitz.

Fig. 17.—NAUTILUS OCCIDENTALUS. 234
A fragment, showing the outer or ventral side, with the two rows of large lateral nodes, and two rows of smaller alternating mesial nodes; nearly natural size.

Fig. 18.—ORTHOCERAS CRIBROSUM. 234
18*a*. Slightly reduced from Professor Geinitz's enlarged figure.
18*b*. Pitted surface of same, enlarged.

Fig. 19.—PLEUROTOMARIA SUBDECUSSATA. 233
A side and aperture view, enlarged about four diameters; after same.

www.ingramcontent.com/pod-product-compliance
Lightning Source LLC
LaVergne TN
LVHW020237110826
845151LV00003B/954

* 9 7 8 1 4 2 5 5 3 0 0 5 1 *